APPLIED MULTIVARIATE STATISTICS
WITH SAS® SOFTWARE

SECOND EDITION

Ravindra Khattree
Dayanand N. Naik

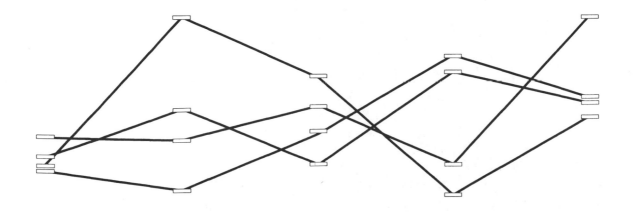

The correct bibliographic citation for this manual is as follows: Khattree, Ravindra, and Dayanand N. Naik. 1999. *Applied Multivariate Statistics with SAS® Software, Second Edition.* Cary, NC: SAS Institute Inc.

Applied Multivariate Statistics with SAS® Software, Second Edition

To the fond memory of my son, Kaushik (R.K.)

To my loving parents (D.N.N.)

Contents

Preface   ix

Commonly Used Notation   xiii

1 Multivariate Analysis Concepts   1
   1.1   Introduction   1
   1.2   Random Vectors, Means, Variances, and Covariances   2
   1.3   Multivariate Normal Distribution   5
   1.4   Sampling from Multivariate Normal Populations   6
   1.5   Some Important Sample Statistics and Their Distributions   8
   1.6   Tests for Multivariate Normality   9
   1.7   Random Vector and Matrix Generation   17

2 Graphical Representation of Multivariate Data   21
   2.1   Introduction   21
   2.2   Scatter Plots   22
   2.3   Profile Plots   31
   2.4   Andrews Function Plots   33
   2.5   Biplots: Plotting Observations and Variables Together   38
   2.6   Q-Q Plots for Assessing Multivariate Normality   45
   2.7   Plots for Detection of Multivariate Outliers   50
   2.8   Bivariate Normal Distribution   53
   2.9   SAS/INSIGHT Software   58
   2.10 Concluding Remarks   59

3 Multivariate Regression   61
   3.1   Introduction   61
   3.2   Statistical Background   62
   3.3   Least Squares Estimation   63
   3.4   ANOVA Partitioning   64
   3.5   Testing Hypotheses: Linear Hypotheses   66
   3.6   Simultaneous Confidence Intervals   84
   3.7   Multiple Response Surface Modeling   87
   3.8   General Linear Hypotheses   91

3.9   Variance and Bias Analyses for Calibration Problems   98
3.10   Regression Diagnostics   102
3.11   Concluding Remarks   116

4   Multivariate Analysis of Experimental Data   **117**
4.1   Introduction   117
4.2   Balanced and Unbalanced Data   120
4.3   One-Way Classification   123
4.4   Two-Way Classification   129
4.5   Blocking   137
4.6   Fractional Factorial Experiments   139
4.7   Analysis of Covariance   145
4.8   Concluding Remarks   149

5   Analysis of Repeated Measures Data   **151**
5.1   Introduction   151
5.2   Single Population   152
5.3 k Populations   176
5.4   Factorial Designs   195
5.5   Analysis in the Presence of Covariates   207
5.6   The Growth Curve Models   219
5.7   Crossover Designs   236
5.8   Concluding Remarks   246

6   Analysis of Repeated Measures Using Mixed Models   **247**
6.1   Introduction   247
6.2   The Mixed Effects Linear Model   248
6.3   An Overview of the MIXED Procedure   252
6.4   Statistical Tests for Covariance Structures   255
6.5   Models with Only Fixed Effects   265
6.6   Analysis in the Presence of Covariates   274
6.7   A Random Coefficient Model   288
6.8   Multivariate Repeated Measures Data   294
6.9   Concluding Remarks   297

References   **299**

Appendix A   A Brief Introduction to the IML Procedure   **305**
A.1   The First SAS Statement   305
A.2   Scalars   305
A.3   Matrices   305
A.4   Printing of Matrices   306
A.5   Algebra of Matrices   306
A.6   Transpose   306
A.7   Inverse   306
A.8   Finding the Number of Rows and Columns   307

A.9   Trace and Determinant   307

A.10   Eigenvalues and Eigenvectors   307

A.11   Square Root of a Symmetric Nonnegative Definite Matrix   308

A.12   Generalized Inverse of a Matrix   308

A.13   Singular Value Decomposition   309

A.14   Symmetric Square Root of a Symmetric Nonnegative Definite Matrix   309

A.15   Kronecker Product   309

A.16   Augmenting Two or More Matrices   310

A.17   Construction of a Design Matrix   310

A.18   Checking the Estimability of a Linear Function $\mathbf{p}'\boldsymbol{\beta}$ 311

A.19   Creating a Matrix from a SAS Data Set   312

A.20   Creating a SAS Data Set from a Matrix   312

A.21   Generation of Normal Random Numbers   312

A.22   Computation of Cumulative Probabilities   313

A.23   Computation of Percentiles and Cut Off Points   313

Appendix B   Data Sets   **315**

Index   **327**

Preface

It was a pleasant surprise when we were asked by our original publisher to come up with a revised second edition within three years of publication of the first edition. The first edition was warmly received by the intended audiences. However, over the last three years we ourselves had critically examined the book and had come to realize that in view of certain newer approaches to the analyses (such as the mixed models approach), a revision could make the book even more useful. It is our pleasure to present to our audiences a second revised edition, which is now copublished by SAS Institute Inc. and John Wiley & Sons.

Applied multivariate techniques are routinely used in a variety of disciplines such as agriculture, anthropology, applied statistics, biological sciences, business, chemistry, econometrics, education, engineering, marketing, medicine, psychology, quality control, and sociology. With such a diverse readership, we thought it essential to present multivariate techniques in a general context while at the same time keeping the mathematical and statistical requirements to a bare minimum. We have sincerely attempted to achieve this objective.

Audience

The book is written both as a handy reference for researchers and practitioners as well as a supplementary college text. Researchers and practitioners can also adapt the material for a self-taught tutorial. Students and their instructors in senior undergraduate or beginning graduate classes in applied statistics will find the book useful as an accompanying computational supplement to a more advanced book on applied multivariate statistics. The book can also be adapted for a statistics service course for graduate students from the nonstatistical disciplines.

Approach

Primary emphasis is on statistical methodology as applied to various scientific disciplines. SAS software is used as the crucial computational aid to carry out various intensive calculations which so naturally occur in any typical multivariate analysis application. Discussion in this volume is limited to only the normal theory-based multivariate analysis.

We believe that those who use multivariate methods should not only understand appropriate statistical techniques useful in their particular situation but should also be able to discern the appropriate approach and distinguish it from an approach that *seems* correct but is completely inappropriate in a particular context. Quite often, these differences are subtle, and there are scenarios where the presumably *best approach* may be completely invalid due to one reason or the other. The problem is further compounded by the understandable temptation to take the shortest route by choosing the analysis that can be readily performed using a particular software package or a canned computer program, regardless of its appropriateness, over a more appropriate analysis not so readily available. This book attempts to demonstrate this process of discernment, problem definition, selection of an appropriate analysis or a combination of many, while providing both the needed SAS code to achieve these goals and the subsequent interpretation of the SAS output.

This approach largely eliminates the need for two books, one for learning multivariate techniques and another for mastering the software usage. Instead of taking various multivariate procedures in SAS one at a time and demonstrating their potential to solve a large number of different problems, we have chosen to discuss various multivariate situations one by one and then identify the most appropriate SAS analyses for them. Many of these analyses may occasionally result from the combined applications of two or more SAS procedures. All multivariate methods are illustrated by appropriate examples. In most cases, the data sets considered are real and are adapted from the published literature from a variety of disciplines.

Prerequisites

A course in applied statistics dealing with the essentials of the (univariate) experimental designs and regression theory and some familiarity with matrix algebra (just enough to interpret the notationally presented statistical models and linear hypotheses) provide an adequate preparation to read this book. Some familiarity with SAS programming (the DATA step and the basic rules of the SAS language) will also be helpful. See the References for a list of SAS documentation.

Overview of Chapters

Chapter 1 provides a summary of important multivariate results. In Chapter 2, various graphical methods for the exploratory multivariate analysis are presented. In Chapter 3, a brief review of the theory of multivariate regression models is provided, which is followed by a number of applications. Chapter 4 deals with the analysis of experimental data. Since the underlying theory, though a bit more complex, is essentially parallel to that presented in Chapter 3, we have largely confined our discussion here to modeling and applications in a variety of experimental designs.

Chapter 5, "Analysis of Repeated Measures Data" and Chapter 6, "Analysis of Repeated Measures Using Mixed Models," occupy a relatively larger space than other chapters in the book. This emphasis requires some further explanation. The repeated measures data are multivariate in nature but are often analyzed using some of the univariate techniques. Both the univariate and multivariate approaches have their own advantages and shortcomings and both are important in their own rights. Both of these approaches are discussed in these chapters. Complexity of models is inherent in the repeated measures data; variety in terms of models is plentiful, and many of these models are commonly used in different disciplines. As a result, we have decided to provide a careful systematic discussion of some of the most commonly used models with an appropriate explanation of the analyses performed by various SAS procedures. However, our coverage, though extensive, is still by no means exhaustive.

The book also contains two appendices. The first of these contains some of the commonly needed and useful multivariate matrix manipulation statements from the SAS IML procedure. It is included so that researchers who wish to perform some further nonstandard analyses of the data should be able to do so with minimal effort using PROC IML. Of course, no attempt is made to be exhaustive, and we readily admit that our selection of items here is purely due to our personal preference and our own exposure and experience with similar analyses. The second appendix contains all the data sets used in the book but not included as part of the corresponding SAS codes due to their large sizes.

Several errors of the first edition have been fixed in the second edition. However, in a work of this size, integrating various aspects of statistical methods and data analysis, there are bound to be some errors and gaps which we may have unintentionally introduced while adding the new material. We will greatly appreciate any comments, suggestions or criticisms which will help us improve this work further.

Acknowledgments

A number of people have contributed to this project in a number of ways. Our sincere thanks are due to outside peer reviewers and internal reviewers at SAS Institute who worked on either of the two editions and to Mr. Jim Ashton (SAS Institute Inc.), Professor Naveen Bansal (Marquette University), Professor Robert Ling (Clemson University) Professor Kenneth Portier (University of Florida), Professor David Scott (Rice University) for critically reading the manuscript for the second edition and for their helpful suggestions. Whole or parts of the manuscript were initially read by Professors A. M. Kshirsagar (University of Michigan), T. K. Nayak (George Washington University), S. D. Peddada (University of Virginia), and N. H. Timm (University of Pittsburgh). We greatly acknowledge their interest in this project. Special thanks are due to Professor Kshirsagar for a number of discussion sessions at various occasions. We thank Professor Robert Kushler (Oakland University) for many helpful criticisms while revising the book and Professor Hans-Peter Piepho (Universitaet Kassel) and Professor Choudary Hanumara (University of Rhode Island) for pointing out certain errors. We also thank Professor N. R. Chaganty (Old Dominion University) for many helpful discussions. Our students Ms. Shobha Prabhala and Ms. Karen Meldrum read parts of an earlier draft of the manuscript. Ms. Raja Vishnubhotla, a local SAS expert at Oakland University, answered our numerous inquiries about SAS. Parts of the book were typed by Ms. Kathy Jegla, Ms. Barbara Jeffrey, and Ms. Sujatha Naik. We kindly thank them for their assistance.

People at SAS Institute, especially Mr. David Baggett, Ms. Caroline Brickley, Ms. Julie Platt, and Ms. Judy Whatley, were most helpful and generous with their suggestions and time. We very much appreciate their understanding and their willingness to extend many of the deadlines.

We would also like to acknowledge the Oakland University Research Foundation for partially supporting some of the travel of R. Khattree and the Old Dominion University for approving the sabbatical of D. N. Naik during Fall 1994 while working on the first edition. We also thank our two departments in the respective universities for playing the host during our numerous visits to each other's institutions.

Last, but not least, our sincere thanks go to our wives, Nidhi and Sujatha, and our little ones Vaidehee and Navin for allowing us to work during late, odd hours and over the weekends. We thank them for their understanding that this book could not have been completed just by working during regular hours.

R. KHATTREE

Rochester, Michigan

D. N. NAIK

Norfolk, Virginia

Commonly Used Notation

$\mathbf{I}_n$	The n by n identity matrix
$\mathbf{1}_n$	The n by 1 vector of unit elements
$\mathbf{O}$	A matrix of appropriate order with all zero entries
λ_i	The ith largest eigenvalue of the matrix under consideration
$|\mathbf{A}|$	The determinant of the square matrix $\mathbf{A}$
$tr(\mathbf{A})$	The trace of the square matrix $\mathbf{A}$
$\mathbf{A}^{-1}$	The inverse of the matrix $\mathbf{A}$
$\mathbf{A}^{1/2}$	The symmetric square root of the matrix $\mathbf{A}$
$\mathbf{A}^-$	A generalized inverse of the matrix $\mathbf{A}$
$E(\mathbf{y})$	Expected value of a random variable or vector $\mathbf{y}$
$v(y), var(y)$	Variance of a random variable y
$cov(\mathbf{x}, \mathbf{y})$	Covariance *of* random variable (vector) $\mathbf{x}$ *with* random variable (or vector) $\mathbf{y}$
$D(\mathbf{y})$	The variance covariance or the dispersion matrix of $\mathbf{y}$
$N_p(\boldsymbol{\mu}, \boldsymbol{\Sigma})$	A p-dimensional normal distribution with mean $\boldsymbol{\mu}$ and the variance covariance matrix $\boldsymbol{\Sigma}$
$W_p(f, \boldsymbol{\Sigma})$	A p-(matrix) variate Wishart distribution with f degrees of freedom and parameter $\boldsymbol{\Sigma}$ (that is, with expected value $f\boldsymbol{\Sigma}$)
$\boldsymbol{\epsilon}$	Error vector
$\mathcal{E}$	Error matrix
$\mathbf{Y}$	n by p matrix of data on dependent variables
$\mathbf{X}$	Regression/Design matrix in the linear model
$\boldsymbol{\beta}$	Regression/Design parameter vector
$\mathbf{B}$	Regression/Design parameter matrix
$\boldsymbol{\Sigma}$	(usually) The Dispersion matrix of errors
df	Degrees of freedom
SS&CP Matrix	Matrix of the sums of squares and crossproducts
$\mathbf{E}$	Error SS&CP matrix
$\mathbf{H}$	Hypothesis SS&CP matrix
$\bar{\mathbf{y}}$	The sample mean vector
$\mathbf{S}$	Sample dispersion matrix (with df as denominator)
$\mathbf{S}_n$	Sample dispersion matrix (with sample size as denominator)
$\mathbf{P}$	The Projection or Hat matrix
T^2	Hotelling's T^2

Λ	Wilks' Lambda
$\beta_{1,p}$	Coefficient of multivariate skewness
$\beta_{2,p}$	Coefficient of multivariate kurtosis
$\otimes$	Kronecker product
AIC	Akaike's information criterion
BIC	Swartz's Bayesian information criterion

Multivariate Analysis Concepts

Chapter
1

1.1 Introduction   1
1.2 Random Vectors, Means, Variances, and Covariances   2
1.3 Multivariate Normal Distribution   5
1.4 Sampling from Multivariate Normal Populations   6
1.5 Some Important Sample Statistics and Their Distributions   8
1.6 Tests for Multivariate Normality   9
1.7 Random Vector and Matrix Generation   17

1.1 Introduction

The subject of multivariate analysis deals with the statistical analysis of the data collected on more than one (response) variable. These variables may be correlated with each other, and their statistical dependence is often taken into account when analyzing such data. In fact, this consideration of statistical dependence makes multivariate analysis somewhat different in approach and considerably more complex than the corresponding univariate analysis, when there is only one response variable under consideration.

Response variables under consideration are often described as random variables and since their dependence is one of the things to be accounted for in the analyses, these response variables are often described by their joint probability distribution. This consideration makes the modeling issue relatively manageable and provides a convenient framework for scientific analysis of the data. Multivariate normal distribution is one of the most frequently made distributional assumptions for the analysis of multivariate data. However, if possible, any such consideration should ideally be dictated by the particular context. Also, in many cases, such as when the data are collected on a nominal or ordinal scales, multivariate normality may not be an appropriate or even viable assumption.

In the real world, most data collection schemes or designed experiments will result in multivariate data. A few examples of such situations are given below.

- During a survey of households, several measurements on *each* household are taken. These measurements, being taken on the same household, will be dependent. For example, the education level of the head of the household and the annual income of the family are related.

- During a production process, a number of different measurements such as the tensile strength, brittleness, diameter, etc. are taken on the same unit. Collectively such data are viewed as multivariate data.

- On a sample of 100 cars, various measurements such as the average gas mileage, number of major repairs, noise level, etc. are taken. Also each car is followed for the first 50,000 miles and these measurements are taken after every 10,000 miles. Measurements taken on the same car at the same mileage and those taken at different mileage are going to be correlated. In fact, these data represent a very complex multivariate analysis problem.

- An engineer wishes to set up a control chart to identify the instances when the production process may have gone out of control. Since an out of control process may produce an excessively large number of out of specification items, detection at an early stage is important. In order to do so, she may wish to monitor several process characteristics on the same units. However, since these characteristics are functions of process parameters (conditions), they are likely to be correlated leading to a set of multivariate data. Thus many times, it is appropriate to set up a single (or only a few) multivariate control chart(s) to detect the occurrence of any out of control conditions. On the other hand, if several univariate control charts are separately set up and individually monitored, one may witness too many false alarms, which is clearly an undesirable situation.

- A new drug is to be compared with a control for its effectiveness. Two different groups of patients are assigned to each of the two treatments and they are observed weekly for next two months. The periodic measurements on the same patient will exhibit dependence and thus the basic problem is multivariate in nature. Additionally, if the measurements on various possible side-effects of the drugs are also considered, the subsequent analysis will have to be done under several carefully chosen models.

- In a designed experiment conducted in a research and development center, various factors are set up at desired levels and a number of response variables are measured for each of these treatment combinations. The problem is to find a combination of the levels of these factors where all the responses are at their 'optimum'. Since a treatment combination which optimizes one response variable may not result in the optimum for the other response variable, one has a problem of conflicting objectives especially when the problem is treated as collection of several univariate optimization problems. Due to dependence among responses, it may be more meaningful to analyze response variables simultaneously.

- In many situations, it is more economical to collect a large number of measurements on the same unit but such measurements are made only on a few units. Such a situation is quite common in many remote sensing data collection plans. Obviously, it is practically impossible to collectively interpret hundreds of univariate analyses to come up with some definite conclusions. A better approach may be that of data reduction by using some meaningful approach. One may eliminate some of the variables which are deemed redundant in the presence of others. Better yet, one may eliminate some of the linear combinations of all variables which contain little or no information and then concentrate only on a few important ones. Which linear combinations of the variables should be retained can be decided using certain multivariate methods such as principal component analysis. Such methods are not discussed in this book, however.

Most of the problems stated above require (at least for the convenience of modeling and for performing statistical tests) the assumption of multivariate normality. There are however, several other aspects of multivariate analysis such as factor analysis, cluster analysis, etc. which are largely distribution free in nature. In this volume, we will only consider the problems of the former class, where multivariate normality assumption may be needed. Therefore, in the next few sections, we will briefly review the theory of multivariate normal and other related distributions. This theory is essential for a proper understanding of various multivariate statistical techniques, notation, and nomenclature. The material presented here is meant to be only a refresher and is far from complete. A more complete discussion of this topic can be found in Kshirsagar (1972), Seber (1984) or Rencher (1995).

1.2 Random Vectors, Means, Variances, and Covariances

Suppose $y_1, \ldots, y_p$ are p possibly correlated random variables with respective means (expected values) $\mu_1, \ldots, \mu_p$. Let us arrange these random variables as a column vector denoted by $\mathbf{y}$, that is, let

$$\mathbf{y} = \begin{bmatrix} y_1 \\ y_2 \\ \vdots \\ y_p \end{bmatrix}.$$

We do the same for $\mu_1, \mu_2, \ldots, \mu_p$ and denote the corresponding vector by $\boldsymbol{\mu}$. Then we say that the vector $\mathbf{y}$ has the mean $\boldsymbol{\mu}$ or in notation $E(\mathbf{y}) = \boldsymbol{\mu}$.

Let us denote the covariance between y_i and y_j by σ_{ij}, $i, j = 1, \ldots, p$, that is

$$\sigma_{ij} = \text{cov}(y_i, y_j) = E[(y_i - \mu_i)(y_j - \mu_j)] = E[(y_i - \mu_i)y_j] = E(y_i y_j) - \mu_i \mu_j$$

and let

$$\boldsymbol{\Sigma} = (\sigma_{ij}) = \begin{bmatrix} \sigma_{11} & \sigma_{12} & \ldots & \sigma_{1p} \\ \sigma_{21} & \sigma_{22} & \ldots & \sigma_{2p} \\ & & & \\ \sigma_{p1} & \sigma_{p2} & \ldots & \sigma_{pp} \end{bmatrix}.$$

Since $\text{cov}(y_i, y_j) = \text{cov}(y_j, y_i)$, we have $\sigma_{ij} = \sigma_{ji}$. Therefore, $\boldsymbol{\Sigma}$ is symmetric with $(i, j)^{th}$ and $(j, i)^{th}$ elements representing the covariance between y_i and y_j. Further, since $\text{var}(y_i) = \text{cov}(y_i, y_i) = \sigma_{ii}$, the ith diagonal place of $\boldsymbol{\Sigma}$ contains the variance of y_i. The matrix $\boldsymbol{\Sigma}$ is called the dispersion or the variance-covariance matrix of $\mathbf{y}$. In notation, we write this fact as $D(\mathbf{y}) = \boldsymbol{\Sigma}$. Various books follow alternative notations for $D(\mathbf{y})$ such as $\text{cov}(\mathbf{y})$ or $\text{var}(\mathbf{y})$. However, we adopt the less ambiguous notation of $D(\mathbf{y})$.

Thus,

$$\boldsymbol{\Sigma} = D(\mathbf{y}) = E[(\mathbf{y} - \boldsymbol{\mu})(\mathbf{y} - \boldsymbol{\mu})'] = E[(\mathbf{y} - \boldsymbol{\mu})\mathbf{y}'] = E(\mathbf{y}\mathbf{y}') - \boldsymbol{\mu}\boldsymbol{\mu}',$$

where for any matrix (vector) $\mathbf{A}$, the notation $\mathbf{A}'$ represents its transpose.

The quantity $tr(\boldsymbol{\Sigma}) = \sum_{i=1}^{p} \sigma_{ii}$ is called *total variance* and a determinant of $\boldsymbol{\Sigma}$, denoted by $|\boldsymbol{\Sigma}|$, is often referred to as the *generalized variance*. The two are often taken as the overall measures of the variability of the random vector $\mathbf{y}$. However, both of these two measures suffer from certain shortcomings. For example, the total variance $tr(\boldsymbol{\Sigma})$ being the sum of only diagonal elements, essentially ignores all covariance terms. On the other hand, the generalized variance $|\boldsymbol{\Sigma}|$ can be misleading since two very different variance covariance structures can sometimes result in the same value of generalized variance. Johnson and Wichern (1998) provide certain interesting illustrations of such situations.

Let $\mathbf{u}_{p \times 1}$ and $\mathbf{z}_{q \times 1}$ be two random vectors, with respective means $\boldsymbol{\mu}_u$ and $\boldsymbol{\mu}_z$. Then the covariance of $\mathbf{u}$ *with* $\mathbf{z}$ is defined as

$$\boldsymbol{\Sigma}_{uz} = \text{cov}(\mathbf{u}, \mathbf{z}) = E[(\mathbf{u} - \boldsymbol{\mu}_u)(\mathbf{z} - \boldsymbol{\mu}_z)'] = E[(\mathbf{u} - \boldsymbol{\mu}_u)\mathbf{z}'] = E(\mathbf{u}\mathbf{z}') - \boldsymbol{\mu}_u \boldsymbol{\mu}_z'.$$

Note that as matrices, the p by q matrix $\boldsymbol{\Sigma}_{uz} = \text{cov}(\mathbf{u}, \mathbf{z})$ is *not* the same as the q by p matrix $\boldsymbol{\Sigma}_{zu} = \text{cov}(\mathbf{z}, \mathbf{u})$, the covariance of $\mathbf{z}$ *with* $\mathbf{u}$. They are, however, related in that

$$\boldsymbol{\Sigma}_{uz} = \boldsymbol{\Sigma}_{zu}'.$$

Notice that for a vector $\mathbf{y}$, $\text{cov}(\mathbf{y}, \mathbf{y}) = D(\mathbf{y})$. Thus, when there is no possibility of confusion, we interchangeably use $D(\mathbf{y})$ and $\text{cov}(\mathbf{y})(= \text{cov}(\mathbf{y}, \mathbf{y}))$ to represent the variance-covariance matrix of $\mathbf{y}$.

A variance-covariance matrix is always positive semidefinite (that is, all its eigenvalues are nonnegative). However, in most of the discussion in this text we encounter dispersion matrices which are positive definite, a condition stronger than positive semidefiniteness in that all eigenvalues are strictly positive. Consequently, such dispersion matrices would also admit an inverse. In the subsequent discussion, we assume our dispersion matrix to be positive definite.

Let us partition the vector $\mathbf{y}$ into two subvectors as

$$\mathbf{y} = \begin{bmatrix} \mathbf{y}_{1_{p_1 \times 1}} \\ \mathbf{y}_{2_{(p-p_1) \times 1}} \end{bmatrix}$$

and partition $\boldsymbol{\Sigma}$ as

$$\boldsymbol{\Sigma} = \begin{bmatrix} \boldsymbol{\Sigma}_{11_{p_1 \times p_1}} & \boldsymbol{\Sigma}_{12_{p_1 \times (p-p_1)}} \\ \boldsymbol{\Sigma}_{21_{(p-p_1) \times p_1}} & \boldsymbol{\Sigma}_{22_{(p-p_1) \times (p-p_1)}} \end{bmatrix}.$$

Then, $E(\mathbf{y}_1) = \boldsymbol{\mu}_1$, $E(\mathbf{y}_2) = \boldsymbol{\mu}_2$, $D(\mathbf{y}_1) = \boldsymbol{\Sigma}_{11}$, $D(\mathbf{y}_2) = \boldsymbol{\Sigma}_{22}$, $\text{cov}(\mathbf{y}_1, \mathbf{y}_2) = \boldsymbol{\Sigma}_{12}$, $\text{cov}(\mathbf{y}_2, \mathbf{y}_1) = \boldsymbol{\Sigma}_{21}$. We also observe that $\boldsymbol{\Sigma}_{12} = \boldsymbol{\Sigma}'_{21}$.

The Pearson's *correlation coefficient* between y_i and y_j, denoted by ρ_{ij}, is defined by

$$\rho_{ij} = \frac{\text{cov}(y_i, y_j)}{\sqrt{\text{var}(y_i)\,\text{var}(y_j)}} = \frac{\sigma_{ij}}{\sqrt{\sigma_{ii}\sigma_{jj}}},$$

and accordingly, we define the *correlation coefficient matrix* of $\mathbf{y}$ as

$$\mathbf{R} = \begin{bmatrix} \rho_{11} & \rho_{12} & \cdots & \rho_{1p} \\ \rho_{21} & \rho_{22} & \cdots & \rho_{2p} \\ & & & \\ \rho_{p1} & \rho_{p2} & \cdots & \rho_{pp} \end{bmatrix}$$

It is easy to verify that the correlation coefficient matrix $\mathbf{R}$ is a symmetric positive definite matrix in which all the diagonal elements are unity. The matrix $\mathbf{R}$ can be written, in terms of matrix $\boldsymbol{\Sigma}$, as

$$\mathbf{R} = [\text{diag}\,(\boldsymbol{\Sigma})]^{-1/2}\,\boldsymbol{\Sigma}\,[\text{diag}\,(\boldsymbol{\Sigma})]^{-1/2},$$

where $\text{diag}\,(\boldsymbol{\Sigma})$ is the diagonal matrix obtained by retaining the diagonal elements of $\boldsymbol{\Sigma}$ and by replacing all the nondiagonal elements by zero. Further, the *square root* of any matrix $\mathbf{A}$, denoted by $\mathbf{A}^{\frac{1}{2}}$, is a symmetric matrix satisfying the condition, $\mathbf{A} = \mathbf{A}^{\frac{1}{2}}\mathbf{A}^{\frac{1}{2}}$.

The probability distribution (density) of a vector $\mathbf{y}$, denoted by $f(\mathbf{y})$, is the same as the joint probability distribution of $y_1, \ldots, y_p$. The marginal distribution $f_1(\mathbf{y}_1)$ of $\mathbf{y}_1 = (y_1, \ldots, y_{p_1})'$, a subvector of $\mathbf{y}$, is obtained by integrating out $\mathbf{y}_2 = (y_{p_1+1}, \ldots, y_p)'$ from the density $f(\mathbf{y})$. The conditional distribution of $\mathbf{y}_2$, when $\mathbf{y}_1$ has been held fixed, is denoted by $g(\mathbf{y}_2|\mathbf{y}_1)$ and is given by

$$g(\mathbf{y}_2|\mathbf{y}_1) = f(\mathbf{y})/f_1(\mathbf{y}_1).$$

An important concept arising from conditional distribution is the *partial correlation coefficient*. If we partition $\mathbf{y}$ as $(\mathbf{y}'_1, \mathbf{y}'_2)'$ where $\mathbf{y}_1$ is a p_1 by 1 vector and $\mathbf{y}_2$ is a $(p - p_1)$ by 1 vector, then the partial correlation coefficient between two components of $\mathbf{y}_1$, say y_i and y_j, is defined as the Pearson's correlation coefficient between y_i and y_j conditional on $\mathbf{y}_2$ (that is, for a given $\mathbf{y}_2$). If $\boldsymbol{\Sigma}_{11\cdot2} = (a_{ij})$ is the p_1 by p_1 variance-covariance matrix of $\mathbf{y}_1$ given $\mathbf{y}_2$, then the population partial correlation coefficient between y_i and y_j, $i, j = 1, \ldots, p_1$ is given by

$$\rho_{ij \cdot p_1+1,\ldots,p} = a_{ij}/\sqrt{a_{ii}a_{jj}}.$$

The matrix of all partial correlation coefficients $\rho_{ij,p_1+1,\ldots,p}$, $i, j = 1, \ldots, p_1$ is denoted by $\mathbf{R}_{11\cdot2}$. More simply, using the matrix notations, $\mathbf{R}_{11\cdot2}$ can be computed as

$$[\text{diag}\,(\boldsymbol{\Sigma}_{11.2})]^{-\frac{1}{2}}\,\boldsymbol{\Sigma}_{11.2}\,[\text{diag}\,(\boldsymbol{\Sigma}_{11.2})]^{-\frac{1}{2}},$$

where $\text{diag}\,(\boldsymbol{\Sigma}_{11.2})$ is a diagonal matrix with respective diagonal entries the same as those in $\boldsymbol{\Sigma}_{11.2}$.

Many times it is of interest to find the correlation coefficients between y_i and y_j, $i, j = 1, \ldots, p$, conditional on all $y_k, k = 1, \ldots, p, k \neq i, k \neq j$. In this case, the partial correlation between y_i and y_j can be interpreted as the strength of correlation between the two variables after eliminating the effects of all the remaining variables.

In many linear model situations, we would like to examine the overall association of a set of variables with a given variable. This is often done by finding the correlation between the variable and a particular linear combination of other variables. The *Multiple correlation coefficient* is an index measuring the association between a random variable y_1 and the set of remaining variables represented by a $(p - 1)$ by 1 vector $\mathbf{y}_2$. It is defined as the maximum correlation between y_1 and $\mathbf{c}'\mathbf{y}_2$, a linear combination of $\mathbf{y}_2$, where the maximum is taken over all possible nonzero vectors $\mathbf{c}$. This maximum value, representing the multiple correlation coefficient between y_1 and $\mathbf{y}_2$, is given by

$$\left(\mathbf{\Sigma}_{12} \mathbf{\Sigma}_{22}^{-1} \mathbf{\Sigma}_{21} \right)^{\frac{1}{2}} / \Sigma_{11}^{\frac{1}{2}}$$

where

$$D \begin{bmatrix} y_1 \\ \mathbf{y}_2 \end{bmatrix} = \begin{bmatrix} \mathbf{\Sigma}_{11} & \mathbf{\Sigma}_{12} \\ \mathbf{\Sigma}_{21} & \mathbf{\Sigma}_{22} \end{bmatrix},$$

and the maximum is attained for the choice $\mathbf{c} = \mathbf{\Sigma}_{22}^{-1} \mathbf{\Sigma}_{21}$. The multiple correlation coefficient always lies between zero and one. The square of the multiple correlation coefficient, often referred to as the *population coefficient of determination*, is generally used to indicate the power of prediction or the effect of regression.

The concept of multiple correlation can be extended to the case in which the random variable y_1 is replaced by a random vector. This leads to what are called *canonical correlation coefficients*.

1.3 Multivariate Normal Distribution

A probability distribution that plays a pivotal role in multivariate analysis is *multivariate normal distribution*. We say that $\mathbf{y}$ has a multivariate normal distribution (with a mean vector $\boldsymbol{\mu}$ and the variance-covariance matrix $\mathbf{\Sigma}$) if its density is given by

$$f(\mathbf{y}) = \frac{1}{(2\pi)^{p/2}|\mathbf{\Sigma}|^{1/2}} \cdot \exp\left(-\frac{1}{2}(\mathbf{y} - \boldsymbol{\mu})'\mathbf{\Sigma}^{-1}(\mathbf{y} - \boldsymbol{\mu}) \right).$$

In notation, we state this fact as $\mathbf{y} \sim N_p(\boldsymbol{\mu}, \mathbf{\Sigma})$. Observe that the above density is a straightforward extension of the univariate normal density to which it will reduce when $p = 1$.

Important properties of the multivariate normal distribution include some of the following:

- Let $\mathbf{A}_{r \times p}$ be a fixed matrix, then $\mathbf{Ay} \sim N_r(\mathbf{A}\boldsymbol{\mu}, \mathbf{A}\mathbf{\Sigma}\mathbf{A}')(r \leq p)$. It may be added that $\mathbf{Ay}$ will admit the density if $\mathbf{A}\mathbf{\Sigma}\mathbf{A}'$ is nonsingular, which will happen if and only if all rows of $\mathbf{A}$ are linearly independent. Further, in principle, r can also be greater than p. However, in that case, the matrix $\mathbf{A}\mathbf{\Sigma}\mathbf{A}'$ will not be nonsingular. Consequently, the vector $\mathbf{Ay}$ will not admit a density function.
- Let $\mathbf{G}$ be such that $\mathbf{\Sigma}^{-1} = \mathbf{GG}'$, then $\mathbf{G}'\mathbf{y} \sim N_p(\mathbf{G}'\boldsymbol{\mu}, I)$ and $\mathbf{G}'(\mathbf{y} - \boldsymbol{\mu}) \sim N_p(\mathbf{0}, I)$.
- Any fixed linear combination of $y_1, \ldots, y_p$, say $\mathbf{c}'\mathbf{y}, \mathbf{c}_{p \times 1} \neq \mathbf{0}$ is also normally distributed. Specifically, $\mathbf{c}'\mathbf{y} \sim N_1(\mathbf{c}'\boldsymbol{\mu}, \mathbf{c}'\mathbf{\Sigma}\mathbf{c})$.
- The subvectors $\mathbf{y}_1$ and $\mathbf{y}_2$ are also normally distributed, specifically, $\mathbf{y}_1 \sim N_{p_1}(\boldsymbol{\mu}_1, \mathbf{\Sigma}_{11})$ and $\mathbf{y}_2 \sim N_{p-p_1}(\boldsymbol{\mu}_2, \mathbf{\Sigma}_{22})$.

- Individual components $y_1, \ldots, y_p$ are all normally distributed. That is, $y_i \sim N_1(\mu_i, \sigma_{ii})$, $i = 1, \ldots, p$.

- The conditional distribution of $\mathbf{y}_1$ given $\mathbf{y}_2$, written as $\mathbf{y}_1|\mathbf{y}_2$, is also normal. Specifically,

$$\mathbf{y}_1|\mathbf{y}_2 \sim N_{p_1}\left(\boldsymbol{\mu}_1 + \boldsymbol{\Sigma}_{12}\boldsymbol{\Sigma}_{22}^{-1}(\mathbf{y}_2 - \boldsymbol{\mu}_2),\ \boldsymbol{\Sigma}_{11} - \boldsymbol{\Sigma}_{12}\boldsymbol{\Sigma}_{22}^{-1}\boldsymbol{\Sigma}_{21}\right).$$

Let $\boldsymbol{\mu}_1 + \boldsymbol{\Sigma}_{12}\boldsymbol{\Sigma}_{22}^{-1}(\mathbf{y}_2 - \boldsymbol{\mu}_2) = \boldsymbol{\mu}_1 - \boldsymbol{\Sigma}_{12}\boldsymbol{\Sigma}_{22}^{-1}\boldsymbol{\mu}_2 + \boldsymbol{\Sigma}_{12}\boldsymbol{\Sigma}_{22}^{-1}\mathbf{y}_2 = \mathbf{B}_0 + \mathbf{B}_1\mathbf{y}_2$, and $\boldsymbol{\Sigma}_{11.2} = \boldsymbol{\Sigma}_{11} - \boldsymbol{\Sigma}_{12}\boldsymbol{\Sigma}_{22}^{-1}\boldsymbol{\Sigma}_{21}$. The conditional expectation of $\mathbf{y}_1$ for given values of $\mathbf{y}_2$ or the regression function of $\mathbf{y}_1$ on $\mathbf{y}_2$ is $\mathbf{B}_0 + \mathbf{B}_1\mathbf{y}_2$, which is linear in $\mathbf{y}_2$. This is a key fact for multivariate multiple linear regression modeling. The matrix $\boldsymbol{\Sigma}_{11.2}$ is usually represented by the variance-covariance matrix of error components in these models. An analogous result (and the interpretation) can be stated for the conditional distribution of $\mathbf{y}_2$ given $\mathbf{y}_1$.

- Let $\boldsymbol{\delta}$ be a fixed $p \times 1$ vector, then

$$\mathbf{y} + \boldsymbol{\delta} \sim N_p(\boldsymbol{\mu} + \boldsymbol{\delta}, \boldsymbol{\Sigma}).$$

- The random components $y_1, \ldots, y_p$ are all independent if and only if $\boldsymbol{\Sigma}$ is a diagonal matrix; that is, when all the covariances (or correlations) are zero.

- Let $\mathbf{u}_1$ and $\mathbf{u}_2$ be respectively distributed as $N_p(\boldsymbol{\mu}_{u_1}, \boldsymbol{\Sigma}_{u_1})$ and $N_p(\boldsymbol{\mu}_{u_2}, \boldsymbol{\Sigma}_{u_2})$, then

$$\mathbf{u}_1 \pm \mathbf{u}_2 \sim N_p(\boldsymbol{\mu}_{u_1} \pm \boldsymbol{\mu}_{u_2}, \boldsymbol{\Sigma}_{u_1} + \boldsymbol{\Sigma}_{u_2} \pm (\text{cov}(\mathbf{u}_1, \mathbf{u}_2) + \text{cov}(\mathbf{u}_2, \mathbf{u}_1))).$$

Note that if $\mathbf{u}_1$ and $\mathbf{u}_2$ were independent, the last two covariance terms would drop out.

There is a vast amount of literature available on multivariate normal distribution, its properties, and the evaluations of multivariate normal probabilities. See Kshirsagar (1972), Rao (1973), and Tong (1990) among many others for further details.

1.4   Sampling from Multivariate Normal Populations

Suppose we have a random sample of size n, say $\mathbf{y}_1, \ldots, \mathbf{y}_n$, from the p dimensional multivariate normal population $N_p(\boldsymbol{\mu}, \boldsymbol{\Sigma})$. Since $\mathbf{y}_1, \ldots, \mathbf{y}_n$ are independently and identically distributed (*iid*), their sample mean

$$\bar{\mathbf{y}} = \frac{1}{n}[\mathbf{y}_1 + \cdots + \mathbf{y}_n] = \frac{1}{n}\sum_{i=1}^{n}\mathbf{y}_i \tag{1.1}$$

is also normally distributed as $N_p(\boldsymbol{\mu}, \boldsymbol{\Sigma}/n)$. Thus, $\bar{\mathbf{y}}$ is an unbiased estimator of $\boldsymbol{\mu}$. Also, observe that $\bar{\mathbf{y}}$ has a dispersion matrix which is a $\frac{1}{n}$ multiple of the original population variance-covariance matrix. These results are straightforward generalizations of the corresponding well known univariate results.

The sample variance of the univariate normal theory is generalized to the sample variance-covariance matrix in the multivariate context. Accordingly, the chi-square distribution is generalized to a matrix distribution known as the *Wishart distribution*.

The p by p sample variance-covariance matrix is obtained as

$$\mathbf{S} = \frac{1}{n-1}\sum_{i=1}^{n}(\mathbf{y}_i - \bar{\mathbf{y}})(\mathbf{y}_i - \bar{\mathbf{y}})' = \frac{1}{n-1}\left\{\sum_{i=1}^{n}\mathbf{y}_i\mathbf{y}_i' - n\bar{\mathbf{y}}\bar{\mathbf{y}}'\right\}. \tag{1.2}$$

The matrix $\mathbf{S}$ is an unbiased estimator of $\boldsymbol{\Sigma}$. Note that $\mathbf{S}$ is a p by p symmetric matrix. Thus, it contains only $\frac{p(p+1)}{2}$ different random variables.

Let

$$
\mathbf{Y} = \begin{bmatrix} \mathbf{y}_1' \\ \mathbf{y}_2' \\ \vdots \\ \mathbf{y}_n' \end{bmatrix}
$$

be the n by p data matrix obtained by stacking $\mathbf{y}_1', \ldots, \mathbf{y}_n'$ one atop the other. Let $\mathbf{I}_n$ stand for an n by n identity matrix and $\mathbf{1}_n$ be an n by 1 column vector with all elements as 1. Then, in terms of $\mathbf{Y}$, the sample mean $\bar{\mathbf{y}}$ can be written as

$$
\bar{\mathbf{y}} = \frac{1}{n}\mathbf{Y}'\mathbf{1}_n
$$

and the sample variance-covariance matrix can be written as

$$
\mathbf{S} = \frac{1}{n-1}\left\{ \mathbf{Y}'\left(\mathbf{I}_n - \frac{1}{n}\mathbf{1}_n\mathbf{1}_n' \right)\mathbf{Y} \right\} = \frac{1}{n-1}\left\{ \mathbf{Y}'\mathbf{Y} - \frac{1}{n}\mathbf{Y}'\mathbf{1}_n\mathbf{1}_n'\mathbf{Y} \right\} = \frac{1}{n-1}\{\mathbf{Y}'\mathbf{Y} - n\bar{\mathbf{y}}\bar{\mathbf{y}}'\}.
$$

It is known that $(n-1)\mathbf{S}$ follows a p-(matrix) variate Wishart distribution with $(n-1)$ degrees of freedom and expectation $(n-1)\boldsymbol{\Sigma}$. We denote this as $(n-1)\mathbf{S} \sim W_p(n-1, \boldsymbol{\Sigma})$. Also, $\mathbf{S}$ is an unbiased estimator of $\boldsymbol{\Sigma}$ (as mentioned earlier, this is always true regardless of the underlying multivariate normality assumption and consequently, without any specific reference to the Wishart distribution).

Since $(n-1)\mathbf{S}$ has a Wishart distribution, the sample variance-covariance matrix $\mathbf{S}$ possesses certain other important properties. Many of these properties are used to obtain the distributions of various estimators and test statistics. Some of these properties are listed as follows.

- $(n-1)s_{ii}/\sigma_{ii} \sim \chi^2(n-1), i = 1, \ldots, p.$
- Let

$$
\mathbf{S} = \begin{bmatrix} \mathbf{S}_{11} & \mathbf{S}_{12} \\ \mathbf{S}_{21} & \mathbf{S}_{22} \end{bmatrix}, \quad \boldsymbol{\Sigma} = \begin{bmatrix} \boldsymbol{\Sigma}_{11} & \boldsymbol{\Sigma}_{12} \\ \boldsymbol{\Sigma}_{21} & \boldsymbol{\Sigma}_{22} \end{bmatrix},
$$

$\mathbf{S}_{11.2} = \mathbf{S}_{11} - \mathbf{S}_{12}\mathbf{S}_{22}^{-1}\mathbf{S}_{21}, \boldsymbol{\Sigma}_{11.2} = \boldsymbol{\Sigma}_{11} - \boldsymbol{\Sigma}_{12}\boldsymbol{\Sigma}_{22}^{-1}\boldsymbol{\Sigma}_{21}, \mathbf{S}_{22.1} = \mathbf{S}_{22} - \mathbf{S}_{21}\mathbf{S}_{11}^{-1}\mathbf{S}_{12}$ and $\boldsymbol{\Sigma}_{22.1} = \boldsymbol{\Sigma}_{22} - \boldsymbol{\Sigma}_{21}\boldsymbol{\Sigma}_{11}^{-1}\boldsymbol{\Sigma}_{12}$, then
  (a) $(n-1)\mathbf{S}_{11} \sim W_{p_1}((n-1), \boldsymbol{\Sigma}_{11}).$
  (b) $(n-1)\mathbf{S}_{22} \sim W_{p_2}((n-1), \boldsymbol{\Sigma}_{22}).$
  (c) $(n-1)\mathbf{S}_{11.2} \sim W_{p_1}((n-p+p_1-1), \boldsymbol{\Sigma}_{11.2}).$
  (d) $(n-1)\mathbf{S}_{22.1} \sim W_{p_2}((n-p_1-1), \boldsymbol{\Sigma}_{22.1}).$
  (e) $\mathbf{S}_{11}$ and $\mathbf{S}_{22.1}$ are independently distributed.
  (f) $\mathbf{S}_{22}$ and $\mathbf{S}_{11.2}$ are independently distributed.
- Let s^{ii} and σ^{ii} be the ith diagonal elements of $\mathbf{S}^{-1}$ and $\boldsymbol{\Sigma}^{-1}$ respectively, then $(n-1)\sigma^{ii}/s^{ii} \sim \chi^2(n-p).$
- Let $\mathbf{c} \neq \mathbf{0}$ be an arbitrary but fixed vector, then

$$
(n-1)\frac{\mathbf{c}'\mathbf{S}\mathbf{c}}{\mathbf{c}'\boldsymbol{\Sigma}\mathbf{c}} \sim \chi^2(n-1),
$$

and $(n-1)\frac{\mathbf{c}'\boldsymbol{\Sigma}^{-1}\mathbf{c}}{\mathbf{c}'\mathbf{S}^{-1}\mathbf{c}} \sim \chi^2(n-p).$

- Let $\mathbf{H}$ be an arbitrary but fixed $k \times p$ matrix ($k \leq p$), then

$$
(n-1)\mathbf{H}\mathbf{S}\mathbf{H}' \sim W_k(n-1, \mathbf{H}\boldsymbol{\Sigma}\mathbf{H}').
$$

In principle, k can also be greater than p but in such a case, the matrix $(n-1)\mathbf{H}\mathbf{S}\mathbf{H}'$ does not admit a probability density.

As a consequence of the above result, if we take $k = p$ and $\mathbf{H} = \mathbf{G}'$ where $\mathbf{\Sigma}^{-1} = \mathbf{GG}'$, then $(n-1)\mathbf{S}^* = (n-1)\mathbf{G}'\mathbf{SG} \sim W_p(n-1, \mathbf{I})$.

In the above discussion, we observed that the Wishart distribution arises naturally in the multivariate normal theory as the distribution of the sample variance-covariance matrix (of course, apart from a scaling by $(n-1)$). Another distribution which is closely related to the Wishart distribution and is useful in various associated hypothesis testing problems is the matrix variate Beta (Type 1) distribution. For example, if $\mathbf{A}_1$ and $\mathbf{A}_2$ are two independent random matrices with $\mathbf{A}_1 \sim W_p(n_1 - 1, \mathbf{\Sigma})$ and $\mathbf{A}_2 \sim W_p(n_2 - 1, \mathbf{\Sigma})$, then $\mathbf{B} = (\mathbf{A}_1 + \mathbf{A}_2)^{-\frac{1}{2}} \mathbf{A}_1 (\mathbf{A}_1 + \mathbf{A}_2)^{-\frac{1}{2}}$ follows a matrix variate Beta Type 1 distribution, denoted by $B_p(\frac{n_1-1}{2}, \frac{n_2-1}{2}, \text{Type 1})$. Similarly, $\mathbf{B}^* = \mathbf{A}_2^{-\frac{1}{2}} \mathbf{A}_1 \mathbf{A}_2^{-\frac{1}{2}}$ follows $B_p(\frac{n_1-1}{2}, \frac{n_2-1}{2}, \text{Type 2})$, a matrix variate Beta Type 2 (or a matrix variate F apart from a constant) distribution. The matrices $\mathbf{A}_2^{-\frac{1}{2}}$ and $(\mathbf{A}_1 + \mathbf{A}_2)^{-\frac{1}{2}}$ respectively are the symmetric "square root" matrices of $\mathbf{A}_2^{-1}$ and $(\mathbf{A}_1 + \mathbf{A}_2)^{-1}$ in the sense that $\mathbf{A}_2^{-1} = (\mathbf{A}_2)^{-\frac{1}{2}}(\mathbf{A}_2)^{-\frac{1}{2}}$ and $(\mathbf{A}_1 + \mathbf{A}_2)^{-1} = (\mathbf{A}_1 + \mathbf{A}_2)^{-\frac{1}{2}}(\mathbf{A}_1 + \mathbf{A}_2)^{-\frac{1}{2}}$. The eigenvalues of the matrices $\mathbf{B}$ and $\mathbf{B}^*$ appear in the expressions of various test statistics used in hypothesis testing problems in multivariate analysis of variance.

Another important fact about the sample mean $\bar{\mathbf{y}}$ and the sample variance-covariance matrix $\mathbf{S}$ is that they are statistically independent under the multivariate normal sampling theory. This fact plays an important role in constructing test statistics for certain statistical hypotheses. For details, see Kshirsagar (1972), Timm (1975), or Muirhead (1982).

1.5 Some Important Sample Statistics and Their Distributions

We have already encountered two important sample statistics in the previous section, namely the sample mean vector $\bar{\mathbf{y}}$ in Equation 1.1 and the sample variance-covariance matrix $\mathbf{S}$ in Equation 1.2. These quantities play a pivotal role in defining the test statistics useful in various hypothesis testing problems. The underlying assumption of multivariate normal population is crucial in obtaining the distribution of these test statistics. Therefore, we will assume that the sample $\mathbf{y}_1, \ldots, \mathbf{y}_n$ of size n is obtained from a multivariate population $N_p(\mu, \mathbf{\Sigma})$.

As we have already indicated, $\bar{\mathbf{y}} \sim N_p(\mu, \mathbf{\Sigma}/n)$ and $(n-1)\mathbf{S} \sim W_p(n-1, \mathbf{\Sigma})$. Consequently, any linear combination of $\bar{\mathbf{y}}$, say $\mathbf{c}'\bar{\mathbf{y}}, \mathbf{c} \neq \mathbf{0}$, follows $N_1(\mathbf{c}'\mu, \mathbf{c}'\mathbf{\Sigma}\mathbf{c}/n)$ and the quadratic form $(n-1)\mathbf{c}'\mathbf{Sc}/\mathbf{c}'\mathbf{\Sigma}\mathbf{c} \sim \chi^2(n-1)$. Further, as pointed out earlier, $\bar{\mathbf{y}}$ and $\mathbf{S}$ are independently distributed and hence the quantity

$$t = \sqrt{n}\mathbf{c}'(\bar{\mathbf{y}} - \mu)/\sqrt{\mathbf{c}'\mathbf{Sc}}$$

follows a t-distribution with $(n-1)$ degrees of freedom. A useful application of this fact is in testing problems for certain contrasts or in testing problems involving a given linear combination of the components of the mean vector.

Often interest may be in testing a hypothesis if the population has its mean vector equal to a given vector, say μ_0. Since $\bar{\mathbf{y}} \sim N_p(\mu, \mathbf{\Sigma}/n)$, it follows that $\mathbf{z} = \sqrt{n}\mathbf{\Sigma}^{-\frac{1}{2}}(\bar{\mathbf{y}} - \mu)$ follows $N_p(\mathbf{0}, \mathbf{I})$. This implies that the components of $\mathbf{z}$ are independent and have the standard normal distribution. As a result, if μ is equal to μ_0 the quantity, $z_1^2 + \cdots + z_p^2 = \mathbf{z}'\mathbf{z} = n(\bar{\mathbf{y}} - \mu_0)'\mathbf{\Sigma}^{-1}(\bar{\mathbf{y}} - \mu_0)$ follows a chi-square distribution with p degrees of freedom. On the other hand, if μ is not equal to μ_0, then this quantity will not have a chi-square distribution. This observation provides a way of testing the hypothesis that the mean of the normal population is equal to a given vector μ_0. However, the assumption of known $\mathbf{\Sigma}$ is needed to actually perform this test. If $\mathbf{\Sigma}$ is unknown, it seems natural to replace it in $n(\bar{\mathbf{y}} - \mu)'\mathbf{\Sigma}^{-1}(\bar{\mathbf{y}} - \mu)$ by its unbiased estimator $\mathbf{S}$, leading to Hotelling's T^2 test statistic defined as

$$T^2 = n(\bar{\mathbf{y}} - \boldsymbol{\mu}_0)' \mathbf{S}^{-1} (\bar{\mathbf{y}} - \boldsymbol{\mu}_0),$$

where we assume that $n \geq p + 1$. This assumption ensures that $\mathbf{S}$ admits an inverse. Under the hypothesis mentioned above, namely $\mu = \mu_0$, the quantity $\frac{n-p}{p(n-1)} T^2$ follows an F distribution with degrees of freedom p and $n - p$.

Assuming normality, the maximum likelihood estimates of $\boldsymbol{\mu}$ and $\boldsymbol{\Sigma}$ are known to be

$$\hat{\boldsymbol{\mu}}_{ml} = \bar{\mathbf{y}}$$

and

$$\hat{\boldsymbol{\Sigma}}_{ml} = \mathbf{S}_n = \frac{1}{n} \mathbf{Y}' \left(\mathbf{I}_n - \frac{1}{n} \mathbf{1}_n \mathbf{1}_n' \right) \mathbf{Y} = \frac{n-1}{n} \mathbf{S}.$$

While $\hat{\boldsymbol{\mu}}_{ml} = \bar{\mathbf{y}}$ is unbiased for $\boldsymbol{\mu}$, $\hat{\boldsymbol{\Sigma}}_{ml} = \mathbf{S}_n$ is a (negatively) biased estimator of $\boldsymbol{\Sigma}$. These quantities are also needed in the process of deriving various maximum likelihood-based tests for the hypothesis testing problems. In general, to test a hypothesis H_0, the likelihood ratio test based on the maximum likelihood estimates is obtained by first maximizing the likelihood within the parameter space restricted by H_0. The next step is maximizing it over the entire parameter space (that is, by evaluating the likelihood at $\hat{\boldsymbol{\mu}}_{ml}$ and $\hat{\boldsymbol{\Sigma}}_{ml}$), and then taking the ratio of the two. Thus, the likelihood ratio test statistic can be written as

$$L = \frac{\max\limits_{H_0} f(\mathbf{Y})}{\max\limits_{\text{unrestricted}} f(\mathbf{Y})} = \frac{\max\limits_{H_0} g(\boldsymbol{\mu}, \boldsymbol{\Sigma} | \mathbf{Y})}{\max\limits_{\text{unrestricted}} g(\boldsymbol{\mu}, \boldsymbol{\Sigma} | \mathbf{Y})},$$

where for optimization purposes the function $g(\boldsymbol{\mu}, \boldsymbol{\Sigma} | \mathbf{Y}) = f(\mathbf{Y})$ is viewed as a function of $\boldsymbol{\mu}$ and $\boldsymbol{\Sigma}$ given data $\mathbf{Y}$. A related test statistic is the Wilks' Λ, which is the $(2/n)^{th}$ power of L. For large n, the quantity $-2 \log L$ approximately follows a chi-square distribution, with degrees of freedom ν, which is a function of the sample size n, the number of parameters estimated, and the number of restrictions imposed by the parameters involved under H_0.

A detailed discussion of various likelihood ratio tests in multivariate analysis context can be found in Kshirsagar (1972), Muirhead (1982) or in Anderson (1984). A brief review of some of the relevant likelihood ratio tests is given in Chapter 6. There are certain other intuitive statistical tests which have been proposed in various contexts and used in applications instead of the likelihood ratio tests. Some of these tests have been discussed in Chapter 3.

1.6   Tests for Multivariate Normality

Often before doing any statistical modeling, it is crucial to verify if the data at hand satisfy the underlying distributional assumptions. Many times such an examination may be needed for the residuals after fitting various models. For most multivariate analyses, it is thus very important that the data indeed follow the multivariate normal, or if not exactly at least approximately. If the answer to such a query is affirmative, it can often reduce the task of searching for procedures which are robust to the departures from multivariate normality. There are many possibilities for departure from multivariate normality and no single procedure is likely to be robust with respect to all such departures from the multivariate normality assumption. Gnanadesikan (1980) and Mardia (1980) provide excellent reviews of various procedures to verify this assumption.

This assumption is often checked by individually examining the univariate normality through various Q-Q plots or some other plots and can at times be very subjective. One of the relatively simpler and mathematically tractable ways to find a support for the assumption of multivariate normality is by using the tests based on Mardia's *multivariate skewness* and *kurtosis* measures. For any general multivariate distribution we define these respectively as

$$\beta_{1,p} = E\left\{ (\mathbf{y} - \boldsymbol{\mu})' \boldsymbol{\Sigma}^{-1} (\mathbf{x} - \boldsymbol{\mu}) \right\}^3,$$

provided that $\mathbf{x}$ is independent of $\mathbf{y}$ but has the same distribution and

$$\beta_{2,p} = E\left\{ (\mathbf{y} - \boldsymbol{\mu})' \boldsymbol{\Sigma}^{-1} (\mathbf{y} - \boldsymbol{\mu}) \right\}^2,$$

provided that the expectations in the expressions of $\beta_{1,p}$ and $\beta_{2,p}$ exist. For the multivariate normal distribution, $\beta_{1,p} = 0$ and $\beta_{2,p} = p(p + 2)$.

For a sample of size n, the estimates of $\beta_{1,p}$ and $\beta_{2,p}$ can be obtained as

$$\hat{\beta}_{1,p} = \frac{1}{n^2} \sum_{i=1}^{n} \sum_{j=1}^{n} g_{ij}^3$$

$$\hat{\beta}_{2,p} = \frac{1}{n} \sum_{i=1}^{n} g_{ii}^2 = \frac{1}{n} \sum_{i=1}^{n} d_i^4$$

where $g_{ij} = (\mathbf{y}_i - \bar{\mathbf{y}})' \mathbf{S}_n^{-1} (\mathbf{y}_j - \bar{\mathbf{y}})$, and $d_i = \sqrt{g_{ii}}$ is the sample version of the squared *Mahalanobis distance* (Mahalanobis, 1936) between $\mathbf{y}_i$ and ($\boldsymbol{\mu}$ which is approximated by) $\bar{\mathbf{y}}$ (Mardia, 1970).

The quantity $\hat{\beta}_{1,p}$ (which is the same as the square of sample skewness coefficient when $p = 1$) as well as $\hat{\beta}_{2,p}$ (which is the same as the sample kurtosis coefficient when $p = 1$) are nonnegative. For the multivariate normal data, we would expect $\hat{\beta}_{1,p}$ to be close to zero. If there is a departure from the spherical symmetry (that is, zero correlation and equal variance), $\hat{\beta}_{2,p}$ will be large. The quantity $\hat{\beta}_{2,p}$ is also useful in indicating the extreme behavior in the squared Mahalanobis distance of the observations from the sample mean.

Thus, $\hat{\beta}_{1,p}$ and $\hat{\beta}_{2,p}$ can be utilized to detect departure from multivariate normality. Mardia (1970) has shown that for large samples, $\kappa_1 = n\hat{\beta}_{1,p}/6$ follows a chi-square distribution with degrees of freedom $p(p + 1)(p + 2)/6$, and $\kappa_2 = \{\hat{\beta}_{2,p} - p(p + 2)\}/\{8p(p + 2)/n\}^{\frac{1}{2}}$ follows a standard normal distribution. Thus, we can use the quantities κ_1 and κ_2 to test the null hypothesis of multivariate normality. For small n, see the tables for the critical values for these test statistics given by Mardia (1974). He also recommends (Mardia, Kent, and Bibby, 1979, p. 149) that if both the hypotheses are accepted, the normal theory for various tests on the mean vector or the covariance matrix can be used. However, in the presence of nonnormality, the normal theory tests on the mean are sensitive to $\beta_{1,p}$, whereas tests on the covariance matrix are influenced by $\beta_{2,p}$.

For a given data set, the multivariate kurtosis can be computed using the CALIS procedure in SAS/STAT software. Notice that the quantities reported in the corresponding SAS output are the centered quantity $(\hat{\beta}_{2,p} - p(p + 2))$ (shown in Output 1.1 as Mardia's Multivariate Kurtosis) and κ_2 (shown in Output 1.1 as Normalized Multivariate Kurtosis).

EXAMPLE 1   *Testing Multivariate Normality, Cork Data*   As an illustration, we consider the cork boring data of Rao (1948) given in Table 1.1, and test the hypothesis that this data set can be considered as a random sample from a multivariate normal population. The data set provided in Table 1.1 consists of the weights of cork borings in four directions (north, east, south, and west) for 28 trees in a block of plantations.

E. S. Pearson had pointed out to C. R. Rao, apparently without any formal statistical testing, that the data are exceedingly asymmetrically distributed. It is therefore of interest to formally test if the data can be assumed to have come from an $N_4(\boldsymbol{\mu}, \boldsymbol{\Sigma})$.

TABLE 1.1 Weights of Cork Boring (in Centigrams) in Four Directions for 28 Trees

Tree	N	E	S	W	Tree	N	E	S	W
1	72	66	76	77	15	91	79	100	75
2	60	53	66	63	16	56	68	47	50
3	56	57	64	58	17	79	65	70	61
4	41	29	36	38	18	81	80	68	58
5	32	32	35	36	19	78	55	67	60
6	30	35	34	26	20	46	38	37	38
7	39	39	31	27	21	39	35	34	37
8	42	43	31	25	22	32	30	30	32
9	37	40	31	25	23	60	50	67	54
10	33	29	27	36	24	35	37	48	39
11	32	30	34	28	25	39	36	39	31
12	63	45	74	63	26	50	34	37	40
13	54	46	60	52	27	43	37	39	50
14	47	51	52	43	28	48	54	57	43

The SAS statements required to compute the multivariate kurtosis using PROC CALIS are given in Program 1.1. A part of the output giving the value of Mardia's multivariate kurtosis ($= -1.0431$) and normalized multivariate kurtosis ($= -0.3984$) is shown as Output 1.1. The output also indicates the observations which are most influential. Although the procedure does not provide the value of multivariate skewness, the IML procedure statements given in Program 1.2 perform all the necessary calculations to compute the multivariate skewness and kurtosis. The results are shown in Output 1.2, which also reports Mardia's test statistics κ_1 and κ_2 described above along with the corresponding p values.

In this program, for the 28 by 4 data matrix $\mathbf{Y}$, we first compute the maximum likelihood estimate of the variance-covariance matrix. This estimate is given by $\mathbf{S}_n = \frac{1}{n}\mathbf{Y}'\mathbf{Q}\mathbf{Y}$, where $\mathbf{Q} = \mathbf{I}_n - \frac{1}{n}\mathbf{1}_n\mathbf{1}_n'$. Also, since the quantities $g_{ij}, i, j = 1, \ldots, n$ needed in the expressions of multivariate skewness and kurtosis are the elements of matrix $\mathbf{G} = \mathbf{Q}\mathbf{Y}\mathbf{S}_n^{-1}\mathbf{Y}'\mathbf{Q}$, we compute the matrix $\mathbf{G}$, using this formula. Their p values are then reported as PVALSKEW and PVALKURT in Output 1.2. It may be remarked that in Program 1.2 the raw data are presented as a matrix entity. One can alternatively read the raw data (as done in Program 1.1) as a data set and then convert it to a matrix. In Appendix 1, we have provided the SAS code to perform this conversion.

```
/* Program 1.1 */

options ls=64 ps=45 nodate nonumber;
data cork;
infile 'cork.dat' firstobs = 1;
input north east south west;
proc calis  data = cork kurtosis;
title1 "Output 1.1";
title2 "Computation of Mardia's Kurtosis";
lineqs
north = e1,
east = e2,
south = e3,
west = e4;
std
e1=eps1, e2=eps2, e3=eps3, e4=eps4;
cov
e1=eps1, e2=eps2, e3=eps3, e4=eps4;
run ;
```

Output 1.1

```
                              Output 1.1
                    Computation of Mardia's Kurtosis

    Mardia's Multivariate Kurtosis . . . . . . . .      -1.0431
    Relative Multivariate Kurtosis . . . . . . . .       0.9565
    Normalized Multivariate Kurtosis . . . . . . .      -0.3984
    Mardia Based Kappa (Browne, 1982). . . . . . .      -0.0435
    Mean Scaled Univariate Kurtosis  . . . . . . .      -0.0770
    Adjusted Mean Scaled Univariate Kurtosis . . .      -0.0770
```

```
/* Program 1.2 */

title 'Output 1.2';
options ls = 64 ps=45 nodate nonumber;

/* This program is for testing the multivariate
normality using Mardia's skewness and kurtosis measures.
Application on C. R. Rao's cork data */

proc iml ;
y ={
72 66 76 77,
60 53 66 63,
56 57 64 58,
41 29 36 38,
32 32 35 36,
30 35 34 26,
39 39 31 27,
42 43 31 25,
37 40 31 25,
33 29 27 36,
32 30 34 28,
63 45 74 63,
54 46 60 52,
47 51 52 43,
91 79 100 75,
56 68 47 50,
79 65 70 61,
81 80 68 58,
78 55 67 60,
46 38 37 38,
39 35 34 37,
32 30 30 32,
60 50 67 54,
35 37 48 39,
39 36 39 31,
50 34 37 40,
43 37 39 50,
48 54 57 43} ;
/* Matrix y can be created from a SAS data set as follows:
data cork;
infile 'cork.dat';
input y1 y2 y3 y4;
run;
proc iml;
use cork;
read all into y;
```

See Appendix 1 for details.
*/
/* Here we determine the number of data points and the dimension
of the vector. The variable dfchi is the degrees of freedom for
the chi square approximation of Multivariate skewness. */

```
n = nrow(y) ;
p = ncol(y) ;
dfchi = p*(p+1)*(p+2)/6 ;
```

/* q is projection matrix, s is the maximum likelihood estimate
of the variance covariance matrix, g_matrix is n by n the matrix
of g(i,j) elements, beta1hat and beta2hat are respectively the
Mardia's sample skewness and kurtosis measures, kappa1 and kappa2
are the test statistics based on skewness and kurtosis to test
for normality and pvalskew and pvalkurt are corresponding p
values. */

```
q = i(n) - (1/n)*j(n,n,1);
s = (1/(n))*y'*q*y ; s_inv = inv(s) ;
g_matrix = q*y*s_inv*y'*q;
beta1hat = ( sum(g_matrix#g_matrix#g_matrix) )/(n*n);
beta2hat =trace( g_matrix#g_matrix )/n ;

kappa1 = n*beta1hat/6 ;
kappa2 = (beta2hat - p*(p+2) ) /sqrt(8*p*(p+2)/n) ;

pvalskew = 1 - probchi(kappa1,dfchi) ;
pvalkurt = 2*( 1 - probnorm(abs(kappa2)) );
print s ;
print s_inv ;
print 'TESTS:';
print 'Based on skewness: ' beta1hat kappa1 pvalskew ;
print 'Based on kurtosis: ' beta2hat kappa2 pvalkurt;
```

Output 1.2

```
                          Output 1.2

                          S
            280.03444 215.76148 278.13648 218.19005
            215.76148 212.07526 220.87883 165.25383
            278.13648 220.87883 337.50383 250.27168
            218.19005 165.25383 250.27168  217.9324

                        S_INV
            0.0332462 -0.016361 -0.008139 -0.011533
            -0.016361 0.0228758 -0.005199 0.0050046
            -0.008139 -0.005199 0.0276698 -0.019685
            -0.011533 0.0050046 -0.019685 0.0349464

                         TESTS:

                        BETA1HAT    KAPPA1   PVALSKEW
    Based on skewness:  4.4763816 20.889781 0.4036454

                        BETA2HAT    KAPPA2   PVALKURT
    Based on kurtosis:  22.95687 -0.398352 0.6903709
```

For this particular data set with its large p values, neither skewness is significantly different from zero, nor is the value of kurtosis significantly different from that for the 4-variate multivariate normal distribution. Consequently, we may assume multivariate normality for testing the various hypotheses on the mean vector and the covariance matrix as far as the present data set is concerned. This particular data set is extensively analyzed in the later chapters under the assumption of normality.

Often we are less interested in the multivariate normality of the original data and more interested in the joint normality of contrasts or any other set of linear combinations of the variables $y_1, \ldots, y_p$. If $\mathbf{C}$ is the corresponding p by r matrix of linear transformations, then the transformed data can be obtained as $\mathbf{Z} = \mathbf{YC}$. Consequently, the only change in Program 1.2 is to replace the earlier definition of $\mathbf{G}$ by $\mathbf{QYC}(\mathbf{C}'\mathbf{S}_n\mathbf{C})^{-1}\mathbf{C}'\mathbf{Y}'\mathbf{Q}$ and replace p by r in the expressions for κ_1, κ_2 and the degrees of freedom corresponding to κ_1.

EXAMPLE 1   *Testing for Contrasts, Cork Data (continued)*   Returning to the cork data, if the interest is in testing if the bark deposit is uniform in all four directions, an appropriate set of transformations would be

$$z_1 = y_1 - y_2 + y_3 - y_4, \quad z_2 = y_3 - y_4, \quad z_3 = y_1 - y_3,$$

where y_1, y_2, y_3, y_4 represent the deposit in four directions listed clockwise and starting from north. The 4 by 3 matrix $\mathbf{C}$ for these transformations will be

$$\mathbf{C} = \begin{bmatrix} 1 & 0 & 1 \\ -1 & 0 & 0 \\ 1 & 1 & -1 \\ -1 & -1 & 0 \end{bmatrix}.$$

It is easy to verify that for these contrasts the assumption of symmetry holds rather more strongly, since the p values corresponding to the skewness are relatively larger. Specifically for these contrasts

$$\hat{\beta}_1 = 1.1770, \qquad \hat{\beta}_2 = 13.5584, \qquad \kappa_1 = 5.4928, \qquad \kappa_2 = -0.6964$$

and the respective p values for skewness and kurtosis tests are 0.8559 and 0.4862. As Rao (1948) points out, this symmetry is not surprising since these are linear combinations, and the contrasts are likely to fit the multivariate normality better than the original data. Since one can easily modify Program 1.1 or Program 1.2 to perform the above analysis on the contrasts z_1, z_2, and z_3, we have not provided the corresponding SAS code or the output.

Mudholkar, McDermott and Srivastava (1992) suggest another simple test of multivariate normality. The idea is based on the facts that (i) the cube root of a chi-square random variable can be approximated by a normal random variable and (ii) the sample mean vector and the sample variance covariance matrix are independent if and only if the underlying distribution is multivariate normal. Lin and Mudholkar (1980) had earlier used these ideas to obtain a test for the univariate normality.

To test multivariate normality (of dimension say p) on the population with mean vector $\boldsymbol{\mu}$ and a variance covariance matrix $\boldsymbol{\Sigma}$, let $\mathbf{y}_1, \ldots, \mathbf{y}_n$ be a random sample of size n then the unbiased estimators of $\boldsymbol{\mu}$ and $\boldsymbol{\Sigma}$ are respectively given by $\bar{\mathbf{y}}$ and $\mathbf{S}$. Corresponding to i^{th} observation we define,

$$D_i^2 = (\mathbf{y}_i - \bar{\mathbf{y}})'\mathbf{S}^{-1}(\mathbf{y}_i - \bar{\mathbf{y}}),$$

$$W_i = (D_i^2)^h,$$

and

$$U_i = \left\{ \sum_{j \neq i} W_j^2 - \left[\sum_{j \neq i} W_j \right]^2 / (n-1) \right\}^{1/3}, \quad i = 1, \ldots, n,$$

where

$$h = \frac{1}{3} - \frac{0.11}{p}.$$

Let r be the sample correlation coefficient between (W_i, U_i), $i = 1, \ldots, n$. Under the null hypothesis of multivariate normality of the data, the quantity, $Z_p = tanh^{-1}(r) = \frac{1}{2}ln\{\frac{1+r}{1-r}\}$, is approximately normal with mean $\mu_{n,p} = E(Z_p) = \frac{A_1(p)}{n} - \frac{A_2(p)}{n^2}$, where $A_1(p) = \frac{-1}{p} - .52p$ and $A_2(p) = 0.8p^2$ and variance, $\sigma_{n,p}^2 = var(Z_p) = \frac{B_1(p)}{n} - \frac{B_2(p)}{n^2}$, where $B_1(p) = 3 - \frac{1.67}{p} + \frac{52}{p^2}$ and $B_2(p) = 1.8p - \frac{9.75}{p^2}$. Thus, the test based on Z_p to test the null hypothesis of multivariate normality rejects it at α level of significance if $|z_{n,p}| = \frac{|Z_p - \mu_{n,p}|}{\sigma_{n,p}} \geq z_{\frac{\alpha}{2}}$, where $z_{\frac{\alpha}{2}}$ is the right $\frac{\alpha}{2}$ cutoff point from the standard normal distribution.

EXAMPLE 1   ***Testing Multivariate Normality, Cork Data (continued)***   In Program 1.3, we reconsider the cork data of C. R. Rao (1948) and test the hypothesis of the multivariate normality of the tree population.

```
/* Program 1.3 */

options ls=64 ps=45 nodate nonumber;
title1 'Output 1.3';
title2 'Testing Multivariate Normality (Cube Root Transformation)';

data D1;
infile 'cork.dat';
input t1 t2 t3 t4 ;
/*
t1=north, t2=east, t3=south, t4=west
n is the number of observations
p is the number of variables
*/
data D2(keep=t1 t2 t3 t4 n p);
set D1;
n=28;
p=4;
run;
data D3(keep=n p);
set D2;
if _n_ > 1 then delete;
run;
proc princomp data=D2 cov std out=D4 noprint;
var t1-t4;
data D5(keep=n1 dsq n p);
set D4;
n1=_n_;
dsq=uss(of prin1-prin4);
run;
data D6(keep=dsq1 n1 );
set  D5;
dsq1=dsq**((1.0/3.0)-(0.11/p));
run;

proc iml;
use D3;
read all var {n p};
u=j(n,1,1);
```

```
use D6;
do k=1 to n;
setin D6 point 0;
sum1=0;
sum2=0;
do data;
read next var{dsq1 n1} ;
if n1 = k then dsq1=0;
sum1=sum1+dsq1**2;
sum2=sum2+dsq1;
end;
u[k]=(sum1-((sum2**2)/(n-1)))**(1.0/3);
end;
varnames={y};
create tyy from u (|colname=varnames|);
append from u;
close tyy;
run;
quit;

data D7;
set D6; set tyy;
run;
proc corr data=D7 noprint outp=D8;
var dsq1;
with y;
run;
data D9;
set D8;
if _TYPE_ ^='CORR' then delete;
run;
data D10(keep=zp r tnp pvalue);
set D9(rename=(dsq1=r));
set D3;
zp=0.5*log((1+r)/(1-r));
b1p=3-1.67/p+0.52/(p**2);
a1p=-1.0/p-0.52*p;
a2p=0.8*p**2;
mnp=(a1p/n)-(a2p/(n**2));
b2p=1.8*p-9.75/(p**2);
ssq1=b1p/n-b2p/(n**2);
snp=ssq1**0.5;
tnp=abs(abs(zp-mnp)/snp);
pvalue=2*(1-probnorm(tnp));
run;
proc print data=D10;
run;
```

The SAS Program 1.3 (adopted from Apprey and Naik (1998)) computes the quantities, Z_p, $\mu_{n,p}$, and $\sigma_{n.p}$ using the expressions listed above. Using these, the test statistic $|z_{n,p}|$ and corresponding p value are computed. A run of the program results in a p value of 0.2216. We thus accept the hypothesis of multivariate normality. This conclusion is consistent with our earlier conclusion using the Mardia's tests for the same data set. Output corresponding to Program 1.3 is suppressed in order to save space.

1.7   Random Vector and Matrix Generation

For various simulation or power studies, it is often necessary to generate a set of random vectors or random matrices. It is therefore of interest to generate these quantities for the probability distributions which arise naturally in the multivariate normal theory. The following sections consider the most common multivariate probability distributions.

1.7.1   Random Vector Generation from $N_p(\mu, \Sigma)$

To generate a random vector from $N_p(\mu, \Sigma)$ use the following steps:

1. Find a matrix $\mathbf{G}$ such that $\Sigma = \mathbf{G}'\mathbf{G}$. This is obtained using the Cholesky decomposition of the symmetric matrix Σ. The functions ROOT of Half in PROC IML can perform this decomposition.
2. Generate p independent standard univariate normal random variables $z_1, \ldots, z_p$ and let $\mathbf{z} = (z_1, \ldots, z_p)'$.
3. Let $\mathbf{y} = \mu + \mathbf{G}'\mathbf{z}$.

The resulting vector $\mathbf{y}$ is an observation from a $N_p(\mu, \Sigma)$ population. To obtain a sample of size n, we repeat the above-mentioned steps n times within a loop.

1.7.2   Generation of Wishart Random Matrix

To generate a matrix $\mathbf{A}_1 \sim W_p(f, \Sigma)$, use the following steps:

1. Find a matrix $\mathbf{G}$ such that $\Sigma = \mathbf{G}'\mathbf{G}$.
2. Generate a random sample of size f, say $\mathbf{z}_1, \ldots, \mathbf{z}_f$ from $N_p(\mathbf{0}, \mathbf{I})$. Let $\mathbf{A}_2 = \sum_{i=1}^{f} \mathbf{z}_i \mathbf{z}_i'$.
3. Define $\mathbf{A}_1 = \mathbf{G}'\mathbf{A}_2\mathbf{G}$.

The generation of Beta matrices can easily be done by first generating two independent Wishart matrices with appropriate degrees of freedom and then forming the appropriate products using these matrices as defined in Section 1.4.

EXAMPLE 2   ***Random Samples from Normal and Wishart Distributions***   In the following example we will illustrate the use of PROC IML for generating samples from the multivariate normal and Wishart distributions respectively. These programs are respectively given as Program 1.4 and Program 1.5. The corresponding outputs have been omitted to save space.

As an example, suppose we want to generate four vectors from $N_3(\mu, \Sigma)$ where

$$\mu = (1\ 3\ 0)'$$

and

$$\Sigma = \begin{bmatrix} 4 & 2 & 1 \\ 2 & 3 & 1 \\ 1 & 1 & 5 \end{bmatrix}.$$

Then save these four vectors as the rows of 4 by 3 matrix $\mathbf{Y}$. It is easy to see that

$$E(\mathbf{Y}) = \begin{bmatrix} \mu' \\ \mu' \\ \mu' \\ \mu' \end{bmatrix} = \mathbf{M}.$$

Also, let $\mathbf{G}$ be a matrix such that $\mathbf{\Sigma} = \mathbf{G}'\mathbf{G}$. This matrix is obtained using the ROOT function which performs the Cholesky decomposition of a symmetric matrix.

```
/* Program 1.4 */

options ls = 64 ps=45 nodate nonumber;
title1 'Output 1.4';

/* Generate n random vector from a p dimensional population
with mean mu and the variance covariance matrix sigma */

proc iml ;
seed = 549065467 ;
n = 4 ;
sigma = { 4 2 1,
          2 3 1,
          1 1 5 };

mu = {1, 3, 0};
p = nrow(sigma);
m = repeat(mu`,n,1) ;
        g =root(sigma);
z =normal(repeat(seed,n,p)) ;
y = z*g + m ;
print 'Multivariate Normal Sample';
print y;
```

We first generate a 4 by 3 random matrix $\mathbf{Z}$, with all its entries distributed as $N(0, 1)$. To do this, we use the normal random number generator (NORMAL) repeated for all the entries of $\mathbf{Z}$, through the REPEAT function. Consequently, if we define $\mathbf{Y} = \mathbf{ZG} + \mathbf{M}$, then the ith row of $\mathbf{Y}$, say $\mathbf{y}'_i$, can be written in terms of the ith row of $\mathbf{Z}$, say $\mathbf{z}'_i$, as

$$\mathbf{y}'_i = \mathbf{z}'_i\mathbf{G} + \boldsymbol{\mu}'$$

or when written as a column vector

$$\mathbf{y}_i = \mathbf{G}'\mathbf{z}_i + \boldsymbol{\mu}.$$

Consequently, $\mathbf{y}_i, i = 1, \ldots, n (= 4 \text{ here})$ are normally distributed with the mean $E(\mathbf{y}_i) = \mathbf{G}'E(\mathbf{z}_i) + \boldsymbol{\mu} = \boldsymbol{\mu}$ and the variance covariance matrix $D(\mathbf{y}_i) = \mathbf{G}'D(\mathbf{z}_i)\mathbf{G} + \mathbf{0} = \mathbf{G}'\mathbf{I}\mathbf{G} = \mathbf{G}'\mathbf{G} = \mathbf{\Sigma}$.

Program 1.5 illustrates the generation of $n = 4$ Wishart matrices from $W_p(f, \mathbf{\Sigma})$ with $f = 7$, $p = 3$, and $\mathbf{\Sigma}$ as given in the previous program. After obtaining the matrix $\mathbf{G}$, as earlier, we generate a 7 by 3 matrix $\mathbf{T}$, for which all the elements are distributed as the standard normal. Consequently, the matrix $\mathbf{W} = \mathbf{G}'\mathbf{T}'\mathbf{TG}$, (written as $\mathbf{X}'\mathbf{X}$, where $\mathbf{X} = \mathbf{TG}$) follows $W_3(7, \mathbf{\Sigma})$ distribution. We have used a DO loop to repeat the process $n = 4$ times to obtain four such matrices.

```
/* Program 1.5 */

options ls=64 ps=45 nodate nonumber;
title1 'Output 1.5';
/* Generate n Wishart matrices of order p by p
with degrees of freedom f */

proc iml;
n = 4 ;
f = 7 ;
```

```
seed = 4509049 ;
sigma = {4 2 1,
         2 3 1,
         1 1 5 } ;
         g = root(sigma);
p = nrow(sigma) ;
print 'Wishart Random Matrix';
do i = 1 to n ;
t = normal(repeat(seed,f,p)) ;
x = t*g ;
w = x'*x ;
print w ;
end ;
```

These programs can be easily modified to generate the Beta matrices of either Type 1 or Type 2, as the generation of such matrices essentially amounts to generating the pairs of Wishart matrices with appropriate degrees of freedom and then combining them as per their definitions.

More efficient algorithms, especially for large values of $f - p$ are available in the literature. One such convenient method based on Bartlett's decomposition can be found in Smith and Hocking (1972). Certain other methods are briefly summarized in Kennedy and Gentle (1980, p. 231).

Graphical Representation of Multivariate Data

Chapter
2

2.1   Introduction   21

2.2   Scatter Plots   22

2.3   Profile Plots   31

2.4   Andrews Function Plots   33

2.5   Biplots: Plotting Observations and Variables Together   38

2.6   Q-Q Plots for Assessing Multivariate Normality   45

2.7   Plots for Detection of Multivariate Outliers   50

2.8   Bivariate Normal Distribution   53

2.9   SAS/INSIGHT Software   58

2.10   Concluding Remarks   59

2.1   Introduction

Graphical techniques have become an integral part of any data analysis, especially now due to a tremendous increase in the accessibility to computing facilities. In general it is easy to use graphical methods for data with one, two, or even three variables. However, for multivariate data in dimensions higher than three, data reduction to two or three variables is needed before it is possible to plot them. Several methods to represent multivariate data are available in the literature. This chapter covers four of these methods in Sections 2.2 through 2.5. It may be mentioned that Section 2.5 may require a slightly higher level of familiarity with matrix decompositions and may be skipped at first reading. See Friendly (1991) for details on various other graphical methods.

The *multivariate normal distribution* is a basis for most of the theory on testing of hypotheses in multivariate analysis. Often graphical methods are used to assess the multivariate normality and to detect multivariate outliers. These methods are covered in Sections 2.6 and 2.7 respectively.

The *probability density function* of a bivariate normal distribution and the contours of the probability density function drawn graphically give information about the magnitude of the variances and correlation between the two variables. Section 2.8 discusses these graphs briefly. SAS/INSIGHT software, an interactive tool for graphical data analysis, is briefly discussed at the end of the chapter.

For illustration purposes, we have confined ourselves to the data set from Rao (1948). This data set, given in Table 1.1, consists of weights of cork boring taken from the north (N), east (E), south (S), and west (W) directions of the trunks of 28 trees in a block of plantations.

2.2 Scatter Plots

Scatter plots are among the most basic and useful of graphical representation techniques. A scatter plot of two sets of variables is simply a two-dimensional representation of the points in a plane to show the relationship between two variables. The scatter plot is most useful in identifying the type of relationship (linear or nonlinear) between two sets of variables. Further, if the relationship is linear they help determine the negative or positive relationship between the two variables. This section uses various SAS procedures to plot scatter plots in two and three dimensions. When there are more than two variables, scatter plots of two variables at a time are displayed in a matrix of plots.

2.2.1 Two-Dimensional Scatter Plots

Two-dimensional scatter plots can be drawn using the PLOT or GPLOT procedures. The SAS code shown in the first two parts of Program 2.1 produces two scatter plots using the GPLOT procedure. The first of several optional statements of the GPLOT procedure in the program specifies the file name where the graphics will be stored as a postscript file (PROG21a.GRAPH in our program). The second statement is essentially used to specify the device name (DEV=PSLMONO in our program). The DEV=PSLMONO specification in that statement instructs SAS to store the graph in black and white in postscript form. The choice of DEV=PS can be used for a color graph. Of course if the PLOT procedure is used to produce the plots then there is no need to include any of the GOPTIONS statements in the program.

```
/* Program 2.1 */

filename gsasfile "prog21a.graph";
goptions reset=all gaccess=gsasfile autofeed dev=pslmono;
goptions horigin=1in vorigin=2in;
goptions hsize=6in vsize=8in;
options ls=64 ps=40 nodate nonumber;
title1 h=1.5 'Two Dimensional Scatter Plot ';
title2 j=l 'Output 2.1';
title3 'Cork Data: Source C.R. Rao (1948)';
data cork;
infile 'cork.dat';
input n e s w; * n:north,e:east,s:south,w:west;
run;
proc gplot data=cork;
plot n*e='star';
label n='Direction: North'
      e='Direction: East';
run;

data d1;
set cork;
y1=n-s;
y2=e-w;
run;
filename gsasfile "prog21b.graph";
title1 h=1.5 'Two Dimensional Scatter Plot of Contrasts';
title2 j=l 'Output 2.1';
title3 'Cork Data: Source C.R. Rao (1948)';
proc gplot data=d1;
plot y1*y2='star';
```

```
label y1='Contrast: North-South'
      y2='Contrast: East-West';
run;

data d2 d3;
set cork;
proc sort data=d2;
by n;
data d3;
set d3;
n_decr=n;
drop n;
proc sort data=d3;
by descending n_decr;
data both;
merge d2 d3;
filename gsasfile "prog21c.graph";
title1 h=1.5 'Testing Symmetry of Data on North Direction';
title2 j=l 'Output 2.1';
title3 'Cork Data: Source C.R. Rao (1948)';
proc gplot data=both;
plot n_decr*n='star';
label n_decr='Descending Ordered Data'
        n='Ascending Ordered Data';
run;
```

The first part of the program plots the data corresponding to the directions of north (N) and east (E), and the second part plots the contrasts of the directions north (N) and south (S) (Y1=N-S) against those of the directions east (E) and west (W) (Y2=E-W). These are shown in Output 2.1. The statement PLOT Y1*Y2 in the program plots the variable Y1 versus variable Y2. That is, the variable listed first in the PLOT statement is plotted on the vertical axis and the other variable is plotted on the horizontal axis. The code

```
proc gplot;
plot y1*y2;
```

uses the default symbols +, in the plot. A statement of the form

```
plot y1*y2='char';
```

can be used to specify a plotting symbol where 'CHAR' stands for the user-specified characters or symbols. The choice of 'star' for 'CHAR' is used in the Output 2.1. The appropriate size of the plot can be determined by the PAGESIZE= (or simply, PS=) and LINESIZE= (or LS=) options.

In a scatter plot, if the points follow an increasing straight line pattern then there may be a positive correlation between the two variables. This pattern indicates that as one variable increases the other increases also. On the other hand, if the points follow a decreasing straight line pattern then there may be a negative correlation indicating that one variable is decreasing as the other variable is increasing. If the points are randomly scattered in the plane then there may be only a weak or no correlation between the two variables. The first scatter plot in Output 2.1 indicates that there is a positive correlation between the cork weights in the directions of north and east. On the other hand, the second scatter plot in Output 2.1 suggests the possibility of a weak or no correlation between the two contrasts, Y1=N-S and Y2=E-W.

There are various variations of scatter plots for a variety of special purposes. For example, scatter plots have been used to examine the symmetry of distribution of the univariate data (Gnanadesikan, 1997). We will briefly discuss this approach. Suppose $y_1, \ldots, y_n$ are n observations on a variable y. To examine if the distribution of y is sym-

Output 2.1

Output 2.1

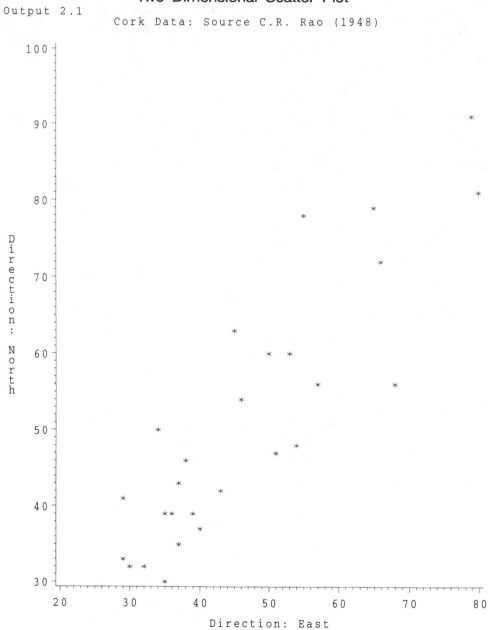

Two Dimensional Scatter Plot

Cork Data: Source C.R. Rao (1948)

metric, we order the data from smallest to largest as $y_{(1)} \leq, \ldots, \leq y_{(n)}$. Then, if the distribution of y is symmetric about a number, say μ, then the scatter plot of the paired data, $(y_{(1)}, y_{(n)}), (y_{(2)}, y_{(n-1)}), \ldots, (y_{(n)}, y_{(1)})$ should approximately fall around a line with slope -1 and intercept 2μ.

In Program 2.1, using the last few SAS statements we examine the symmetry of the cork data in the north direction (N) only. Using the sorted data sets D2 and D3 which arrange the observations on N in increasing and decreasing orders, we create a data set termed BOTH, which pairs the observations as $(y_{(1)}, y_{(n)}), (y_{(2)}, y_{(n-1)}), \ldots, (y_{(n)}, y_{(1)})$. These are then

Output 2.1
continued

Output 2.1

Two Dimensional Scatter Plot of Contrasts

Cork Data: Source C.R. Rao (1948)

plotted using the GPLOT procedure. The scatter plot shows a certain degree of departure from symmetry.

Gnanadesikan also suggests another scatter plot for symmetry where the pairs are defined not in terms of the original observations but in terms of deviations from the median, say m, of the data. Specifically, the paired values $(m - y_{(1)}, y_{(n)} - m)$, $(m - y_{(2)}, y_{(n-1)} - m)$, ..., $(y_{(n)} - m, m - y_{(1)})$, are plotted. If the original distribution is symmetric, the points should form a linear pattern along a line with slope 1 and intercept zero. The SAS code of Program 2.1 can be easily modified for this plot.

Output 2.1
continued

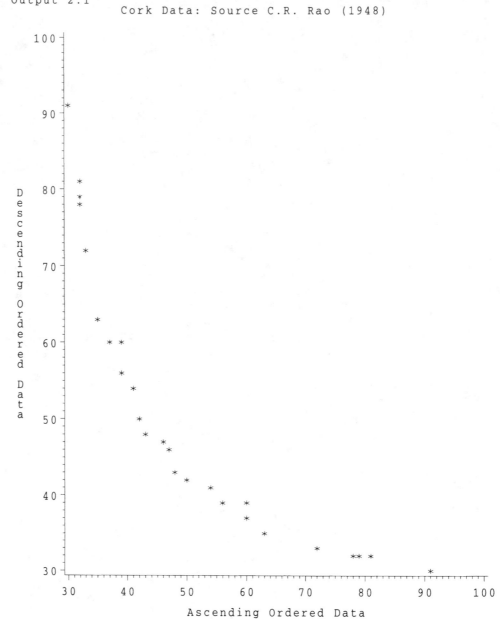

Output 2.1

Testing Symmetry of Data on North Direction

Cork Data: Source C.R. Rao (1948)

2.2.2   Three-Dimensional Scatter Plots

A three-dimensional scatter plot is needed to simultaneously display the relationships between three variables. The SCATTER statement in the G3D procedure can be used to draw
a three-dimensional scatter plot. The code given in Program 2.2 produces a scatter plot of
the variables N, E, and S by taking the variables N and S on the horizontal plane and E on
the axis perpendicular to the plane as displayed in Output 2.2.

```
/* Program 2.2 */

filename gsasfile "prog22.graph";
goptions reset=all gaccess=gsasfile autofeed dev=pslmono;
goptions horigin=1in vorigin=2in;
goptions hsize=6in vsize=8in;
options ls=64 ps=45 nodate nonumber;
data cork;
infile 'cork.dat';
input n e s w;
title1 h=1.5 'Three-D Scatter Plot for Cork Data';
title2 j=l 'Output 2.2';
```

Output 2.2

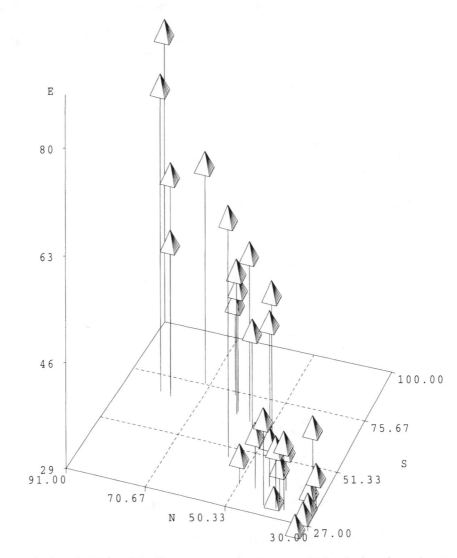

Output 2.2

Three-D Scatter Plot for Cork Data

by weight of cork boring
Source: C.R. Rao (1948)

N:Cork boring in North   E:Cork boring in East
S:Cork boring in South   W:West boring is not shown

```
title3 'by weight of cork boring';
title4 'Source: C.R. Rao (1948)';
footnote1 j=l 'N:Cork boring in North'
          j=r 'E:Cork boring in East';
footnote2 j=l 'S:Cork boring in South'
          j=r 'W:West boring is not shown';
proc g3d data=cork;
scatter n*s=e;
run;
```

Notice the SCATTER statement in Program 2.2 that plots the values of variables N and S on the horizontal plane and those of E on the axis perpendicular to that plane. The options J=L and J=R in the FOOTNOTE and TITLE statements indicate that the footnote or the title should be written on the left and on the right side of the page, respectively.

As in the two-dimensional scatter plot, if the points follow a pattern in the space then there may be correlations between any two or all three variables. If the points are scattered in the space then there is a weak or no correlation between any of the three variables. For example, the scatter plot of the three variables N, S, and E indicates that the points have an increasing pattern not only in the horizontal plane but also in the perpendicular direction. This seems to indicate that there is a positive correlation between the variables (N,S), between (S,E), and between (N, E).

Program 2.3 generates a three-dimensional scatter plot for the three contrasts C1=N-E-W+S, C2=N-S, and C3=E-W shown in Output 2.3.

```
/* Program 2.3 */

filename gsasfile "prog23.graph";
goptions gaccess=gsasfile autofeed dev=pslmono;
goptions horigin=1in vorigin=2in;
goptions hsize=6in vsize=8in;
options ls=64 ps=40 nodate nonumber;
data cork;
infile 'cork.dat';
input n e s w;
c1=n-e-w+s;
c2=n-s;
c3=e-w;
title1 h=1.5 'Three-Dimensional Scatter Plot for Cork Data';
title2 j=l 'Output 2.3';
title3 'Contrasts of weights of cork boring';
title4 'Source: C.R. Rao (1948)';
footnote1 j=l 'C1:Contrast N-E-W+S'
          j=r 'C2:Contrast N-S';
footnote2 j=r 'C3:Contrast E-W';
proc g3d data=cork;
scatter c1*c2=c3;
run;
```

This scatter plot seems to show weak or no correlation among the three contrasts except perhaps between C2 and C3.

2.2.3 Scatter Plot Matrix

For multivariate data with p variables, $y_1, \ldots, y_p$, a scatter plot of each pair of variables can be displayed in a p by p matrix of scatter plots. In this matrix the scatter plot of two different variables y_i and y_j is in the $(i, j)^{th}$ position of the matrix. The diagonal positions

Output 2.3   Three – Dimensional Scatter Plot for Cork Data

Output 2.3
                    Contrasts of weights of cork boring
                         Source: C.R. Rao (1948)

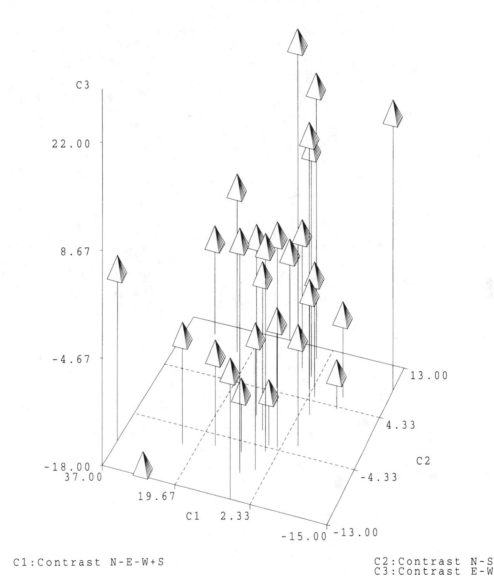

C1:Contrast N-E-W+S   C2:Contrast N-S
                                             C3:Contrast E-W

are usually used for writing descriptive comments. The scatter plot matrix is a useful way of representing multivariate data on a single two-dimensional display. It simultaneously identifies the relationships between various variables. In this sense it is a graphic analog of a correlation matrix. However, it may sometimes be more effective in that apart from the strength of linear relationships, any nonlinearities can also be easily spotted.

A macro for drawing a scatter plot matrix is given in Friendly (1991). However, a version of a scatter plot matrix can also be drawn very easily using SAS/INSIGHT software. See Section 2.9 for a brief description of the software. Program 2.4 produces a scatter plot matrix in a compact lower triangular form presented in Output 2.4.

```
/* Program 2.4 */

filename gsasfile "prog24.graph";
goptions reset=all gaccess=gsasfile autofeed dev=pslmono;
goptions horigin=1in vorigin=2in;
goptions hsize=6in vsize=8in;
options ls=64 ps=40 nodate nonumber;
title1 h=1.5 'Scatter Plot Matrix for Cork Data';
title2 'Output 2.4';
data cork;
infile 'cork.dat';
input n e s w;
proc insight data=cork;
scatter n e s w * n e s w;
run;
```

Output 2.4   Scatter Plot Matrix for Cork Data
                                      Output 2.4

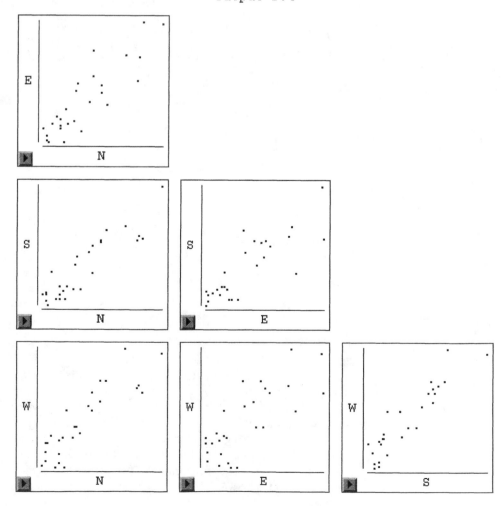

The plot indicates that there is a positive correlation between every pair of variables in the four directions. The correlation seems to be strongest between the variables S and W, but weakest between the variables E and W.

It may sometimes be cumbersome to represent all the variables on a matrix plot, especially if the number of variables is large. In order to visually extract the maximum information possible from these plots it may be necessary to restrict the choice to a moderate number of variables (say 5 or 6) at a time.

2.3   Profile Plots

One of the simplest ways of representing p-dimensional measurements is by using profile plots. These plots are the polygonal representations of p-dimensional observation vectors. Each p-dimensional observation vector is represented by p points with the vertical coordinate of each point proportional to the value of the corresponding variable. The successive points are joined using straight line segments. The resulting curve is called the *profile* of that observation. These plots can be very helpful in identifying clusters of the observations and outliers. Many times it may be more meaningful to plot the standardized variables in order to have a uniform scale for each variable. The standardization of variables can easily be achieved using the STANDARD procedure. See the *SAS Procedures Guide* for details on PROC STANDARD. Hartigan (1975) has suggested more effective displays of profile by optimally smoothing (linearizing) each profile as much as possible.

Program 2.5 produces a profile plot, as shown in Output 2.5. In the program, the features of the TRANSPOSE procedure have been utilized for data manipulation. The new variable DIRECTN placed in the first column is used in the PLOT statement to plot the tree profiles for each of the four directions. An alternative set of SAS code for drawing the profiles, using the ARRAY statement instead of PROC TRANSPOSE, is commented out in Program 2.5. By removing the comment delimiters, here as well as in all the programs to come, it is possible to use this alternative. In Output 2.5 the profile plot of the fifteenth tree (denoted by the letter M) stands out. This tree may possibly be an outlier. An examination of the data indicates that this tree has bark deposit measurements that are unusually large in magnitude compared to the rest. The profile plots also seem to indicate that there is a cluster of 12 to 14 trees, with relatively smaller measurements. A profile plot of standardized values also depicts similar conclusions about the data.

```
/* Program 2.5 */

filename gsasfile "prog25.graph";
goptions reset=all gaccess=gsasfile autofeed dev=pslmono;
options ls=64 ps=45 nodate nonumber;
data cork;
infile 'cork.dat';
input y1 y2 y3 y4; /*y1=north, y2=east, y3=south, y4=west*/
tree=_n_;
proc transpose data=cork
out=cork2 name=directn;
by tree;
proc gplot data=cork2(rename=(col1=weight));
/*
   data plot;
   set cork;
   array y{4} y1 y2 y3 y4;
   do directn=1 to 4;
   weight =y(directn);
   output;
   end;
```

```
      drop y1 y2 y3 y4;
      proc gplot data=plot;
*/
goptions horigin=1in vorigin=2in;
goptions hsize=6in vsize=8in;
plot weight*directn=tree/
vaxis=axis1 haxis=axis2 legend=legend1;
axis1 label=(a=90 h=1.2 'Standardized Weight of Cork Boring');
axis2 offset=(2) label=(h=1.2 'Direction');

symbol1 i=join v=star;
symbol2 i=join v=+;
symbol3 i=join v=A;
symbol4 i=join v=B;
symbol5 i=join v=C;
symbol6 i=join v=D;
symbol7 i=join v=E;
symbol8 i=join v=F;
symbol9 i=join v=G;
symbol10 i=join v=H;
symbol11 i=join v=I;
symbol12 i=join v=J;
symbol13 i=join v=K;
symbol14 i=join v=L;
symbol15 i=join v=M;
symbol16 i=join v=N;
symbol17 i=join v=O;
symbol18 i=join v=P;
symbol19 i=join v=Q;
symbol20 i=join v=R;
symbol21 i=join v=S;
symbol22 i=join v=T;
symbol23 i=join v=U;
symbol24 i=join v=V;
symbol25 i=join v=W;
symbol26 i=join v=X;
symbol27 i=join v=Y;
symbol28 i=join v=Z;
legend1 across=4;
title1 h=1.5 'Profiles of Standardized Cork Data';
title2 j=1 'Output 2.5';
title3 'Source: C.R. Rao (1948)';
run;
```

Profile plots of a large data set may be too cumbersome to be practically useful. Diggle, Liang and Zeger (1995), in the context of repeated measures data, suggested displaying the profiles of few systematically selected individuals (observations). The observations corresponding to certain quantiles of a meaningful summary statistic (of an observation) may be selected for displaying. For example, if we take the average of cork weights of a tree as the summary statistic, we will have 28 averages for the present data set, corresponding to 28 trees. The idea is to display the profiles of the trees having the minimum average weight, the maximum average weight, the 10th percentile average weight and so on. Such a plot may not be able to determine the clusters in the data set, but should be able to determine the outliers.

Output 2.5

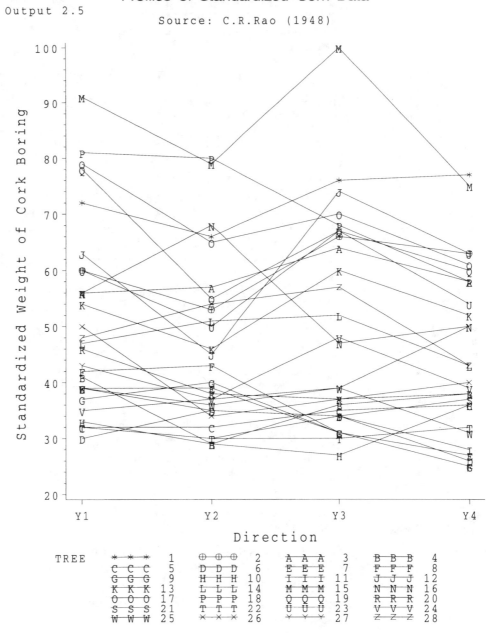

Plotting the profiles of sample mean vectors for different groups helps one to see whether the profiles are parallel. These profile plots serve as convenient graphical tools to explore the data before any formal multivariate statistical analysis techniques, like profile analysis (see Section 5.3.2), are applied to a data set.

2.4 Andrews Function Plots

Andrews (1972) suggests an innovative method to pictorially display the multivariate data points as curves. This pictorial representation can sometimes be very helpful in visually grouping similar objects together and in searching for any striking dissimilarities between the objects or the groups of the objects on which the multivariate data are collected.

Andrews' approach consists of representing a p-dimensional vector $\mathbf{y} = (y_1, \ldots, y_p)'$ by a function $f_{\mathbf{y}}(t)$ of a single variable t. This function when plotted on a two-dimensional space over t leads to a periodic curve. Andrews suggested the function $f_{\mathbf{y}}(t)$ to be the finite Fourier series

$$f_{\mathbf{y}}(t) = y_1/\sqrt{2} + y_2 sin(t) + y_3 cos(t) + y_4 sin(2t) + y_5 cos(2t) + \cdots .$$

By storing the coefficients of $y_1, y_2, \ldots$ in a column vector

$$\mathbf{a}(t) = \left(\frac{1}{\sqrt{2}}, \ sin(t), \ cos(t), \ sin(2t), \ cos(2t), \ldots \right)',$$

we can write $f_{\mathbf{y}} = \mathbf{a}'(t)\mathbf{y}$ as a linear combination of $\mathbf{y}$, for a fixed t. Thus for a given $t = t_0$, $(t_0, f_{\mathbf{y}}(t_0))$ represents a point on the curve, and by varying t between $-\pi$ and π, an *Andrews curve* of $\mathbf{y}$ is obtained as a collection of all such points. Corresponding to n different multivariate observations $\mathbf{y}_1, \ldots, \mathbf{y}_n$, there will be n different Andrews curves. A plot consisting of such curves is called an *Andrews plot*.

The Andrews function, $f_{\mathbf{y}}(t)$, has certain useful properties. Specifically,

(i) If the vector $\bar{\mathbf{y}}$ represents the mean of n multivariate observations, $\mathbf{y}_1, \mathbf{y}_2, \ldots, \mathbf{y}_n$, then

$$f_{\bar{\mathbf{y}}}(t) = \bar{f}_{\mathbf{y}}(t) = \frac{1}{n} \sum_{i=1}^{n} f_{\mathbf{y}_i}(t).$$

Thus, the function $f_{\mathbf{y}}(t)$ preserves the mean and as a result, in the Andrews plots the average of the data will be represented by the average of the corresponding Andrews curves in the plot.

(ii) Apart from a constant, the L_2-*distance* between two curves $f_{\mathbf{y}_i}(t)$ and $f_{\mathbf{y}_{i'}}(t)$, defined as, $\int_{-\pi}^{\pi} (f_{\mathbf{y}_i}(t) - f_{\mathbf{y}_{i'}}(t))^2 dt$ is preserved as the squared *Euclidean distance* between the multidimensional points $\mathbf{y}_i = (y_{i1}, y_{i2}, \ldots)'$ and $\mathbf{y}_{i'} = (y_{i'1}, y_{i'2}, \ldots)'$, that is, $\sum_l (y_{il} - y_{i'l})^2$. Specifically,

$$\int_{-\pi}^{\pi} (f_{\mathbf{y}_i}(t) - f_{\mathbf{y}_{i'}}(t))^2 dt = \pi \sum_l (y_{il} - y_{i'l})^2.$$

Thus, the points which are closer to each other are represented as curves which are nearer to each other and vice versa. Therefore, these plots can be used to detect the clusters and the outliers within the data set.

(iii) Apart from a constant, the curves also (almost) preserve the variances provided the variables are uncorrelated with a common variance σ^2. In particular,

$$Var(f_{\mathbf{y}}(t)) = \sigma^2 \left(\frac{1}{2} + sin^2 t + cos^2 t + sin^2 2t + cos^2 2t + \cdots \right).$$

If the number of components in $\mathbf{y}$ is odd, then the variance of $f_{\mathbf{y}}(t)$ is a constant, $\frac{1}{2} p\sigma^2$. In case of even number of components, say p, the variance satisfies

$$\sigma^2 \left(\frac{p-1}{2} \right) \leq Var(f_{\mathbf{y}}(t)) \leq \sigma^2 \left(\frac{p+1}{2} \right).$$

(iv) The function preserves linear relationships. If $\mathbf{y}$ lies on the line joining $\mathbf{x}$ and $\mathbf{z}$, the curve $f_{\mathbf{y}}(t)$ is also sandwiched between the curves $f_{\mathbf{x}}(t)$ and $f_{\mathbf{z}}(t)$.

(v) The representation yields one-dimensional projections. For a fixed $t = t_0$ the value, $f_{\mathbf{y}}(t_0)$, of the function represents, apart from a constant, the length of the projection of $\mathbf{y}$ on the vector $\mathbf{a}(t_0)$. Such projections may reveal patterns, groupings or outliers in the data, in this particular one-dimensional subspace.

Program 2.6 draws Andrews function plots for the cork data. For each observation $\mathbf{y}_i$, the values of the function $f_{\mathbf{y}_i}(t)$ are generated for various values of t between $-\pi$ and π at the steps of $2\pi/100 = \pi/50$ units. These are denoted by the variable Z in Program 2.6. The OUTPUT statement writes the values of the function for each value of t in the SAS data set ANDREWS. A two-dimensional plot of Z versus t is obtained by using PROC GPLOT. Output 2.6a is the result of this program.

```
/* Program 2.6 */

filename gsasfile "prog26.graph";
goptions reset=all gaccess=gsasfile autofeed dev=pslmono;
goptions horigin=1in vorigin=2in;
goptions hsize=6in vsize=8in;
options ls=64 ps=40 nodate nonumber;
title1 h=1.5 'Andrews Function Plot for Cork Data';
*title1 h=1.5 'Modified Andrews Function Plot for Cork Data';
title2 j=l 'Output 2.6';
data andrews;
infile 'cork.dat';
input y1-y4;
tree=_n_;
pi=3.14159265;
s=1/sqrt(2);
inc=2*pi/100;
/* The function z defines the Andrews function and is used for
plotting Andrews plot.  The function mz defines the Modified Andrews
function and is used for plotting Modified Andrews plot. */

do t=-pi to pi by inc;
z=s*y1+sin(t)*y2+cos(t)*y3+sin(2*t)*y4;
*mz=s*(y1+(sin(t)+cos(t))*y2+(sin(t)-cos(t))*y3+
(sin(2*t)+cos(2*t))*y4);
output;
end;

symbol1 i=join v=star;
symbol2 i=join v=+;
symbol3 i=join v=A;
symbol4 i=join v=B;
symbol5 i=join v=C;
symbol6 i=join v=D;
symbol7 i=join v=E;
symbol8 i=join v=F;
symbol9 i=join v=G;
symbol10 i=join v=H;
symbol11 i=join v=I;
symbol12 i=join v=J;
symbol13 i=join v=K;
symbol14 i=join v=L;
symbol15 i=join v=M;
symbol16 i=join v=N;
symbol17 i=join v=O;
symbol18 i=join v=P;
symbol19 i=join v=Q;
symbol20 i=join v=R;
symbol21 i=join v=S;
symbol22 i=join v=T;
symbol23 i=join v=U;
symbol24 i=join v=V;
```

```
symbol25 i=join v=W;
symbol26 i=join v=X;
symbol27 i=join v=Y;
symbol28 i=join v=Z;
legend1 across=4;
proc gplot data=andrews;
plot z*t=tree/vaxis=axis1 haxis=axis2 legend=legend1;
*plot mz*t=tree/vaxis=axis1 haxis=axis2 legend=legend1;
axis1 label=(a=90 h=1.5 f=duplex 'f(t)');
axis2 label=(h=1.5 f=duplex 't')offset=(2);
run;
```

Output 2.6a

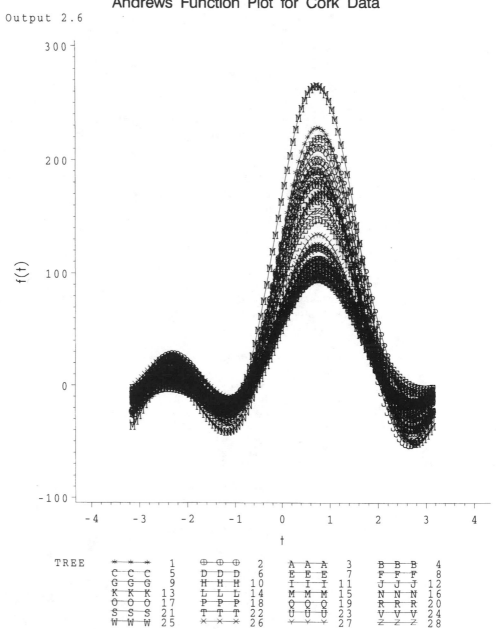

Output 2.6

Andrews Function Plot for Cork Data

Output 2.6
continued

Output 2.6

Modified Andrews Function Plot for Cork Data

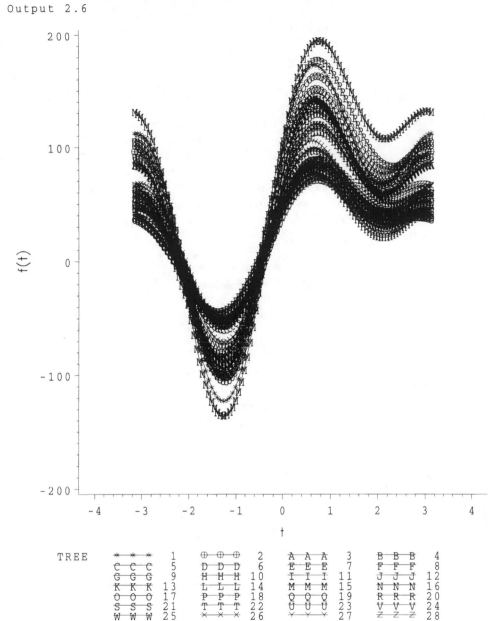

An examination of the Andrews plot indicates that the fifteenth tree (M) again stands out from the rest. As noted in the profile plot from the last section, this tree may be an outlier. As earlier, there is also a group of 12 to 14 trees that are clumped together.

One of the major shortcomings of an Andrews plot is that while it is able to preserve the distance and the average, it does not preserve order. Consequently, its shapes, patterns and clusterings, etc., may be affected by interchanging the coefficients of the terms in the Fourier series. Specifically, if $\mathbf{y}^{(p)}$ is a permutation of $\mathbf{y}$, then $f_{\mathbf{y}(p)}(t) = \mathbf{a}'(t)\mathbf{y}^{(p)}$ and $f_{\mathbf{y}}(t) = \mathbf{a}'(t)\mathbf{y}$ will represent different curves and their shapes and/or specific patterns may be drastically altered or get hidden.

Since low frequencies (that is, high periodicities) in any curve are more readily caught by the human eye than the high frequencies, the effects of the variables listed first in the multivariate vector will be more prominently displayed in the graph. Hence it is useful to associate the most important variables with the low frequencies and thus arrange the variables in a multivariate vector **y** in the decreasing order of their importance. Such an arrangement will also minimize to some extent, the effects of the above mentioned short-comings of these plots. Another shortcoming of the Andrews' functions is the fact that at $t = 0$, the terms indexed with the even numbers vanish. Thus, around $t = 0$, which constitutes the visual center of the plot, the behavior of only odd numbered terms is illustrated. The same can also be said about the left and right ends of the plots—that is, around $t = -\pi$ and $t = \pi$.

Several variations of Andrews plots are available in the literature. Khattree and Naik (1998) give a list of these and also provide a new alternative to Andrews plot. This alternative modification of the Andrews plot is more descriptive and informative. Since in Andrews functions, variables y_j are used as the coefficients of the trigonometric functions, the statistical variation of the data gets intermixed with the periodic variation of the sine and cosine waves, thereby making the plots sometimes harder to interpret. This is an important issue since for multivariate data, there may be several additional complex problems such as dealing with scaling (or lack of it) of variables, correlated variables, unequal variances, etc. These problems make the interpretation difficult. Since the trigonometric functions form a natural choice, they cannot be completely discarded. Khattree and Naik (1998) suggest the modified Andrews' functions,

$$g_{\mathbf{y}}(t) = \frac{1}{\sqrt{2}}\{y_1 + y_2(sin(t) + cos(t)) + y_3(sin(t) - cos(t)) +$$

$$y_4(sin(2t) + cos(2t)) + y_5(sin(2t) - cos(2t)) + \cdots\}, -\pi \leq t \leq \pi.$$

The coefficient $\frac{1}{\sqrt{2}}$ in the above expression is really not needed. We will however include it to provide the plot (Output 2.6b, following) which is essentially on the same scale as the Andrews plot and hence the more readily comparable. Note that the addition and difference of the functions $sin(jt)$ and $cos(jt)$ result in each y_j being exposed to a sine function as well as a cosine function. Consequently, from a statistical point of view, these plots may be more informative. Also, unlike the case of the Andrews original function, the trigonometric terms in $g_{\mathbf{y}}(t)$ do not simultaneously vanish at any given t. Moreover, this function still retains all five properties of the Andrews function listed above.

In Program 2.6 the values of the function $g_{\mathbf{y}_i}(t)$ are also computed and plotted. The second Output 2.6 contains the modified Andrews plot as well. A similar structure of the data is observed in this plot as well. However, this plot seems to display these findings more prominently.

Although we have not illustrated it here, like in the profile plots, one can use the standardized variables for computing and plotting either the Andrews plots or the modified Andrews plots.

2.5   Biplots: Plotting Observations and Variables Together

The biplot, introduced by Gabriel (1971) and discussed extensively in Gower and Hand (1996), is a graphical representation of a data matrix by means of two sets of markers representing its rows and columns respectively. These graphs describe relationships between the observations, by helping to form groups and clusters, as well as between the variables.

Suppose we have a set of n observations on p variables. A biplot describes the relationships among the p variables and the n observations. It is based on the fact (commonly referred to as a QR decomposition) that any n by p matrix **Y** of rank r can be expressed as

$$\mathbf{Y} = \mathbf{GH}', \tag{2.1}$$

where **G** and **H** respectively are n by r and p by r matrices of rank r. Each y_{ij} the $(i, j)^{th}$ element of **Y** is thus expressed as $y_{ij} = \mathbf{g}_i'\mathbf{h}_j$, where $\mathbf{g}_i'$ is the ith row of **G**, and $\mathbf{h}_j$ is the

j^{th} column of $\mathbf{H}'$. Thus each element y_{ij} of $\mathbf{Y}$ is represented by two r-dimensional vectors, $\mathbf{g}_i'$, corresponding to the ith row and $\mathbf{h}_j$, corresponding to the jth column of the matrix $\mathbf{Y}$. It may be remarked that in applications the matrix $\mathbf{Y}$ is often corrected for the means.

When the rank of the data matrix $\mathbf{Y}$ is $r = 2$, the vectors $\mathbf{g}_i$ and $\mathbf{h}_j$ are all of size 2 by 1. Therefore, the $n + p$ points, $\mathbf{g}_1, \ldots, \mathbf{g}_n$ and $\mathbf{h}_1, \ldots, \mathbf{h}_p$ can be plotted on the plane to get the biplot. The same procedure can be adopted if $r = 3$, and in that case the corresponding biplot will be in a three-dimensional space. For the data matrix $\mathbf{Y}$ with the rank $r > 3$, an approximation matrix of the lower rank, say of rank 2 or 3, can be constructed, and it can be used for plotting the biplot yielding an approximate biplot for $\mathbf{Y}$.

To obtain meaningful properties for the vectors $\mathbf{g}_i$ and $\mathbf{h}_j$ Gabriel (1971) suggests that the *singular value decomposition* (SVD) (Rao, 1973, p. 42) of the data matrix be used for the representation (Equation 2.1). That is, write $\mathbf{Y}$ as

$$\mathbf{Y} = \mathbf{U}\Lambda\mathbf{V}' = \sum_{i=1}^{r} \lambda_i \mathbf{u}_i \mathbf{v}_i',$$

where Λ is an r by r diagonal matrix with positive diagonal elements in sorted order $\lambda_1 \geq \lambda_2 \geq \cdots \geq \lambda_r > 0$, $\mathbf{U}$, an n by r matrix with columns $\mathbf{u}_1, \ldots, \mathbf{u}_r$, is such that $\mathbf{U}'\mathbf{U} = \mathbf{I}_r$ and $\mathbf{V}$, a p by r matrix with columns $\mathbf{v}_1, \ldots, \mathbf{v}_r$, is such that $\mathbf{V}'\mathbf{V} = \mathbf{I}_r$. The values $\lambda_1, \ldots, \lambda_r$ are called the singular values of $\mathbf{Y}$. In fact, $\lambda_1^2, \ldots, \lambda_r^2$ are the nonzero eigenvalues of the matrix $\mathbf{Y}\mathbf{Y}'$ or $\mathbf{Y}'\mathbf{Y}$, the vectors $\mathbf{u}_1, \ldots, \mathbf{u}_r$ are the corresponding eigenvectors of $\mathbf{Y}\mathbf{Y}'$, and $\mathbf{v}_1, \ldots, \mathbf{v}_r$ are those corresponding to $\mathbf{Y}'\mathbf{Y}$.

For the rest of the discussions in this section suppose that the data matrix $\mathbf{Y}$ is already centered to have zero column means. Then $\lambda_1^2, \ldots, \lambda_r^2$ respectively represent the portion of the total variation accounted for by the dimensions $1, \ldots, r$. If an approximation of dimension two is used for $\mathbf{Y}$, that is, if

$$\mathbf{Y} \approx \lambda_1 \mathbf{u}_1 \mathbf{v}_1' + \lambda_2 \mathbf{u}_2 \mathbf{v}_2', \tag{2.2}$$

then we get an approximate biplot for $\mathbf{Y}$ and the corresponding goodness of fit is measured by

$$\eta = \frac{\lambda_1^2 + \lambda_2^2}{\sum_{i=1}^{r} \lambda_i^2}.$$

If the actual dimension of $\mathbf{Y}$ is 2, then $\eta = 1$. If $r \geq 3$, $\eta < 1$. Thus if η is near one, the two-dimensional biplot will give a good visual approximation of the data matrix $\mathbf{Y}$ which has dimension r.

Gabriel (1971) suggested these choices for the coordinates of the biplot. Although some of these choices have better statistical interpretations, the interpretation of biplots is essentially the same except that the plotting coordinates are scaled differently.

a.

$$\mathbf{g}_i' = \left(\sqrt{\lambda_1} u_{1i}, \sqrt{\lambda_2} u_{2i} \right), i = 1, \ldots, n,$$

$$\mathbf{h}_j' = \left(\sqrt{\lambda_1} v_{1j}, \sqrt{\lambda_2} v_{2j} \right), j = 1, \ldots, p.$$

This perhaps is the most common representation that is used in practice as it seemingly provides an equal division of weights through λ_1 and λ_2.

b.

$$\mathbf{g}_i' = (u_{1i}, u_{2i}), i = 1, \ldots, n,$$

$$\mathbf{h}_j' = (\lambda_1 v_{1j}, \lambda_2 v_{2j}), j = 1, \ldots, p.$$

This representation has some interesting interpretations. For example, the distance between any two $\mathbf{g}_i$'s, say $\mathbf{g}_i$ and $\mathbf{g}_{i'}$, approximates the squared Mahalanobis distance between the observation vectors (that is, the ith and the i'^{th} rows of the data matrix $\mathbf{Y}$). Further, since $\mathbf{Y'Y} = \mathbf{HH'}$ is n times the variance covariance matrix, the inner product $\mathbf{h}_{j'}\mathbf{h}_{j'}$ between the jth and j'^{th} rows is the covariance between the corresponding variables y_j and $y_{j'}$ and the squared length of the jth row of $\mathbf{H}$ is the variance of variable y_j. Also the angle (cosine of the angle) between any two $\mathbf{h}_j$'s, say $\mathbf{h}_j$ and $\mathbf{h}_{j'}$, approximates the angle between the corresponding columns of the data matrix $\mathbf{Y}$ (approximates to the correlation between variables y_j and $y_{j'}$).

c.

$$\mathbf{g}'_i = (\lambda_1 u_{1i}, \lambda_2 u_{2i}), i = 1, \ldots, n, \tag{2.3}$$

$$\mathbf{h}'_j = (v_{1j}, v_{2j}), j = 1, \ldots, p. \tag{2.4}$$

In this representation the usual Euclidean distance between $\mathbf{g}_i$ and $\mathbf{g}_{i'}$ approximates the Euclidean distance between the ith and the i'^{th} rows of the data matrix $\mathbf{Y}$. This is also the same as the principal components' representation of the data. The $\mathbf{g}_i$'s are the same as the principal component scores and the $\mathbf{h}_j$'s are the principal component loadings or weights (see the PRINCOMP procedure in *SAS/STAT User's Guide*).

The macro BIPLOT in Friendly (1991) is designed to draw a biplot of any of the above three types **(a)**, **(b)**, or **(c)**. For completeness' sake we have given this macro below, with some minor notational changes. Program 2.7 is used to draw a biplot of cork data after correcting for the corresponding means using choice **(a)**, as shown in Output 2.7. It gives the biplot coordinates and the plot using PROC GPLOT.

```
/* Program 2.7 */

filename gsasfile "prog27.graph";
goptions reset=all gaccess=gsasfile autofeed dev=pslmono;
goptions horigin=1in vorigin=2in;
goptions hsize=7in vsize=8in;
options ls=64 ps=45 nodate nonumber;
title1 'Output 2.7';
data cork;
infile 'newcork.dat';
input tree$ north east south west;
%include biplot; /* Include the macro "biplot.sas" */
%biplot( data = cork, var = North East South West,
        id = TREE, factype=SYM, std =STD   );
proc gplot data=biplot;
plot dim2 * dim1  /anno=bianno frame
        href=0 vref=0 lvref=3 lhref=3
         vaxis=axis2 haxis=axis1 vminor=1 hminor=1;
axis1 length=6 in order=(-.8 to .8 by .2)
        offset=(2) label = (h=1.3 'Dimension 1');
axis2 length=6 in order =(-.8 to .8 by .2)
        offset=(2) label=(h=1.3 a=90 r=0  'Dimension 2');
symbol v=none;
title1 h=1.5 'Biplot of Cork Data ';
title2 j=l 'Output 2.7';
title3 'Observations are points, Variables are vectors';
run;


/* The BIPLOT Macro: biplot.sas */

%macro BIPLOT(
        data=_LAST_,
        var =_NUMERIC_,
```

```
           id = ID,
          dim = 2,
      factype=SYM,
        scale=1,
          out=BIPLOT,
         anno=BIANNO,
          std=MEAN,
        pplot=YES);

%let factype=%upcase(&factype);
        %if &factype=GH  %then %let p=0;
%else   %if &factype=SYM %then %let p=.5;
%else   %if &factype=JK  %then %let p=1;
%else %do;
   %put  BIPLOT:  FACTYPE must be GH, SYM, or JK.
"&factype" is not valid.;
   %goto done;
   %end;


Proc IML;
Start BIPLOT(Y,ID,VARS,OUT,power,scale);
   N = nrow(Y);
   P = ncol(Y);
   %if &std = NONE
       %then Y = Y - Y[:] %str(;);   /*remove grand mean */
       %else Y = Y - J(N,1,1)*Y[:,] %str(;); /*remove column means*/
   %if &std = STD %then  %do;
       S = sqrt(Y[##,]);
       Y = Y * diag (1/S);
   %end;


*_ _ Singular value decomposition:
      Y is expressed as U diag(Q) V prime
      Q contains singular values in descending order;
   call svd(u,q,v,y);

   reset fw=8 noname;
   percent = 100*q##2 / q[##];
     *__ cumulate by multiplying by lower
triangular matrix of 1s;

   j = nrow(q);
   tri = (1:j)' * repeat(1,1,j) >= repeat(1,j,1)*(1:j);
   cum = tri*percent;
   Print "Singular values and variance accounted for",,
         q        [colname={'Singular Values'} format=9.4]
         percent [colname={'Percent'} format=8.2]
         cum      [colname={'cum % '} format = 8.2];

    d = &dim;
   *__extract  first d columns of U & V,and first d elements of Q;

     U=U[,1:d];
     V=V[,1:d];
     Q=Q[1:d];

    *__ scale the vectors by QL ,QR;

    QL= diag(Q ## power);
```

```
QR= diag(Q ## (1-power));
A = U * QL;
B = V * QR # scale;
OUT=A // B;

*__ Create observation labels;
id = id // vars';
type = repeat({"OBS "},n,1) //  repeat({"VAR "},p,1);
 id  = concat(type,id);

factype = {"GH" "Symmetric" "JK"}[1+2#power];
print "Biplot Factor Type",factype;
cvar = concat(shape({"DIM"},1,d),char(1:d,1.));
print "Biplot coordinates",
        out[rowname=id colname=cvar];
%if &pplot = YES %then
call pgraf(out,substr(id,5),'Dimension 1','Dimension 2','Biplot');
;
create &out from out[rowname=id colname=cvar];
append from out[rowname=id];
finish;

  use &data;
  read all var{&var} into y[colname=vars rowname=&id];
  power=&p;
  scale=&scale;
  run biplot(y,&id,vars,out,power,scale);
  quit;

 /*__ split id into _type_ and _Name_*/

  data &out;
      set &out;
      drop id;
      length _type_  $3 _name_ $16;
      _type_ = scan(id,1);
      _name_ = scan(id,2);

  /*Annotate  observation labels and variable vectors */
  data &anno;
      set &out;
      length function text $8;
      xsys='2'; ysys='2';
      text=_name_;

  if _type_='OBS' then do;
      color = 'BLACK';
      x = dim1;y = dim2;
      position='5';
      function='LABEL     ';output;
      end;

    if _type_ ='VAR' then do;    /*Draw  line from*/
      color='RED   ';
      x=0; y=0;                  /*the origin to*/
      function ='MOVE'    ;output;
      x=dim1;y=dim2;             /* the variable point*/
      function ='DRAW'    ;output;
      if dim1>=0
```

```
            then position ='6';           /*left justify*/
            else position ='2';           /*right justify*/
        function='LABEL       ';output;  /* variable name */
        end;


%done:
%mend BIPLOT;
run;
```

Output 2.7

Output 2.7

Singular values and variance accounted for

Singular Values	Percent	cum %
1.8965	89.92	89.92
0.5036	6.34	96.26
0.2830	2.00	98.27
0.2634	1.73	100.00

Biplot Factor Type
Symmetric

Biplot coordinates

		DIM1	DIM2
OBS	T1	0.427188	-0.11806
OBS	T2	0.215056	-0.13767
OBS	T3	0.185794	-0.03076
OBS	T4	-0.20326	-0.1258
OBS	T5	-0.24035	-0.07762
OBS	T6	-0.2851	0.065699
OBS	T7	-0.23596	0.142421
OBS	T8	-0.21443	0.224351
OBS	T9	-0.24903	0.172549
OBS	T10	-0.28019	-0.08246
OBS	T11	-0.29024	-0.02005
OBS	T12	0.221587	-0.27774
OBS	T13	0.084425	-0.10957
OBS	T14	0.005872	0.071865
OBS	T15	0.64783	0.016692
OBS	T16	0.134329	0.277319
OBS	T17	0.355213	0.070339
OBS	T18	0.410399	0.322066
OBS	T19	0.289498	-0.04754
OBS	T20	-0.13755	0.005078
OBS	T21	-0.19654	-0.02756
OBS	T22	-0.28693	-0.04313
OBS	T23	0.16359	-0.0926
OBS	T24	-0.1418	-0.08908
OBS	T25	-0.20075	0.025326
OBS	T26	-0.12975	-0.06274
OBS	T27	-0.0916	-0.14513
OBS	T28	0.042699	0.093809
VAR	NORTH	0.703585	0.089931
VAR	EAST	0.665153	0.539686
VAR	SOUTH	0.699975	-0.21337
VAR	WEST	0.684921	-0.39843

Output 2.7
continued   Output 2.7

Biplot of Cork Data

Observations are points, Variables are vectors

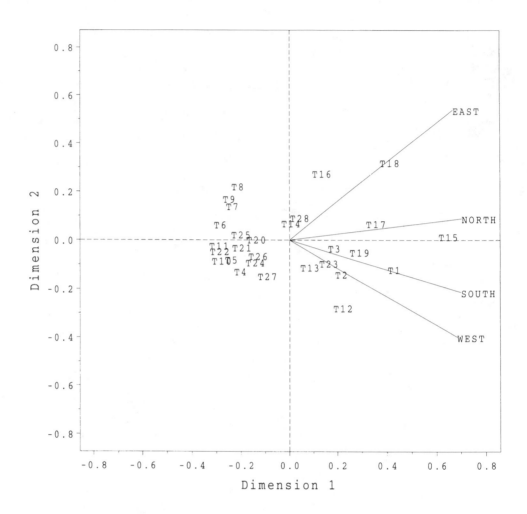

The first two dimensions together account for more than 96% of the variation in the data (see Output 2.7, column corresponding to cum %). Dimension 1 (the horizontal axis) in Output 2.7 is interpreted as the overall score (weighted average), corrected for the mean, of an observation vector corresponding to a tree. This is so since the coefficients of the linear combination of the variables that are used to form the value corresponding to Dimension 1 are all positive and are approximately of the same magnitude (see the first coordinates of the four variables). Hence the trees whose corresponding values fall at the right in the positive direction have an overall larger score. For example, the fifteenth tree (T15) seems to have the largest weighted average after correcting for the mean. The trees T1 and T18 have the next largest averages. There is a group of trees (as noted previously) that have smaller weighted averages than the overall mean and are clustered together. Dimension 2 (the vertical axis) represents a contrast between the cork weight by direction, that is, the contrast N+E-S-W. This is so since the second coordinates of the four variables have positive signs for north and east directions, but negative signs for south and west directions. The trees T12 and T18 seem to have the greatest difference in the cork weights in those directions.

The vectors in the biplot represent various variables (four directions in the present example). The cosines of the angles between these vectors indicate the degree of correlations between the variables. Variables corresponding to the pair of vectors with small angles between them are highly positively correlated. Variables corresponding to the vectors at right angles are uncorrelated, and those with angles more than $90°$ are negatively correlated. An examination of these vectors in the biplot indicates that the correlation between the measurements in the directions of west and south is the highest followed by the correlation between south and north, and then between west and north. The weakest correlation is observed between the measurements in the east and west directions.

2.6   Q-Q Plots for Assessing Multivariate Normality

We present a simple method of assessing multivariate normality using a Q-Q plot. A Q-Q plot is a quantile-quantile comparison of two distributions either or both of which may be empirical or theoretical. When we compare the probability distributions of two random variables, the Q-Q plot will result in a straight line if the two variables are linearly related. These plots are especially good for discriminating in the tail areas of the distributions since the quantiles change more rapidly there and hence the larger distances will occur between consecutive quantiles. Often to assess the multivariate normality of a set of multivariate data, the marginal univariate normality of each component is assessed using the normal Q-Q plots. However, marginal normality of each component does not imply their joint multivariate normality.

In the other approach to assess multivariate normality, Q-Q plots are used rather indirectly. Since for multivariate normal data certain quantiles (as described below) approximately follow a chi-square distribution, the empirical quantiles obtained from the data are therefore plotted against the theoretical quantiles of certain chi-square distributions. The details are described below.

Let $\mathbf{y}_i$, $i = 1, \ldots, n$, be a random sample presumably from a multivariate normal distribution $N_p(\boldsymbol{\mu}, \boldsymbol{\Sigma})$. Then $\mathbf{z}_i = \boldsymbol{\Sigma}^{-1/2}(\mathbf{y}_i - \boldsymbol{\mu})$, $i = 1, \ldots, n$ are *iid* $N_p(\mathbf{0}, \mathbf{I})$ and hence

$$\delta_i^2 = \mathbf{z}_i'\mathbf{z}_i = (\mathbf{y}_i - \boldsymbol{\mu})'\boldsymbol{\Sigma}^{-1}(\mathbf{y}_i - \boldsymbol{\mu}), i = 1, \ldots, n,$$

follows a chi-square distribution with p degrees of freedom. The quantity δ_i^2 is the squared Mahalanobis distance (Mahalanobis, 1936) between $\mathbf{y}_i$ and its expectation $\boldsymbol{\mu}$. If the observations, $\mathbf{y}_i$'s, are indeed from an $N_p(\boldsymbol{\mu}, \boldsymbol{\Sigma})$ then the distances (the sample versions of squared Mahalanobis distances)

$$d_i^2 = (\mathbf{y}_i - \bar{\mathbf{y}})'\mathbf{S}^{-1}(\mathbf{y}_i - \bar{\mathbf{y}}), i = 1, \ldots, n,$$

where $\bar{\mathbf{y}} = \frac{1}{n} \sum_{i=1}^{n} \mathbf{y}_i$ and $\mathbf{S} = \frac{1}{n-1} \sum_{i=1}^{n} (\mathbf{y}_i - \bar{\mathbf{y}})(\mathbf{y}_i - \bar{\mathbf{y}})'$ will approximately be distributed as a chi-square on p degrees of freedom. Hence the suggestion is to plot ordered d_i^2 values against the quantiles of chi-square distribution on p degrees of freedom. If the assumed normality holds, then the plot should approximately resemble a straight line passing through the origin at a $45°$ angle with the horizontal axis.

An efficient program using the IML procedure to plot the above Q-Q plot is provided by Friendly (1991). Program 2.8 uses the PRINCOMP procedure to compute d_i^2, and then produces the Q-Q plot shown in Output 2.8. The cork data has been used for illustration.

```
/* Program 2.8 */

filename gsasfile "prog28.graph";
goptions reset=all gaccess=gsasfile autofeed dev=pslmono;
goptions horigin=1in vorigin=2in;
goptions hsize=6in vsize=8in;
options ls=64 ps=45 nodate nonumber;
title1 h=1.5 'Q-Q Plot for Assessing Normality';
```

```
title2 j=l 'Output 2.8';
title3 'Cork Data';
data a;
infile 'cork.dat';
input y1-y4;
proc princomp data=a cov std out=b noprint;
var y1-y4;
data chiq;
set b;
dsq=uss(of prin1-prin4);
proc sort;
by dsq;
proc means noprint;
var dsq;
output out=chiqn n=totn;
data chiqq;
if(_n_=1) then set chiqn;
set chiq;
novar=4; /* novar=number of variables. */
chisq=cinv(((_n_-.5)/ totn),novar);
if mod(_n_,2)=0 then chiline=chisq;
proc gplot;
plot dsq*chisq chiline*chisq/overlay;
label dsq='Mahalanobis D Square'
    chisq='Chi-Square Quantile';
symbol1 v=star;
symbol2 i=join v=+;
run;
```

In order to explain the computations performed in Program 2.8, let $u_1, \ldots, u_p$ be the p sample principal components with estimated means $\bar{u}_1, \ldots, \bar{u}_p$ and the estimated variances $l_1, \ldots, l_p$ respectively. Then for each $i = 1, \ldots, n$ the standardized variables $(u_{ij} - \bar{u}_j)/\sqrt{l_j}$, $j = 1, \ldots, p$ approximately follow independent standard normal distributions. This yields the approximate distribution of

$$d_i^2 = (\mathbf{y}_i - \bar{\mathbf{y}})'\mathbf{S}^{-1}(\mathbf{y}_i - \bar{\mathbf{y}}) = \sum_{j=1}^{p} \left(\frac{u_{ij} - \bar{u}_j}{\sqrt{l_j}} \right)^2,$$

$i = 1, \ldots, n$ as a chi-square on p degrees of freedom. The values of d_i^2 can be easily computed using the right-most expression in the equation given above. In SAS/STAT software, this can be achieved by using the STD option (to standardize the principal components) in the PROC PRINCOMP statement. Then the SAS function USS can be used to compute their (uncorrected) sums of squares. The automatic variable $\_N\_$ created by SAS is used in the process of computing the probabilities at which the quantiles of the chi-square variable (on $p = 4$ degrees of freedom) are generated. The function CINV computes these quantiles for the given probabilities. For this program these probability values are chosen as $(i - 0.5)/28$, $i = 1, \ldots, 28$ and these are specified as $(\_N\_ - 0.5)/TOTN$ within the function CINV.

A 45° angle line passing through the origin on the same graph using the OVERLAY option in the PLOT statement has also been included. These points denoted by a plus sign (+) in Output 2.8 are joined to form a line. In addition, to avoid counting the number of observations and explicitly specifying the number of observations as TOTN=28.0, we have used the MEANS procedure to calculate the total number of observations (TOTN) in the program. Examination of the plot in Output 2.8 indicates that most of the points are around the 45° angle line passing through the origin. Hence it can be assumed that the observations are coming from a multivariate normal population. This is not surprising since

Output 2.8

Output 2.8

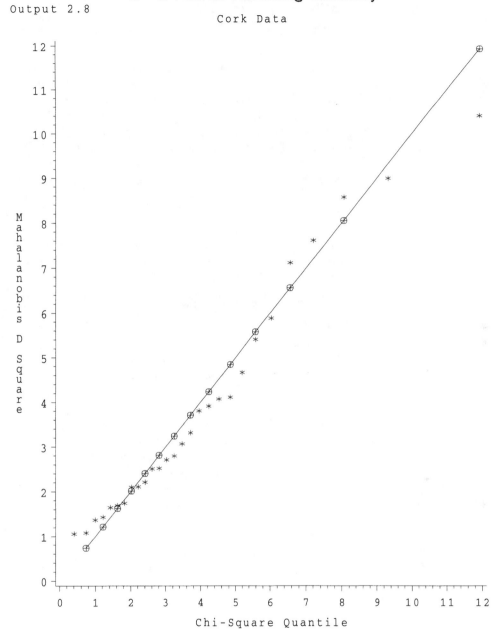

Q−Q Plot for Assessing Normality

Cork Data

the two tests for multivariate normality for the data described in Chapter 1 also resulted in the acceptance of multivariate normality assumption.

The above procedure requires the use of only SAS/STAT and SAS/GRAPH software. If SAS/QC software is available, as an alternative, the QQPLOT statement in the CAPABILITY procedure can be used to draw a Q-Q plot. Since the chi-square distribution on p degrees of freedom is the same as the gamma distribution with a shape parameter $\alpha = p/2$, a scale parameter $\sigma = 2$, and a shift parameter $\theta = 0$, we can use the Q-Q plot corresponding to the gamma distribution in the CAPABILITY procedure. The appropriate SAS statements for Q-Q plot of DSQ, which has an approximate chi-square distribution with $p = 4$ degrees of freedom, are

```
proc capability graphics noprint;
qqplot dsq/gamma(alpha=2 sigma=2 theta=0);
run;
```

The GRAPHICS option results in a graphic plot instead of a simpler line printer plot. The quantiles of data are plotted against the theoretical quantiles of gamma distribution with the specified shape parameter α and shift parameter θ, but with the standardized scale parameter $\sigma = 1$.

As an alternative to a Q-Q plot one may also use the probability plot (using the PROBPLOT statement), which is a plot of quantiles of data against the corresponding theoretical cumulative probabilities for the distribution under consideration. The plot is made on a probability paper where the points on the horizontal axis are not equally spaced, but are spaced in such a way that if the data indeed come from the assumed probability distribution then a straight line pattern will be observed in the probability plot.

Another choice for the same purpose is a probability-probability or simply P-P plot, which is a plot of empirical cumulative probabilities against the theoretical cumulative probabilities under the assumed distribution, corresponding to the respective quantiles computed from the data. If the data indeed come from the assumed probability distribution, the P-P plot will show a linear pattern of the points. This plot can be obtained using the PPPLOT statement of the CAPABILITY procedure.

For the cork data, all three plots have been presented in Output 2.9 and all seem to support (graphically) the hypothesis of multivariate normality of the data. However, a natural question is: which one of the above three plots is preferred? Gnanadesikan (1997) favors the Q-Q plot because it is invariant of any location shift or any scaling of the data and because it is able to detect any departure from the theoretical distribution in the tail areas more effectively than the other two plots.

To assess the multivariate normality, in addition to Q-Q plots of d_i^2, Gnanadesikan (1997) suggests Q-Q plots of certain other beta distributed quantities as well. For details about these and other graphical and numerical methods for assessing or testing for multivariate normality see Gnanadesikan (1997).

```
/* Program 2.9 */

filename gsasfile "prog29a.graph";
goptions reset=all gaccess=gsasfile autofeed dev=pslmono;
goptions horigin=1in vorigin=2in;
goptions hsize=6in vsize=8in;
options ls=64 ps=45 nodate nonumber;
data a;
infile 'cork.dat';
input y1-y4;
run;
proc princomp data=a cov std out=b noprint;
var y1-y4;
run;
data qq;
set b;
dsq=uss(of prin1-prin4);
run;
title1 h=1.5 'Q-Q Plot for Assessing Normality';
title2 j=l 'Output 2.9';
title3 'Cork Data';
proc capability data=qq noprint graphics;
qqplot dsq/gamma(alpha=2 sigma=2 theta=0);
run;
```

```
filename gsasfile "prog29b.graph";
title1 h=1.5 'PROB Plot for Assessing Normality';
title2 j=l 'Output 2.9';
title3 'Cork Data';
proc capability data=qq noprint graphics;
probplot dsq/gamma(alpha=2 sigma=2 theta=0);
run;

filename gsasfile "prog29c.graph";
title1 h=1.5 'P-P Plot for Assessing Normality';
title2 j=l 'Output 2.9';
title3 'Cork Data';
proc capability data=qq noprint graphics;
ppplot dsq/gamma(alpha=2 sigma=2 theta=0);
run;
```

Output 2.9

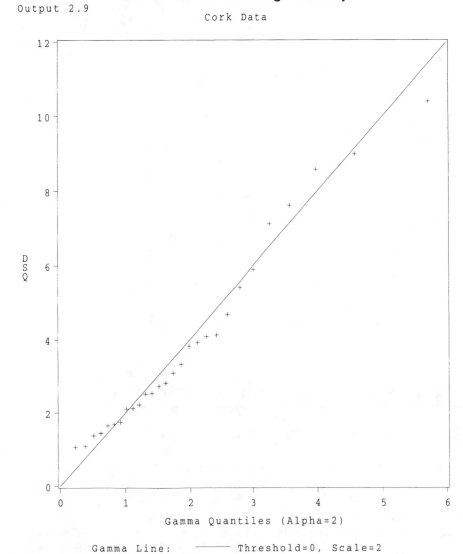

Output 2.9

Q—Q Plot for Assessing Normality

Cork Data

Gamma Quantiles (Alpha=2)

Gamma Line: ———— Threshold=0, Scale=2

Output 2.9
continued   Output 2.9

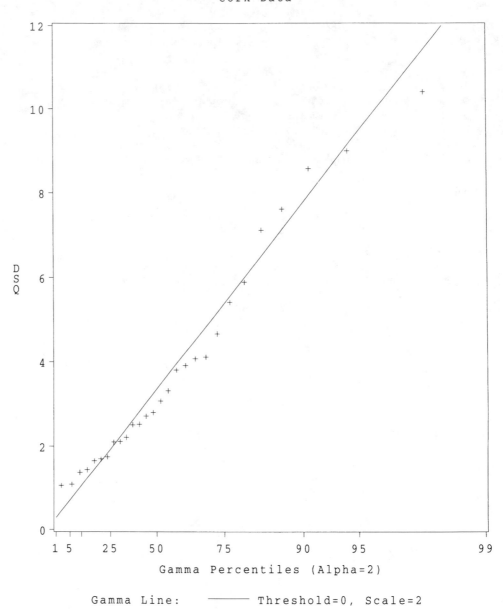

PROB Plot for Assessing Normality

Cork Data

Gamma Line: ——— Threshold=0, Scale=2

2.7   Plots for Detection of Multivariate Outliers

Sometimes a set of observations may violate certain model assumptions (e.g., data follows multivariate normal distribution). These observations are called *outliers*. The plots explained in the previous section can be used to detect possible outliers in the multivariate data. If one or more points fall outside the majority of the points on the Q-Q plot, then those points are suspected to be outliers. However, it is known that the statistics $\bar{\mathbf{y}}$ and $\mathbf{S}$ are both sensitive to the presence of outliers. Hence the squared Mahalanobis distance d_i^2 calculated using the formula $d_i^2 = (\mathbf{y}_i - \bar{\mathbf{y}})'\mathbf{S}^{-1}(\mathbf{y}_i - \bar{\mathbf{y}})$ may not indicate $\mathbf{y}_i$ to be an outlier even when it actually is. An alternative is to use other more robust estimators of $\boldsymbol{\mu}$ and $\boldsymbol{\Sigma}$

Output 2.9
continued   Output 2.9

P−P Plot for Assessing Normality

Cork Data

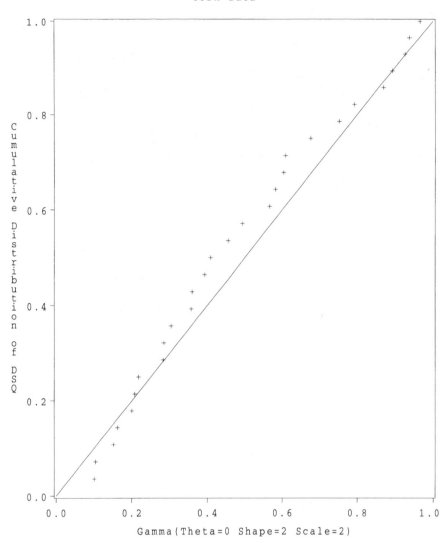

in place of $\bar{\mathbf{y}}$ and $\mathbf{S}$ to compute d_i^2. One such procedure is to use $\bar{\mathbf{y}}_{(i)}$ and $\mathbf{S}_{(i)}$ in place of $\bar{\mathbf{y}}$ and $\mathbf{S}$ in the definition of d_i^2, where $\bar{\mathbf{y}}_{(i)}$ and $\mathbf{S}_{(i)}$ are the values of the sample mean vector and the sample variance covariance matrix without using the ith observation vector. Thus for every $i = 1, \ldots, n$ we compute the robust squared distances

$$D_i^2 = (\mathbf{y}_i - \bar{\mathbf{y}}_{(i)})'\mathbf{S}_{(i)}^{-1}(\mathbf{y}_i - \bar{\mathbf{y}}_{(i)}).$$

To get a Q-Q plot similar to that in the previous section, plot the values of D_i^2 against the quantiles of chi-square distribution with p degrees of freedom. Fortunately, the quantities d_i^2 and D_i^2 are functionally related as

$$D_i^2 = \left(\frac{n}{n-1}\right)^2 \frac{(n-2)\frac{d_i^2}{n-1}}{1 - \left(\frac{n}{n-1}\right)\frac{d_i^2}{n-1}}.$$

This relationship between d_i^2 and D_i^2 can be established using the functional relationship that exists between $\bar{\mathbf{y}}_{(i)}$ and $\bar{\mathbf{y}}$ and that between $\mathbf{S}_{(i)}$ and $\mathbf{S}$. See Cook and Weisberg (1982) for these functional relationships in an univariate context. Program 2.8 can be suitably modified, as shown in Program 2.10, by adding a few additional commands to compute D_i^2 from d_i^2. As earlier, the CAPABILITY procedure can also be used as an alternative.

The robust squared distances (RDSQ), the chi-square quantiles (CHISQ), and the plot are presented in Output 2.10.

```
/* Program 2.10 */

filename gsasfile "prog210.graph";
goptions reset=all gaccess=gsasfile autofeed dev=pslmono;
goptions horigin=1in vorigin=2in;
goptions hsize=6in vsize=8in;
options ls=64 ps=45 nodate nonumber;
title1 'Output 2.10';
data a;
infile 'cork.dat';
input y1-y4;
totn=28.0; * totn is the no. of observations;
novar=4.0; * novar=number of variables;
proc princomp data=a cov std out=b noprint;
var y1-y4;
data chiq;
set b;
tree=_n_;
dsq=uss(of prin1-prin4);
rdsq=(totn/(totn-1))**2*(((totn-2)*dsq/(totn-1))/
(1-(totn*dsq/(totn-1)**2)));
proc sort;
by rdsq;
data chiq;
set chiq;
chisq=cinv(((_n_-.5)/ totn),novar);
if mod(_n_,2)=0 then chiline=chisq;
run;
proc print data=chiq;
var tree rdsq chisq;
run;
proc gplot;
plot rdsq*chisq chiline*chisq/overlay;
label rdsq='Robust Mahalanobis D Square'
      chisq='Chi-Square Quantile';
symbol1 v=star;
symbol2 i=join v=+;
title1 h=1.5 'Chi-square Q-Q Plot of Robust Squared Distances';
title2 j=l 'Output 2.10';
run;
```

Output 2.10

Output 2.10

OBS	TREE	RDSQ	CHISQ
1	13	1.1471	0.4041
2	21	1.1740	0.7390
3	22	1.5008	0.9939
4	20	1.5727	1.2188

5	25	1.8211	1.4282
6	14	1.8729	1.6290
7	5	1.9360	1.8253
8	11	2.3652	2.0197
9	23	2.3789	2.2142
10	7	2.5047	2.4106
11	2	2.8762	2.6103
12	4	2.8910	2.8148
13	3	3.1381	3.0255
14	9	3.2502	3.2440
15	6	3.6033	3.4720
16	10	3.9346	3.7117
17	28	4.6177	3.9654
18	24	4.7694	4.2361
19	17	5.0041	4.5276
20	8	5.0601	4.8450
21	26	5.8892	5.1951
22	27	7.0757	5.5875
23	12	7.8851	6.0366
24	19	10.1572	6.5654
25	1	11.1555	7.2140
26	18	13.2461	8.0633
27	15	14.2371	9.3204
28	16	17.9606	11.9329

The plot indicates that there is definitely one point, and possibly more, that stands out from the rest. The points that stand separate are those with high robust distance. As shown by the RDSQ values presented in Output 2.10, the highest distance turns out to be for observation 16 (not 15, as we may have expected). The squared distance D^2_{16} for this observation is 17.32 which is considerably larger compared to the corresponding chi-square value of 11.93. Observation 15 with a D^2_i value of 13.73, observation 18 with 12.77, and observation 1 with 10.76 also stand apart from the majority of the data vectors. As we have previously seen, observation 15 is different from the rest because of the magnitudes of its individual components. A closer look at observation 16 reveals that this particular tree is unique in the sense that its measurement in the direction of south is unusually low compared to those in the other directions. This phenomenon is markedly different from the majority of the trees for which the measurement in the direction of south is higher. See the profile plot in Output 2.5. Hence the sixteenth tree may be classified as an outlier. Trees 15, 18, and 1 may be classified as outliers also.

Detection of outliers from a set of multivariate data is a difficult problem. Rao (1964) has suggested another method for detection of outliers using a distance measure based on the last few principal components. It has also been observed that Mardia's multivariate kurtosis can be used as a measure to detect any outliers in the data that are supposedly distributed as the multivariate normal. See Schwager and Margolin (1982), Das and Sinha (1986), and Naik (1989) for details. Further elaborate discussions of outlier detection methods may be found in Barnett and Lewis (1994).

2.8   Bivariate Normal Distribution

One of the most commonly used distributions in multivariate data analysis is multivariate normal distribution. This distribution has been briefly discussed in Chapter 1. The p-variate normal distribution with $p = 2$ is often referred to as a *bivariate normal distribution*. For a bivariate normal distribution, it is possible to present much of the information about

Output 2.10
continued

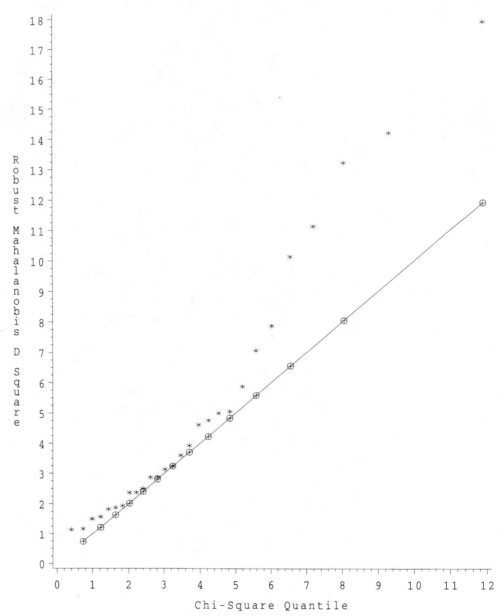

the distribution very effectively in a graph. In this section we consider a bivariate normal distribution and give plots for its probability density function (pdf) as well as its contours. *Contours* of a function on the higher dimension are the graphs of the projections of the function on a plane at the fixed values of the function. In the present context, these plots can help us visualize the shape of the pdf of the multivariate normal distribution by helping us to examine its various bivariate marginal pdfs and their contour plots.

2.8.1   Probability Density Function Plotting

The pdf of a p-variate normal distribution with mean vector $\boldsymbol{\mu}$ and variance covariance matrix $\boldsymbol{\Sigma}$ is given by

$$f(\mathbf{y}) = \frac{1}{(2\pi)^{p/2}|\boldsymbol{\Sigma}|^{1/2}} exp \left\{ -\frac{1}{2}(\mathbf{y} - \boldsymbol{\mu})'\boldsymbol{\Sigma}^{-1}(\mathbf{y} - \boldsymbol{\mu}) \right\}.$$

When $p = 2$, the mean vector of $\mathbf{y} = (y_1, y_2)'$ is $\boldsymbol{\mu} = (\mu_1, \mu_2)'$ and the dispersion matrix is a 2 by 2 matrix

$$\boldsymbol{\Sigma} = \begin{bmatrix} \sigma_1^2 & \rho\sigma_1\sigma_2 \\ \rho\sigma_1\sigma_2 & \sigma_2^2 \end{bmatrix},$$

where $\sigma_1^2 = var(y_1)$, $\sigma_2^2 = var(y_2)$ and ρ is the correlation coefficient between y_1 and y_2. In Program 2.11 we use PROC G3D to plot the pdf of the bivariate normal distribution for specific values of $\mu_1 = 0.0$, $\mu_2 = 0.0$, $\sigma_1^2 = 2.0$, $\sigma_2^2 = 1.0$, and $\rho = 0.5$. The KEEP statement in the program saves the variables that are listed in that statement. Alternatively, a DROP statement could be used to drop the variables not needed. The output of Program 2.11 is shown in Output 2.11.

```
/* Program 2.11 */

filename gsasfile "prog211.graph";
goptions reset=all gaccess=gsasfile  autofeed dev=pslmono;
goptions horigin=1in vorigin=2in;
goptions hsize=6in vsize=8in;
options ls=64 ps=45 nodate nonumber;
title1 h=1.5 'PDF of Bivariate Normal Distribution';
title2 j=l 'Output 2.11';
title3 'Mu_1=0, Mu_2=0, Sigma_1^2=2, Sigma_2^2=1 and Rho=0.5';
data normal;
mu_1=0.0;
mu_2=0.0;
vy1=2;
vy2=1;
rho=.5;
keep y1 y2 z;
label z='Density';
con=1/(2*3.141592654*sqrt(vy1*vy2*(1-rho*rho)));
do y1=-4 to 4 by 0.10;
do y2=-3 to 3 by 0.10;
zy1=(y1-mu_1)/sqrt(vy1);
zy2=(y2-mu_2)/sqrt(vy2);
hy=zy1**2+zy2**2-2*rho*zy1*zy2;
z=con*exp(-hy/(2*(1-rho**2)));
if z>.001 then output;
end;
end;
proc g3d data=normal;
plot y1*y2=z;
*plot y1*y2=z/ rotate=30;
run;
```

An examination of the pdf plot in Output 2.11 shows how the variance of y_1 being larger than that of y_2 affects the density plot. That is, the spread of the plot on the axis representing the variable y_1 is more than that on the axis representing y_2. Further, the effect of positive correlation between these two variables on the density plot can be seen from the shape of the density surface which is concentrated along the line $y_1 = y_2$ in the horizontal plane.

Output 2.11

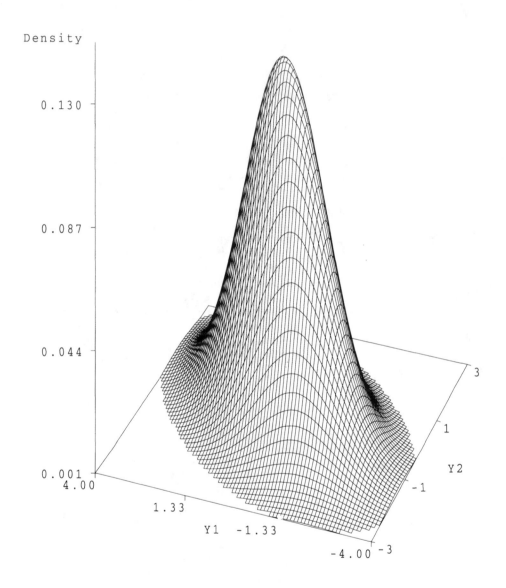

Output 2.11

PDF of Bivariate Normal Distribution

Mu_1=0, Mu_2=0, Sigma_1^2=2, Sigma_2^2=1 and Rho=0.5

2.8.2   Contour Plot of Density

The contour plots of a bivariate probability density function show the degrees of association between the two random variables. For the same data as Program 2.11 we draw the contours of the pdf using the GCONTOUR procedure. By adding a few more SAS statements to Program 2.11 we have Program 2.12 which achieves the desired objective. The output is shown in Output 2.12.

```
/* Program 2.12 */

filename gsasfile "prog212.graph";
goptions reset=all gaccess=gsasfile  autofeed dev=pslmono;
goptions horigin=1in vorigin=2in;
goptions hsize=6in vsize=8in;
options ls=64 ps=45 nodate nonumber;
title1 h=1.5 'Contours of Bivariate Normal Distribution';
title2 j=l 'Output 2.12';
title3 'Mu_1=0, Mu_2=0, Sigma_1^2=2, Sigma_2^2=1 and Rho=0.5';
data normal;
vy1=2;
vy2=1;
rho=.5;
keep y1 y2 z;
label z='Density';
con=1/(2*3.141592654*sqrt(vy1*vy2*(1-rho*rho)));
do y1=-4 to 4 by 0.3;
do y2=-3 to 3 by 0.10;
zy1=y1/sqrt(vy1);
zy2=y2/sqrt(vy2);
hy=zy1**2+zy2**2-2*rho*zy1*zy2;
z=con*exp(-hy/(2*(1-rho**2)));
if z>.001 then output;
end;
end;
proc gcontour data=normal;
plot y2*y1=z/levels=.02 .03 .04 .05 .06 .07 .08;
run;
```

The LEVELS option in the PLOT statement of the program is used to specify the fixed values of the pdf for which the contours are to be drawn. These values should be the plausible values of the function and hence should be between zero and the maximum possible value of the pdf. Noting that the maximum value of the pdf of a bivariate normal distribution corresponds to $y_1 = \mu_1$ and $y_2 = \mu_2$, we can determine the maximum value that can be given in the LEVELS option, for the given values of σ_1^2, σ_2^2, and ρ. For example, the maximum value of the pdf is $\frac{1}{2\pi\sqrt{1.5}}$ for the choices $\mu_1 = \mu_2 = 0$, $\sigma_1^2 = 2.0$, $\sigma_2^2 = 1.0$, and $\rho = 0.5$.

The contours of a bivariate probability density function have the following interpretations.

- For a zero correlation between the variables and equal variances, the contours are circles centered at (μ_1, μ_2).

- For zero correlation and the variance of y_1 greater than that of y_2, the contours are ellipses whose major axes are parallel to the horizontal axis. (If the variance of y_2 is greater than that of y_1 then the major axis will be parallel to vertical axis.)

- If the correlation between the variables is nonzero, then the contours are ellipses.

- Additionally if the two variances are equal then for any contour, the major axis is at an angle (with the horizontal axis) whose cosine is same as the correlation coefficient between the two variables.

The contours in Output 2.12 indicate the positive correlation between the two variables y_1 and y_2.

Output 2.12

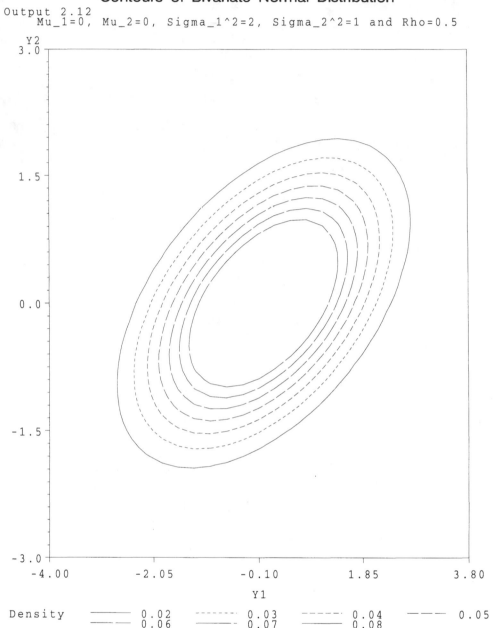

Contours of Bivariate Normal Distribution

Output 2.12
    Mu_1=0, Mu_2=0, Sigma_1^2=2, Sigma_2^2=1 and Rho=0.5

2.9   SAS/INSIGHT Software

SAS/INSIGHT software is an interactive tool for data exploration and analysis. We can use it to explore data through a variety of interactive graphs and analyses linked across multiple windows. In addition, it can be used to analyze univariate distributions, investigate multivariate distributions, and fit explanatory models using analysis of variance, regression, and the generalized linear model. The following summary of features is adapted from the *SAS/INSIGHT User's Guide, Version 6, First Edition* published by SAS Institute.

SAS/INSIGHT software offers a comprehensive set of graphical tools. For example, it can rotate data in three-dimensional plots; create Q-Q (quantile-quantile) plots; apply transformation to data; fit curves including polynomials, provide kernel density estimates, and smoothing techniques using spline curves; and create residual and leverage plots. Because it is a part of the SAS System, SAS/INSIGHT software can explore results from any SAS procedure.

The statements given in Program 2.4 invoke SAS/INSIGHT software. Once SAS/INSIGHT is invoked, there are several options for extensive data analysis to choose from just by clicking the button on the mouse. The scatter plot matrix given in Output 2.4 was obtained using SAS/INSIGHT.

2.10   Concluding Remarks

Finally it may be remarked that multivariate graphical methodology is an area which is still evolving. A variety of other techniques such as Chernoff faces, star plots, and Wegman plots also exist. Further variations of techniques provided in this chapter can be developed to fit the specific type of problems. Books by Chambers, Cleveland, Kleiner, and Tukey (1983) and Gnanadesikan (1997) provide elaborate discussions of various other methods and Friendly (1991) gives SAS macros for many of these methods.

Multivariate Regression

3.1   Introduction   61

3.2   Statistical Background   62

3.3   Least Squares Estimation   63

3.4   ANOVA Partitioning   64

3.5   Testing Hypotheses: Linear Hypotheses   66

3.6   Simultaneous Confidence Intervals   84

3.7   Multiple Response Surface Modeling   87

3.8   General Linear Hypotheses   91

3.9   Variance and Bias Analyses for Calibration Problems   98

3.10 Regression Diagnostics   102

3.11 Concluding Remarks   116

3.1   Introduction

Regression analysis primarily deals with the issues related to estimating or predicting the expected value of the dependent or response variable using the known values of one or more independent or predictor variables. Usually a model is postulated relating the response variable to the predictor variables with certain unknown coefficients. A model that is linear in these coefficients is often referred to as a *linear regression model* or simply a linear model.

The multivariate linear regression model is a natural generalization of a (univariate) linear regression model. That is, two or more possibly correlated dependent variables are simultaneously modeled as the linear functions of the same set of predictor variables. For reasons of mathematical convenience in developing an appropriate theory, it is required that the particular model for each response variable be in exactly the same functional form. For example, if y_1 and y_2 are the response variables and x_1 and x_2 are the predictors, then the univariate models $y_1 = a_0 + a_1 x_1 + a_2 x_2 + \epsilon_1$ and $y_2 = b_0 + b_1 x_1 + b_2 x_2 + \epsilon_2$ have the same functional forms, whereas the models $y_1 = a_0 + a_1 x_1 + a_2 x_2 + \epsilon_1$ and $y_2 = b_0 + b_1 x_1 + \epsilon_2$ do not. It is possible to argue that the last model (for y_2) has the same functional form as the model for y_1 with the choice $b_2 = 0$. However, this suggests that b_2 is completely known and hence its estimation is irrelevant. But multivariate regression theory assumes that all the coefficients in the model are unknown and are to be estimated.

In the univariate regression models, we assume that there are n observations available on a response variable y as well as on predictors $x_1, \ldots, x_k$. Suppose these n data values on y are stored in an n by 1 column vector $\mathbf{y}$ and values on x_i, $i = 1, \ldots, k$ are stacked, in the same order, in an n by 1 vector $\mathbf{x}_i$. Then the complete linear model for the data can be expressed as the linear relation between these column vectors

$$\mathbf{y} = \beta_0 \mathbf{1}_n + \beta_1 \mathbf{x}_1 + \cdots + \beta_k \mathbf{x}_k + \boldsymbol{\epsilon}.$$

In multivariate situations, that is, when there are two or more response variables, the functional form of the linear model for each of these response variables is assumed to be the same as above. However, each model will have a different set of unknown coefficients $\beta_0, \ldots, \beta_k$ and a different error vector.

3.2 Statistical Background

A multivariate linear model in p (possibly correlated) response variables $y_1, \ldots, y_p$ and k independent or predictor variables $x_1, \ldots, x_k$ is represented by a system of p univariate linear models

$$
\left.
\begin{aligned}
\mathbf{y}_1 &= \beta_{01}\mathbf{1}_n + \beta_{11}\mathbf{x}_1 + \beta_{21}\mathbf{x}_2 + \cdots + \beta_{k1}\mathbf{x}_k + \boldsymbol{\epsilon}_1 \\
\mathbf{y}_2 &= \beta_{02}\mathbf{1}_n + \beta_{12}\mathbf{x}_1 + \beta_{22}\mathbf{x}_2 + \cdots + \beta_{k2}\mathbf{x}_k + \boldsymbol{\epsilon}_2 \\
&\;\;\vdots \\
\mathbf{y}_p &= \beta_{0p}\mathbf{1}_n + \beta_{1p}\mathbf{x}_1 + \beta_{2p}\mathbf{x}_2 + \cdots + \beta_{kp}\mathbf{x}_k + \boldsymbol{\epsilon}_p
\end{aligned}
\right\},
\tag{3.1}
$$

where $\mathbf{y}_i$, $\mathbf{x}_i$, and $\boldsymbol{\epsilon}_i$ are all n by 1 vectors. The vectors $\mathbf{y}_i$ and $\mathbf{x}_i$ are respectively the data vectors on the variables y_i and x_i. These equations can compactly be represented using matrix notation as

$$
\mathbf{Y} = \mathbf{XB} + \mathcal{E},
\tag{3.2}
$$

where

$$
\begin{aligned}
\mathbf{Y}_{n \times p} &= (\mathbf{y}_1 : \mathbf{y}_2 : \ldots : \mathbf{y}_p), \\
\mathbf{X}_{n \times (k+1)} &= (\mathbf{1}_n : \mathbf{x}_1 : \mathbf{x}_2 : \ldots : \mathbf{x}_k),
\end{aligned}
$$

$$
\mathbf{B} =
\begin{bmatrix}
\beta_{01} & \beta_{02} & \cdots & \beta_{0p} \\
\beta_{11} & \beta_{12} & \cdots & \beta_{1p} \\
\cdot & & & \\
\cdot & & & \\
\beta_{k1} & \beta_{k2} & \cdots & \beta_{kp}
\end{bmatrix}
$$

and $\mathcal{E} = (\boldsymbol{\epsilon}_1 : \boldsymbol{\epsilon}_2 : \ldots : \boldsymbol{\epsilon}_p)$. The vector $\mathbf{1}_n$ here represents an n by 1 column vector with all elements as unity. Assume that $n > (k + 1)$.

In case we wish to deal with a model without an intercept term, we could still write the corresponding model as in Equation 3.2 by omitting the first columns of the matrix $\mathbf{X}$ and the first row of matrix $\mathbf{B}$ defined above. The only additional change would be to replace $k + 1$ by k in what follows. One such situation where models without intercept terms are encountered is the analysis of mixture experiment data.

A typical equation in Equation 3.1, say the ith one,

$$
\mathbf{y}_i = \beta_{0i}\mathbf{1}_n + \beta_{1i}\mathbf{x}_1 + \cdots + \beta_{ki}\mathbf{x}_k + \boldsymbol{\epsilon}_i
$$

represents a univariate regression model, which can be written as $\mathbf{y}_i = \mathbf{X}\boldsymbol{\beta}_i + \boldsymbol{\epsilon}_i$ with $\boldsymbol{\beta}_i = (\beta_{0i}, \beta_{1i}, \ldots, \beta_{ki})'$ and hence could be analyzed independently of the other regression models in Equation 3.1. However, since the response variables may themselves be correlated with each other, this dependence should also be taken into account when drawing the statistical conclusions using the inferential methods. This suggests the need for using the model in Equation 3.2, where the collection of all the dependent variables is analyzed as a single data set.

Similarly, because observations or responses on different units are assumed to be independent, each $\boldsymbol{\epsilon}_i$ is assumed to be distributed with zero mean and the variance-covariance matrix $D(\boldsymbol{\epsilon}_i) = \sigma_{ii}\mathbf{I}_n$, where $\mathbf{I}_n$ is an n by n identity matrix. However, to incorporate the dependence between the response variables, it is assumed that for $i, j = 1, \ldots, p$, $cov(\mathbf{y}_i, \mathbf{y}_j|\mathbf{X}) = cov(\boldsymbol{\epsilon}_i, \boldsymbol{\epsilon}_j) = \sigma_{ij}\mathbf{I}_n$. In other words, while the ith column of $\mathcal{E}$ has the variance-covariance matrix $\sigma_{ii}\mathbf{I}_n$, a typical row of $\mathcal{E}$ has that as $\boldsymbol{\Sigma} = (\sigma_{ij})$, a symmetric p by p matrix that is assumed to be positive definite.

As in the univariate linear regression setup, assume that the regression matrix $\mathbf{X}$ is of full column rank (that is all columns of $\mathbf{X}$ are linearly independent), thereby implying that $Rank(\mathbf{X}'\mathbf{X}) = Rank(\mathbf{X}) = k + 1$. This, in turn, ensures the existence of the inverse of $\mathbf{X}'\mathbf{X}$. If $\mathbf{X}$ is not of full rank, $(\mathbf{X}'\mathbf{X})^-$, a *generalized inverse* (Rao, 1973) of $\mathbf{X}'\mathbf{X}$, will replace $(\mathbf{X}'\mathbf{X})^{-1}$ in most situations. However, extra care is needed in interpreting the results of the data analysis in such instances.

An assumption of multivariate normality for error vectors is needed for hypothesis testing problems and construction of confidence regions, even though it is not needed for the linear estimation problems. Most of the resulting exact and approximate statistical tests for the multivariate linear regression models are the consequences of this assumption.

3.3   Least Squares Estimation

A natural criterion to obtain some meaningful estimators of $\mathbf{B}$ is to minimize $\sum_{i=1}^{p} \boldsymbol{\epsilon}_i'\boldsymbol{\epsilon}_i = \sum_{i=1}^{p}(\mathbf{y}_i - \mathbf{X}\boldsymbol{\beta}_i)'(\mathbf{y}_i - \mathbf{X}\boldsymbol{\beta}_i)$ with respect to the matrix $\mathbf{B} = (\boldsymbol{\beta}_1 : \boldsymbol{\beta}_2 : \ldots : \boldsymbol{\beta}_p)$. This is merely the sum of the squared deviations from the corresponding means over all observations and over all responses or dependent variables. This criterion is the same as that of minimizing $tr(\mathbf{Y} - \mathbf{XB})'(\mathbf{Y} - \mathbf{XB})$, the *trace* (the sum of the diagonal elements) of the p by p matrix $(\mathbf{Y} - \mathbf{XB})'(\mathbf{Y} - \mathbf{XB})$, resulting in the system of normal (matrix) equations

$$(\mathbf{X}'\mathbf{X})\hat{\mathbf{B}} = \mathbf{X}'\mathbf{Y} \tag{3.3}$$

and yielding

$$\hat{\mathbf{B}} = (\mathbf{X}'\mathbf{X})^{-1}\mathbf{X}'\mathbf{Y} \tag{3.4}$$

as the least squares estimator of matrix $\mathbf{B}$. It means that $\hat{\boldsymbol{\beta}}_i$, the ith column of $\hat{\mathbf{B}}$, which estimates $\boldsymbol{\beta}_i$, the ith column of $\mathbf{B}$, is given by

$$\hat{\boldsymbol{\beta}}_i = (\mathbf{X}'\mathbf{X})^{-1}\mathbf{X}'\mathbf{y}_i, \ i = 1, \ldots, p.$$

It is easy to demonstrate that $\hat{\mathbf{B}}$ given by Equation 3.4 is unbiased for $\mathbf{B}$, that is, $E(\hat{\mathbf{B}}) = \mathbf{B}$. Further, $cov(\hat{\boldsymbol{\beta}}_i, \hat{\boldsymbol{\beta}}_j) = \sigma_{ij}(\mathbf{X}'\mathbf{X})^{-1}$ for all $i, j = 1, \ldots, p$. In addition, under the assumption of the model in Equation 3.2, $\hat{\mathbf{B}}$ is the Best Linear Unbiased Estimator (BLUE) of $\mathbf{B}$ in the sense that it has the smallest total variance among all linear unbiased estimators. By *total variance* we mean the sum of the variances of all elements of the matrix used as the estimator.

When the matrix $\mathbf{X}$ is not of full rank, a least squares solution to the system of normal equations, Equation 3.3, is given by

$$\hat{\mathbf{B}}^{(g)} = (\mathbf{X}'\mathbf{X})^-\mathbf{X}'\mathbf{Y} \tag{3.5}$$

and

$$\hat{\boldsymbol{\beta}}_i^{(g)} = (\mathbf{X}'\mathbf{X})^-\mathbf{X}'\mathbf{y}_i, \ i = 1, \ldots, p.$$

The matrix $\hat{\mathbf{B}}^{(g)}$ defined in Equation 3.5

- is not unique,
- depends on the particular choice of the generalized inverse used, and
- merely represents one of the many solutions to a *singular* (that is where at least one linear equation is redundant and is implied by the others) system of normal equations.

In this sense, $\hat{\mathbf{B}}^{(g)}$ is really not an estimator and one or more components of $\hat{\mathbf{B}}^{(g)}$ may be biased for their counterparts in $\mathbf{B}$. However, as indicated in Searle (1971), certain linear functions of $\mathbf{B}$ can still be uniquely estimated. Such functions are called the *estimable functions*. Specifically, as shown by Bose (1951), and Searle (1971), a linear function $\mathbf{c}'\mathbf{B}$, where $\mathbf{c} \neq \mathbf{0}$ is a nonrandom $(k+1)$ by 1 vector, is *estimable* if and only if

$$(\mathbf{X}'\mathbf{X})(\mathbf{X}'\mathbf{X})^{-}\mathbf{c} = \mathbf{c}.$$

Accordingly, a linear hypothesis on the regression parameters will be a "testable hypothesis" if and only if it involves only the estimable functions of $\mathbf{B}$.

3.4   ANOVA Partitioning

In the multivariate context, the role of the total sum of squares is played by the p by p positive definite matrix of (corrected) total sums of squares and crossproducts (SS&CP) defined as

$$\mathbf{T} = \mathbf{Y}'\mathbf{Y} - \frac{1}{n}\mathbf{Y}'\mathbf{1}_n\mathbf{1}_n'\mathbf{Y} = \mathbf{Y}'\left(\mathbf{I}_n - \frac{1}{n}\mathbf{1}_n\mathbf{1}_n'\right)\mathbf{Y}. \qquad (3.6)$$

Apart from the dividing factor, a typical element of the matrix in Equation 3.6, say the one corresponding to the ith row and j^{th} column, is the same as the sample covariance between the ith and j^{th} dependent variables. Consequently, the diagonal elements of $\mathbf{T}$ are the (corrected) total sums of squares for the respective dependent variables.

Assuming that $Rank(\mathbf{X}) = k + 1$, this matrix can be partitioned as the sum of the two p by p positive definite matrices

$$\mathbf{T} = \mathbf{R} + \mathbf{E},$$

where

$$\mathbf{R} = \mathbf{Y}'\left[\mathbf{X}(\mathbf{X}'\mathbf{X})^{-1}\mathbf{X}' - \frac{1}{n}\mathbf{1}_n\mathbf{1}_n'\right]\mathbf{Y} = \mathbf{Y}'\mathbf{X}\hat{\mathbf{B}} - \frac{1}{n}\mathbf{Y}'\mathbf{1}_n\mathbf{1}_n'\mathbf{Y}, \qquad (3.7)$$

$$\mathbf{E} = \mathbf{Y}'[\mathbf{I} - \mathbf{X}(\mathbf{X}'\mathbf{X})^{-1}\mathbf{X}']\mathbf{Y} = \mathbf{Y}'\mathbf{Y} - \mathbf{Y}'\mathbf{X}(\mathbf{X}'\mathbf{X})^{-1}\mathbf{X}'\mathbf{Y}$$

$$= \mathbf{Y}'\mathbf{Y} - \mathbf{Y}'\mathbf{X}\hat{\mathbf{B}} = (\mathbf{Y} - \hat{\mathbf{Y}})'(\mathbf{Y} - \hat{\mathbf{Y}}), \qquad (3.8)$$

and $\hat{\mathbf{Y}} = \mathbf{X}\hat{\mathbf{B}}$ is the matrix of the predicted values of matrix $\mathbf{Y}$. The matrix $\mathbf{R}$ represents the matrix of model or regression sums of squares and crossproducts, while the matrix $\mathbf{E}$ represents that corresponding to error. Note that the diagonal elements of these matrices respectively represent the usual regression and error sums of squares for the corresponding dependent variables in the univariate linear regression setup.

Table 3.1 summarizes the partitioning explained above, along with a similar partitioning for the degrees of freedom (df).

An unbiased estimator of $\boldsymbol{\Sigma}$ is given by

$$\hat{\boldsymbol{\Sigma}} = \mathbf{E}/(n - k - 1). \qquad (3.9)$$

TABLE 3.1 Multivariate Analysis of Variance (Manova)

Source	df	SS & CP	E(SS & CP)
Regression	k	**R**	$k\Sigma + \mathbf{B}'\mathbf{X}'(\mathbf{I_n} - \frac{1}{n}\mathbf{11}')\mathbf{XB}$
Error	$n - k - 1$	**E**	$(n - k - 1)\Sigma$
Corrected Total	$n - 1$	**T**	

When $\mathbf{X}$ is not of full rank, $(\mathbf{X}'\mathbf{X})^{-1}$ is replaced by $(\mathbf{X}'\mathbf{X})^{-}$, a generalized inverse of $\mathbf{X}'\mathbf{X}$ in the formulas in Equations 3.7 and 3.8, but the matrices $\mathbf{R}$, $\mathbf{E}$ and $\hat{\mathbf{Y}}$ are invariant of the choice of a particular generalized inverse and remain the same regardless of which generalized inverse is used. However, in this case the unbiased estimator of Σ is given by $\mathbf{E}/(n - Rank(\mathbf{X}))$, which differs from Equation 3.9 in its denominator.

Depending on the rank of $\mathbf{R}$, the matrix $\mathbf{R}$ can further be partitioned into two or more positive definite matrices. This fact is useful in developing the tests for various linear hypotheses on $\mathbf{B}$. If needed, for example, as in the lack-of-fit analysis, a further partitioning of matrix $\mathbf{E}$ is also possible.

The matrix $\mathbf{R}$, of regression sums of squares and crossproducts, measures the effect of the part of the model involving the independent variables. By contrast, $\mathbf{E}$, the error sums of squares and product matrix, measures the effect due to random error or the variation not explained by the independent variables. Further partitioning of $\mathbf{R}$ and $\mathbf{E}$ can also be given certain similar interpretations.

In univariate regression models, the coefficient of determination R^2, which is the ratio of regression sum of squares to total sum of squares, is taken as an index to measure the adequacy of the fitted model. Analogously, in the present context, it is possible to define

$$|\mathbf{R}|/|\mathbf{R} + \mathbf{E}|$$

and

$$\frac{1}{p}tr[\mathbf{R}(\mathbf{R} + \mathbf{E})^{-1}]$$

as two possible generalizations of R^2. In the hypothesis testing context, the latter measure is often referred to as *Pillai's trace statistic*. These indices can be interpreted in essentially the same way as the univariate coefficient of determination R^2.

Another useful measure of the strength of the relationship or the adequacy of the model can be defined as $1 - |\mathbf{E}|/|\mathbf{R} + \mathbf{E}| = 1 - \Lambda$, where $\Lambda = |\mathbf{E}|/|\mathbf{R} + \mathbf{E}|$ is called the *Wilks' ratio*. However, this index of association is strongly biased. Jobson (1992) provides two modifications of $1 - \Lambda$, one of which has considerably less bias while another provided by Tatsuoka (1988) is approximately unbiased. These are respectively given by

$$\omega = 1 - \frac{n\Lambda}{\Lambda + n - k - 1}$$

and

$$\omega_c = \omega - \left(\frac{p^2 + k^2}{3n}\right)(1 - \omega)$$

As the value of Λ is produced by several SAS procedures performing multivariate analyses, these two measures are easily computable. It may be remarked that ω and ω_c can be negative.

3.5 Testing Hypotheses: Linear Hypotheses

One important aspect of statistical inference is testing the hypothesis of interest. In the context of multivariate linear models, the hypotheses may be functions of

- matrix **B** only,
- matrix Σ only, or
- both **B** and Σ.

In this section, we will confine the discussion to hypotheses that are functions of **B** only. Certain hypotheses involving the matrix Σ are discussed in Chapters 5 and 6.

A hypothesis on **B** may be of interest in a variety of situations. It may be needed in the context of data reduction, to test the redundancy of certain variables, or in connection with model reduction schemes as in the process of selecting variables. Of course a number of hypotheses are important in their own right, e.g., when a comparison of two or more populations is needed.

Most of the hypotheses of interest on **B** can be expressed as the general linear hypothesis

$$H_0 : \mathbf{LB} = \mathbf{D} \text{ vs. } H_1 : \mathbf{LB} \neq \mathbf{D} \qquad (3.10)$$

with known full row rank matrix **L** of order r by $(k + 1)$ and known **D**. The matrix **D** is usually a zero matrix for most linear hypotheses; in the cases when **D** is not a zero matrix, a suitable transformation of data on dependent variables would provide an equivalent linear hypothesis with zero matrix on the right-hand side in terms of the reparameterization of **B**, as indicated below.

Suppose $\mathbf{D} \neq \mathbf{0}$. Since **L** is of full row rank, $\mathbf{LL}'$ admits the inverse. Consequently, by subtracting $\mathbf{XL}'(\mathbf{LL}')^{-1}\mathbf{D}$ from both sides of Equation 3.2, we have,

$$\mathbf{Y} - \mathbf{XL}'(\mathbf{LL}')^{-1}\mathbf{D} = \mathbf{X}(\mathbf{B} - \mathbf{L}'(\mathbf{LL}')^{-1}\mathbf{D}) + \mathcal{E}$$

or equivalently

$$\mathbf{Y}^* = \mathbf{X\Gamma} + \mathcal{E},$$

where $\mathbf{Y}^* = \mathbf{Y} - \mathbf{XL}'(\mathbf{LL}')^{-1}\mathbf{D}$, $\mathbf{\Gamma} = \mathbf{B} - \mathbf{L}'(\mathbf{LL}')^{-1}\mathbf{D}$, and the hypothesis in Equation 3.10 can be equivalently written as

$$H_0 : \mathbf{L\Gamma} = \mathbf{0} \text{ vs. } H_1 : \mathbf{L\Gamma} \neq \mathbf{0}.$$

Hence, it suffices to consider the hypotheses of type

$$H_0 : \mathbf{LB} = \mathbf{0} \text{ vs. } H_1 : \mathbf{LB} \neq \mathbf{0} \qquad (3.11)$$

for the discussion that follows. In the following subsections, we present some statistical tests for various hypotheses. For this purpose, multivariate normality for the rows of $\mathcal{E}$ is assumed.

3.5.1 Multivariate Tests

To test the null hypothesis in Equation 3.11, various test criteria based on the eigenvalues of certain matrices (which may be the functions of **L** and $\hat{\mathbf{B}}$) are available. Specifically, let

$$\mathbf{H} = \hat{\mathbf{B}}'\mathbf{L}'[\mathbf{L}(\mathbf{X}'\mathbf{X})^{-1}\mathbf{L}']^{-1}\mathbf{L}\hat{\mathbf{B}}, \qquad (3.12)$$

and define $\lambda_1 \geq \lambda_2 \geq \cdots \geq \lambda_p \geq 0$ as the ordered eigenvalues of $\mathbf{E}^{-1}\mathbf{H}$. As **E** is assumed to be positive definite, the existence of $\mathbf{E}^{-1}$ is ensured. Various test statistics for the hypothesis in Equation 3.11 are given below.

Wilks' Λ Criterion

$$\Lambda = |(\mathbf{H} + \mathbf{E})^{-1}\mathbf{E}| = |\mathbf{E}^{-1}\mathbf{H} + \mathbf{I}|^{-1} = \prod_{i=1}^{p}(1 + \lambda_i)^{-1} \qquad (3.13)$$

Pillai's Trace Criterion (Bartlett-Nanda-Pillai's Trace)

$$V = tr[(\mathbf{H} + \mathbf{E})^{-1}\mathbf{H}] = \sum_{i=1}^{p} \lambda_i/(1 + \lambda_i) \qquad (3.14)$$

Hotelling-Lawley Trace Criterion (Bartlett-Hotelling-Lawley Trace)

$$U = tr(\mathbf{E}^{-1}\mathbf{H}) = \sum_{i=1}^{p} \lambda_i \qquad (3.15)$$

Roy's Maximum Root Criterion

$$\lambda_{\max} = \max(\lambda_1, \lambda_2, \ldots, \lambda_p) = \lambda_1. \qquad (3.16)$$

For tables of critical values for these test statistics, see Pillai (1960). However, approximations to F statistics are summarized in Table 3.2.

TABLE 3.2 *F Approximations for Various Tests*

Criterion	F	Approximate Distribution of F under H_0
Wilks (Λ)	$\frac{gt-2u}{rr_t}\frac{1-\Lambda^{1/t}}{\Lambda^{1/t}}$	$F(rr_t, gt - 2u)$
Pillai (V)	$\frac{2m_2+s+1}{2m_1+s+1}\frac{V}{s-V}$	$F(s(2m_1 + s + 1), s(2m_2 + s + 1))$
Hotelling - Lawley (U)	$\frac{2(sm_2+1)}{s^2(2m_1+s+1)}U$	$F(s(2m_1 + s + 1), 2(sm_2 + 1))$
Roy (λ_{max})	$\frac{n-k-h+r-1}{h}\lambda_{max}$	$F(h, n - k - h + r - 1)^*$

* This F statistic is an upper bound on the F statistic that provides a lower bound on the assumed level of significance.

Various quantities used in Table 3.2 are defined below.

$$r_t = \text{Rank } (\mathbf{H} + \mathbf{E})$$

$$r = \text{Rank } (\mathbf{L})$$

$$s = \min(r, r_t)$$

$$h = \max(r, r_t)$$

$$m_1 = [|r - r_t| - 1]/2$$

$$m_2 = (n - k - r_t - 2)/2$$

$$g = (n - k - 1) - \frac{(r_t - r + 1)}{2}$$

$$u = (rr_t - 2)/4$$

$$t = \begin{cases} \sqrt{(r^2r_t^2 - 4)/(r_t^2 + r^2 - 5)} & \text{if } (r_t^2 + r^2 - 5) > 0 \\ 1 & \text{otherwise} \end{cases}$$

The *SAS/STAT User's Guide*, pp. 18–19 provides these formulas with a slightly different notation. Specifically, quantities (r_t, r, h, m_1, m_2, g) defined above are the same as quantities (p, q, r^*, m, n, r) defined in the *User's Guide*. It may be noted that r^* has been referred as just r on page 19 of the *Guide*.

The F approximation to Wilks' Λ is often referred to as Rao's F (Rao, 1951). Table 3.3 (*Reprinted by permission of John Wiley & Sons, Inc.*) adapted from Rao (1973, p. 555) shows that in certain special cases a transformation of Wilks' Λ to F statistic is exact:

TABLE 3.3 The Exact F in Special Cases

Values of r and p	F	Exact F Under H_0
$r = 1$ *for any* p	$\frac{n-k+r-p-1}{p}\frac{1-\Lambda}{\Lambda}$	$F(p, n-k+r-p-1)$
$r = 2$ *for any* p	$\frac{n-k+r-p-2}{p}\frac{1-\sqrt{\Lambda}}{\sqrt{\Lambda}}$	$F(2p, 2(n-k+r-p-2))$
$p = 1$ *for any* r	$\frac{n-k-1}{r}\frac{1-\Lambda}{\Lambda}$	$F(r, n-k-1)$
$p = 2$ *for any* r	$\frac{n-k-1}{r}\frac{1-\sqrt{\Lambda}}{\sqrt{\Lambda}}$	$F(2r, 2(n-k-1))$

Two considerably more accurate approximations than those given in Table 3.2 of the distribution of Wilks' Λ are suggested by Gupta and Richards (1983). The simpler of the two approximates the distribution of $-ln(\Lambda)$ by a chi-square distribution. Specifically,

$$Pr[-ln(\Lambda) < u] = Pr[\chi^2_{pd_{\mathbf{H}}} < u/\zeta]$$

where,

$$\zeta = [2(d_{\mathbf{E}} - p + 1)]^{-1} + [d_{\mathbf{E}} + d_{\mathbf{H}} - 2]^{-1},$$

$$d_{\mathbf{E}} = \text{degrees of freedom of } \mathbf{E},$$

$$d_{\mathbf{H}} = \text{degrees of freedom of } \mathbf{H}$$

and $\chi^2_{pd_{\mathbf{H}}}$ is a chi-square random variable with degrees of freedom $pd_{\mathbf{H}}$. The value of Λ is readily available in the SAS output and $d_{\mathbf{H}}$ and $d_{\mathbf{E}}$ can be calculated from the appropriate partitioning of the total SS&CP matrix. Therefore, the probabilities corresponding to the Wilks' Λ statistic can be more accurately estimated via this approximation.

In general, none of the tests based on the criteria defined in Equations 3.13 through 3.16 is uniformly best. Giri (1977, p. 219) provides a review of various optimality properties of these tests. Based on the power studies by Pillai and Jayachandran (1967, 1968), the following general recommendations can be made (Seber, 1984, p. 415, Muirhead, 1982, p. 484).

- For $p = 2$ and for small departures from H_0 or for large deviations from H_0 with the two nonzero eigenvalues of $\mathbf{HE}^{-1}$ nearly equal, Pillai's test is superior to Wilks' Λ which in turn is better than Hotelling-Lawley's U.

- For $p = 2$ and for large departures from H_0 and when the two nonzero eigenvalues of $\mathbf{HE}^{-1}$ are very different, the order of preference given above is reversed.

- For general p, and for small departures from H_0, the order of preference is the same as that in the first item above.

Asymptotically, all the three tests are equivalent. Specifically, as $n \to \infty$, and under H_0, $\{n - k - \frac{1}{2}(p - r + 3)\}ln\,\Lambda$, $(n - k - 1)U$, and $(n - k - 1)V$ are all asymptotically distributed as central chi-square distributions with degrees of freedom pr. When $p = 1$, these tests and also Roy's λ_{max} test are all identical.

The four multivariate tests described above are available as options in the GLM procedure as well as in the REG procedure. Both of these perform various aspects of multivariate regression analysis. However, on occasion, one may be superior to the other in achieving certain specific tasks. PROC GLM is more general in that it can be used for analyzing the regression as well as the experimental design models and can also be applied with little effort to situations involving the blocking effects or *covariables*. In contrast, PROC REG is more convenient for regression analysis and provides certain options for specialized analyses in the regression context. In the examples that follow, we will utilize both choices depending on the context as well as the convenience and we occasionally also comment on the specific differences as well as their relative merits.

EXAMPLE 1   *Hotelling T^2 Test for Cork Data*   The cork data in Rao (1948) presented in Table 1.1 consist of the weights of cork boring from the north, east, south, and west directions of the trunks for 28 trees. The problem is to test whether the cork deposit varies, in thickness and hence in weight, in the four directions. We therefore set up the null hypothesis as "the cork deposit is uniform along all four directions". The multivariate tests illustrated above will be described for this null hypothesis. If we denote the 28 by 1 column vectors of cork deposits in four directions by $\mathbf{y}_1$, $\mathbf{y}_2$, $\mathbf{y}_3$ and $\mathbf{y}_4$ respectively, then, with $\mathbf{Y} = (\mathbf{y}_1 : \mathbf{y}_2 : \mathbf{y}_3 : \mathbf{y}_4)$, we have the model

$$(\mathbf{y}_1 : \mathbf{y}_2 : \mathbf{y}_3 : \mathbf{y}_4) = (\mu_1\mathbf{1} : \mu_2\mathbf{1} : \mu_3\mathbf{1} : \mu_4\mathbf{1}) + (\epsilon_1 : \epsilon_2 : \epsilon_3 : \epsilon_4),$$

$$\text{i.e. } \mathbf{Y} = \mathbf{1B} + \mathcal{E},$$

where $\mathbf{B} = (\mu_1 : \mu_2 : \mu_3 : \mu_4)$ is a 1 by 4 matrix of mean bark deposit in four directions. The null hypothesis $\mu_1 = \mu_2 = \mu_3 = \mu_4$ can be presented as

$$H_0 : \begin{cases} \mu_1 - \mu_2 = 0 \\ \mu_1 - \mu_3 = 0 \\ \mu_1 - \mu_4 = 0. \end{cases}$$

If we accordingly define three variables

$$DNE = y_1 - y_2 \text{ (difference between north and east measurements)}$$

$$DNS = y_1 - y_3 \text{ (difference between north and south measurements)}$$

and

$$DNW = y_1 - y_4 \text{ (difference between north and west measurements)}$$

then under H_0, the three variables defined above have zero means. Since there are no independent variables, the linear model in the three variables has the matrix of regression coefficients, say $\mathbf{B}_d$, consisting of only one row of intercepts. In this case, our null hypothesis is equivalent to testing the hypothesis that $\mathbf{B}_d = \mathbf{0}$. This hypothesis can be tested by performing multivariate significance tests on the model with only an intercept and no independent variables. The SAS code given in Program 3.1 performs these tests, and the results are shown in Ouput 3.1.

```
/* Program 3.1 */

options ls=64 ps=45 nodate nonumber;
data cork;
infile 'cork.dat' firstobs = 1 ;
input north east south west;
y1=north;
y2=east;
y3=south;
y4=west;
/* Hotelling's T-square by creating the differences */
dne=y1-y2;
dns=y1-y3;
dnw=y1-y4;
proc glm data=cork;
```

```
model dne dns dnw= /nouni;
manova h=intercept;
title1 ' Output 3.1 ';
title2 ' Equality of the Components of the Mean Vector ';
title3 ' Cork Data';
run;
```

First, within the data set named CORK.DAT we create three new variables DNE, DNS, and DNW as defined in Program 3.1. The NOUNI option suppresses the univariate analysis of individual variables. Finally, the statement

```
manova  h=intercept;
```

performs multivariate hypothesis testing, when the hypothesis of interest is on the intercept vector.

Output 3.1

```
                              Output 3.1
            Equality of the Components of the Mean Vector
                              Cork Data

                    General Linear Models Procedure

                Number of observations in data set = 28

                    General Linear Models Procedure
                    Multivariate Analysis of Variance

          Characteristic Roots and Vectors of: E Inverse * H, where
        H = Type III SS&CP Matrix for INTERCEPT   E = Error SS&CP Matrix

        Characteristic   Percent        Characteristic Vector  V'EV=1
            Root
                                            DNE              DNS
                                            DNW

           0.76822288    100.00         0.01256587      -0.01086704
                                         0.02243791

           0.00000000      0.00        -0.00378101       0.02474866
                                        -0.00088456

           0.00000000      0.00        -0.02093912      -0.01046880
                                         0.01870549



                Manova Test Criteria and Exact F Statistics for
                   the Hypothesis of no Overall INTERCEPT Effect
        H = Type III SS&CP Matrix for INTERCEPT   E = Error SS&CP Matrix

                      S=1     M=0.5    N=11.5

        Statistic                 Value      F     Num DF  Den DF  Pr > F

        Wilks' Lambda           0.56554    6.4019     3      25    0.0023
        Pillai's Trace          0.43446    6.4019     3      25    0.0023
        Hotelling-Lawley Trace  0.768223   6.4019     3      25    0.0023
        Roy's Greatest Root     0.768223   6.4019     3      25    0.0023
```

In Output 3.1 note that since the model has only an intercept vector and no other independent variables, the matrix $\mathbf{E}^{-1}\mathbf{H}$ is of rank 1. Consequently, only the first eigenvalue is nonzero and is 0.7682. The four test statistics Λ, V, U and λ_{max} are reported next. All four tests are equivalent in this case, and lead to the same observed value of the F statistic. With $n = 28$, $k = 0$, $p = 3$, and $r = 1$, the observed value of exact F statistic is given by $F = \frac{n-k+r-p-1}{p}(\frac{1-\Lambda}{\Lambda}) = 6.4019$ (see Table 3.3). It follows an F distribution with (3, 25) degrees of freedom, leading to the p value of 0.0023. Consequently, in view of small p value, we reject the null hypothesis of the uniform cork deposits in the four directions.

The value of Hotelling's T^2 can be easily obtained from the value of Wilks' Λ. Specifically,

$$T^2 = (n-1)\left(\frac{1-\Lambda}{\Lambda}\right),$$

which is equal to 20.7420 for this particular data set.

There is another way of testing the same hypothesis, yet without directly creating the data on the differences, and that is to use the SAS option M=. These alternative statements are given in Program 3.2.

```
/* Program 3.2 */

options ls=64 ps=45 nodate nonumber;
data cork;
infile 'cork.dat' firstobs=1;
input north east south west;
y1=north;
y2=east;
y3=south;
y4=west;
/* Hotelling's T-square using m = option*/
proc glm data=cork;
model y1 y2 y3 y4= /nouni;
manova h=intercept
m=y2-y1, y3-y1, y4-y1
mnames=d1 d2 d3;
title1 ' Output 3.2 ' ;
title2 'Use of m=option for Testing Equal Means for Cork Data';
run;
```

Program 3.2 attempts to directly fit the model $\mathbf{Y} = \mathbf{XB} + \mathcal{E}$ with $\mathbf{X} = \mathbf{1}_n$, an n by 1 column vector of ones. Through the MANOVA statement, we specify the null hypothesis of interest; the intercepts of the variables indicated in the M=specification are zero. In the M= specification, we essentially define the variables DNE, DNS, and DNW from Program 3.1, but without any such explicit assignment. This, in turn, defines a matrix

$$\mathbf{M} = \begin{bmatrix} -1 & 1 & 0 & 0 \\ -1 & 0 & 1 & 0 \\ -1 & 0 & 0 & 1 \end{bmatrix}',$$

and with $\mathbf{L} = \mathbf{I} = 1$, the hypothesis of interest is written as $H_0 : \mathbf{LBM} = \mathbf{0}$, the general linear hypothesis which is further described in Section 3.8. The corresponding SAS code is

```
proc glm;
model y1 y2 y3 y4= /nouni;
manova h=intercept m=y2-y1,
                   y3-y1,
                   y4-y1
mnames=d1 d2 d3;
```

Thus in the M= specification, three variables earlier denoted by DNE, DNS, and DNW respectively are defined and are subsequently named $D1$, $D2$, and $D3$ using the MNAMES= specification. The NOUNI option on the MODEL statement suppresses the output corresponding to the univariate models in $y1$, $y2$, $y3$, and $y4$. Output 3.2 first prints the **M** matrix (in fact, its transpose **M'**) indicated above and then produces the output which is essentially identical to Output 3.1. An alternative to the M = specification indicated above is to explicitly specify the **M** matrix column by column. That is, by specifying

```
m = (-1  1  0  0,
     -1  0  1  0,
     -1  0  0  1);
```

Since the specification is by columns, the representation above resembles **M'** and not **M**. The resulting output and the corresponding interpretations are essentially the same as those for Output 3.1.

Output 3.2

Output 3.2
Use of m=option for Testing Equal Means for Cork Data

General Linear Models Procedure
Multivariate Analysis of Variance

M Matrix Describing Transformed Variables

	Y1	Y2	Y3	Y4
D1	-1	1	0	0
D2	-1	0	1	0
D3	-1	0	0	1

Characteristic Roots and Vectors of: E Inverse * H, where
H = Type III SS&CP Matrix for INTERCEPT   E = Error SS&CP Matrix

Variables have been transformed by the M Matrix

Characteristic Root	Percent	Characteristic Vector V'EV=1	
		D1 D3	D2
0.76822288	100.00	0.01256587 0.02243791	-0.01086704
0.00000000	0.00	-0.00378101 -0.00088456	0.02474866
0.00000000	0.00	-0.02093912 0.01870549	-0.01046880

Manova Test Criteria and Exact F Statistics for
the Hypothesis of no Overall INTERCEPT Effect
on the variables defined by the M Matrix Transformation
H = Type III SS&CP Matrix for INTERCEPT   E = Error SS&CP Matrix

S=1   M=0.5   N=11.5

Statistic	Value	F	Num DF	Den DF	Pr > F
Wilks' Lambda	0.56554	6.4019	3	25	0.0023
Pillai's Trace	0.43446	6.4019	3	25	0.0023
Hotelling-Lawley Trace	0.768223	6.4019	3	25	0.0023
Roy's Greatest Root	0.768223	6.4019	3	25	0.0023

EXAMPLE 2   ***Multivariate Regression for Fish Data***   The data for the illustration of multivariate regression are taken from Srivastava and Carter (1983). This toxicity study was conducted to determine the effect of copper on fish mortality. Twenty-five tanks of twenty trout were given various doses (DOSE) of copper in mg. per liter. The average weight of the fish (WT) was also recorded and used as one of the covariables. The proportions of dead fish after 8, 14, 24, 36, and 48 hours were recorded and an arcsine transformation was used to obtain the transformed variables y_i, $i = 1, \ldots, 5$. Such a transformation was used to stabilize the variances. Further, in keeping with the standard practice in many dose response studies, the various doses were converted into logarithmic scale. This transformation in turn made the spacing between the various levels more uniform.

The problems of interest are to

- fit a multivariate model expressing y_i as the linear functions of weights and the natural logarithm of the dose, which are to be treated as the independent variables,

- test the significance of the dose as a variable,

- test the significance of the fish weight as a variable, and

- test the overall significance of the model.

The SAS statements presented in Program 3.3 are used to obtain the appropriate output presented as Output 3.3. Only selected parts of output are presented.

After reading the raw data on proportions p_1 (corresponding to $t =8$), $\ldots$, p_5 (corresponding to $t =48$), we obtain the variables $y_1, \ldots, y_5$ by defining

$$y_i = \sin^{-1}\{\sqrt{p_i}\}, \quad i = 1, \ldots, 5.$$

This is done in the DATA step. We denote the natural logarithm of dose by $X1$ and the weight of the fish by $X2$. In PROC GLM, the MODEL statement fits the model

$$\mathbf{Y}_{25\times 5} = \mathbf{X}_{25\times 3}\mathbf{B}_{3\times 5} + \mathcal{E}_{25\times 5},$$

where $\mathbf{B} = \begin{bmatrix} \beta_0 \\ \beta_1 \\ \beta_2 \end{bmatrix}$. The 1 by 5 row vector β_0 contains the intercepts of the five different models corresponding to the five time points. Similarly, the row vectors β_1 and β_2 respectively contain the slope parameters in these models for the independent variables X1 (logarithms of DOSE) and X2 (WT). To test the significance of the model, we have the null hypothesis $H_0 : \begin{bmatrix} \beta_1 \\ \beta_2 \end{bmatrix} = \mathbf{0}$. Assuming the multivariate normality of the error, this null hypothesis can be tested using Wilks' Λ test. This test statistic cannot be computed using PROC GLM. However, we will describe its computation later using PROC REG. The statement

```
manova h=x1 x2;
```

is used to test the other two hypotheses of interest, namely $H_0^{(1)} : \beta_1 = \mathbf{0}$ (no dose effect) and $H_0^{(2)} : \beta_2 = \mathbf{0}$ (no weight effect). The options PRINTE and PRINTH are used to print the error sums of squares and products matrix $\mathbf{E}$ and the appropriate $\mathbf{H}$ matrix corresponding to the desired hypothesis.

Output 3.3 first presents the error sums of squares and crossproducts matrix $\mathbf{E}$. The sums of squares due to error in the case of individual variables y_i can be obtained as the diagonal elements of this matrix. The output then presents the analysis corresponding to variable $X1$, the natural logarithm of the variable DOSE. By default, Type III sums of squares and crossproducts are used and hence these are adjusted for the other variable $X2$ which represents the initial weight of the fish. Various types of sums of squares are discussed in some detail in Chapter 4 and in relatively greater detail in *SAS/STAT User's Guide*.

All four multivariate tests (all of which are exact in this case) indicate that the variable $X1$ does indeed have a very significant effect on the variables $Y1, \ldots, Y5$. At the same time, a similar test for $X2$ shows that initial weight does not significantly contribute to the death rate. Thus, the null hypothesis corresponding to the significance of dose is rejected while that corresponding to the significance of weight is not rejected. It may be pointed out that the data analyzed here represent repeated measurements. See Chapters 5 and 6 for extensive analyses of repeated measures data.

```
  /* Program 3.3 */

options ls=64 ps=45 nodate nonumber;
title1 'Output 3.3';
data fish;
infile 'fish.dat' firstobs = 1;
input p1 p2 p3 p4 p5 dose wt @@;
y1=arsin(sqrt(p1));
y2=arsin(sqrt(p2));
y3=arsin(sqrt(p3));
y4=arsin(sqrt(p4));
y5=arsin(sqrt(p5));
x1=log(dose);
x2=wt;
proc print data=fish;
var p1 p2 p3 p4 p5 dose x2;
title2 'Data on Proportions of Dead Fish';
run;
proc print data=fish;
var y1 y2 y3 y4 y5 x1 x2;
title2 'Transformed Fish Data';
run;
proc glm data=fish;
model y1 y2 y3 y4 y5=x1 x2/nouni;
manova h=x1 x2/printe printh;
title2 'Multivariate Regression for Fish Data';
run;
/*
mtest option of proc reg can be used instead of manova
option of proc glm to get the same results.  This is done
using the last two statements of the following program.
*/
proc reg data=fish;
model y1 y2 y3 y4 y5=x1 x2;
Model: mtest x1, x2/print;
Onlyx1: mtest x1/print;
Onlyx2: mtest x2/print;
```

The PROC GLM code in Program 3.3 does not provide any test for $H_0^{(3)} : \begin{bmatrix} \beta_1 \\ \beta_2 \end{bmatrix} = \mathbf{0}$. This is an important hypothesis which simultaneously tests that responses neither depend on the dose nor depend on the weights of the fish (and therefore, for any given response, the mean response for all fish is constant and is not affected by these two variables). Since the hypothesis is on the entire deterministic part of the model, the problem is also commonly referred to as *testing for the model*. The above test can be completed by using the MTEST statement in PROC REG. Specifically, for the null hypothesis of the type indicated above, that is, those on all the regression coefficients, the corresponding MTEST statement should include all the independent variables separated by commas. If the intercept is also one of the parameters in the null hypothesis, it is specified using the keyword INTERCEPT in the MTEST statement. In our case the SAS code is

```
proc reg;
y1 y2 y3 y4 y5=x1 x2;
mtest x1, x2/print;
```

The PRINT option prints the corresponding **H** matrix and the error SS&CP matrix **E**. The output contains the value of Wilks' Λ as well as the other three statistics. For our data set, the p value for each of the four tests is quite small, and consequently the null hypothesis can be rejected. Hence we conclude that the death rate does indeed depend on the average weight or the dose or both.

The MTEST statement in PROC REG can also be used to test the individual hypotheses on X1 and X2. The corresponding SAS statements are

```
mtest x1/print;
mtest x2/print;
```

Thus, it is possible to use the MTEST statement in PROC REG as an alternative to the MANOVA statement in PROC GLM. In addition using MTEST is a useful way of performing *partial tests*, the multivariate versions of univariate partial F tests, on the subsets of independent variables; that task is not easy to accomplish with PROC GLM. To perform a partial test, we list these independent variables and separate them by commas in the MTEST statement. On the other hand, unlike PROC GLM, PROC REG does not provide any facility to suppress the accompanying univariate analyses, and hence its use may not always be an optimal way of performing multivariate analysis of the data.

The MTEST statement is also useful for a variety of other multivariate null hypotheses. For example, null hypotheses stating any interrelationships between independent variables can be included as equations in the MTEST statement. For example, a hypothesis of the form $H_0 : \beta_1 - \beta_2 = \mathbf{c}_0$, where $\mathbf{c}_0$ is a specified constant vector, can be tested using the statement

```
mtest x1-x2=c0/print;
```

The dependent variables can also be included in the MTEST statement. It is a very useful feature for situations where two or more univariate regression models are to be compared. We will address these issues in Section 3.8.

Output 3.3

```
                              Output 3.3
                     Data on Proportions of Dead Fish
```

OBS	P1	P2	P3	P4	P5	DOSE	X2
1	0.00	0.00	0.25	0.25	0.25	270	0.6695
2	0.00	0.10	0.30	0.30	0.30	410	0.6405
3	0.00	0.50	0.75	0.90	0.90	610	0.7290
4	0.15	0.65	1.00	1.00	1.00	940	0.7700
5	0.45	1.00	1.00	1.00	1.00	1450	0.5655
6	0.00	0.05	0.20	0.20	0.20	270	0.7820
7	0.05	0.10	0.30	0.30	0.30	410	0.8120

8	0.05	0.45	0.95	1.00	1.00	610	0.8215
9	0.10	0.70	1.00	1.00	1.00	940	0.8690
10	0.20	0.85	1.00	1.00	1.00	1450	0.8395
11	0.00	0.00	0.00	0.00	0.05	270	0.8615
12	0.00	0.05	0.15	0.25	0.30	410	0.9045
13	0.00	0.15	0.95	0.95	0.95	610	1.0280
14	0.00	0.55	0.95	1.00	1.00	940	1.0445
15	0.10	0.85	1.00	1.00	1.00	1450	1.0455
16	0.00	0.00	0.00	0.05	0.10	270	0.6195
17	0.00	0.05	0.15	0.20	0.25	410	0.5305
18	0.10	0.45	0.95	0.95	0.95	610	0.5970
19	0.10	0.70	1.00	1.00	1.00	940	0.6385
20	0.35	0.95	1.00	1.00	1.00	1450	0.6645
21	0.00	0.05	0.20	0.20	0.20	270	0.5685
22	0.00	0.00	0.15	0.25	0.25	410	0.6040
23	0.00	0.40	0.90	1.00	1.00	610	0.6325
24	0.05	0.65	1.00	1.00	1.00	940	0.6845
25	0.30	0.85	1.00	1.00	1.00	1450	0.7230

Output 3.3
continued

Output 3.3
Transformed Fish Data

OBS	Y1	Y2	Y3	Y4	Y5	X1	X2
1	0.00000	0.00000	0.52360	0.52360	0.52360	5.59842	0.6695
2	0.00000	0.32175	0.57964	0.57964	0.57964	6.01616	0.6405
3	0.00000	0.78540	1.04720	1.24905	1.24905	6.41346	0.7290
4	0.39770	0.93774	1.57080	1.57080	1.57080	6.84588	0.7700
5	0.73531	1.57080	1.57080	1.57080	1.57080	7.27932	0.5655
6	0.00000	0.22551	0.46365	0.46365	0.46365	5.59842	0.7820
7	0.22551	0.32175	0.57964	0.57964	0.57964	6.01616	0.8120
8	0.22551	0.73531	1.34528	1.57080	1.57080	6.41346	0.8215
9	0.32175	0.99116	1.57080	1.57080	1.57080	6.84588	0.8690
10	0.46365	1.17310	1.57080	1.57080	1.57080	7.27932	0.8395
11	0.00000	0.00000	0.00000	0.00000	0.22551	5.59842	0.8615
12	0.00000	0.22551	0.39770	0.52360	0.57964	6.01616	0.9045
13	0.00000	0.39770	1.34528	1.34528	1.34528	6.41346	1.0280
14	0.00000	0.83548	1.34528	1.57080	1.57080	6.84588	1.0445
15	0.32175	1.17310	1.57080	1.57080	1.57080	7.27932	1.0455
16	0.00000	0.00000	0.00000	0.22551	0.32175	5.59842	0.6195
17	0.00000	0.22551	0.39770	0.46365	0.52360	6.01616	0.5305
18	0.32175	0.73531	1.34528	1.34528	1.34528	6.41346	0.5970
19	0.32175	0.99116	1.57080	1.57080	1.57080	6.84588	0.6385
20	0.63305	1.34528	1.57080	1.57080	1.57080	7.27932	0.6645
21	0.00000	0.22551	0.46365	0.46365	0.46365	5.59842	0.5685
22	0.00000	0.00000	0.39770	0.52360	0.52360	6.01616	0.6040
23	0.00000	0.68472	1.24905	1.57080	1.57080	6.41346	0.6325
24	0.22551	0.93774	1.57080	1.57080	1.57080	6.84588	0.6845
25	0.57964	1.17310	1.57080	1.57080	1.57080	7.27932	0.7230

Output 3.3
Multivariate Regression for Fish Data
General Linear Models Procedure
Number of observations in data set=25

E = Error SS&CP Matrix

	Y1	Y2	Y3	Y4	Y5
Y1	0.35212077	0.13237233	-0.0815826	-0.2385054	-0.2115503
Y2	0.13237233	0.3986007	0.24790465	0.23529861	0.2252679
Y3	-0.0815826	0.24790465	1.22735578	1.22213318	1.08918722
Y4	-0.2385054	0.23529861	1.22213318	1.45057752	1.3142694
Y5	-0.2115503	0.2252679	1.08918722	1.3142694	1.2243379

H = Type III SS&CP Matrix for X1

	Y1	Y2	Y3	Y4	Y5
Y1	0.94443317	2.13966069	2.40045274	2.31897003	2.20973844
Y2	2.13966069	4.84750856	5.43834602	5.25374286	5.00627322
Y3	2.40045274	5.43834602	6.10119756	5.89409409	5.61646167
Y4	2.31897003	5.25374286	5.89409409	5.69402069	5.42581243
Y5	2.20973844	5.00627322	5.61646167	5.42581243	5.17023771

Manova Test Criteria and Exact F Statistics for
the Hypothesis of no Overall X1 Effect
H = Type III SS&CP Matrix for X1   E = Error SS&CP Matrix

S=1   M=1.5   N=8

Statistic	Value	F	Num DF	Den DF	Pr > F
Wilks' Lambda	0.065585	51.29	5	18	0.0001
Pillai's Trace	0.934415	51.29	5	18	0.0001
Hotelling-Lawley Trace	14.24729	51.29	5	18	0.0001
Roy's Greatest Root	14.24729	51.29	5	18	0.0001

Manova Test Criteria and Exact F Statistics for
the Hypothesis of no Overall X2 Effect
H = Type III SS&CP Matrix for X2   E = Error SS&CP Matrix

S=1   M=1.5   N=8

Statistic	Value	F	Num DF	Den DF	Pr > F
Wilks' Lambda	0.654906	1.897	5	18	0.1449
Pillai's Trace	0.345094	1.897	5	18	0.1449
Hotelling-Lawley Trace	0.526938	1.897	5	18	0.1449
Roy's Greatest Root	0.526938	1.897	5	18	0.1449

Multivariate Test: MODEL

E, the Error Matrix

0.3521207721	0.1323723295	-0.081582595
0.1323723295	0.3986006993	0.2479046457

```
        -0.081582595      0.2479046457      1.2273557756
        -0.238505417      0.2352986084      1.2221331781
        -0.211550264      0.2252678955      1.0891872222
```

```
              -0.238505417     -0.211550264
               0.2352986084     0.2252678955
               1.2221331781     1.0891872222
               1.4505775203     1.3142693993
               1.3142693993     1.2243378999
```

H, the Hypothesis Matrix

```
        0.9538786705      2.1157383555      2.3410489485
        2.1157383555      4.9080959325         5.58879619
        2.3410489485         5.58879619      6.4747944636
        2.2590396342      5.4055267572      6.2710028881
        2.1485124778      5.1613383564      6.0015184245
```

```
               2.2590396342      2.1485124778
               5.4055267572      5.1613383564
               6.2710028881      6.0015184245
               6.0742707424      5.8142826656
               5.8142826656       5.567105838
```

Multivariate Statistics and F Approximations

S=2   M=1   N=8

Statistic	Value	F	Num DF	Den DF	Pr > F
Wilks' Lambda	0.046688	13.061	10	36	0.0001
Pillai's Trace	1.2132	5.8594	10	38	0.0001
Hotelling-Lawley Trace	14.85233	25.249	10	34	0.0001
Roy's Greatest Root	14.46758	54.977	5	19	0.0001

NOTE: F Statistic for Roy's Greatest Root is an upper bound.
NOTE: F Statistic for Wilks' Lambda is exact.

Multivariate Test: ONLYX1

E, the Error Matrix

```
        0.3521207721      0.1323723295     -0.081582595
        0.1323723295      0.3986006993      0.2479046457
       -0.081582595       0.2479046457      1.2273557756
       -0.238505417       0.2352986084      1.2221331781
       -0.211550264       0.2252678955      1.0891872222
```

```
        -0.238505417        -0.211550264
         0.2352986084        0.2252678955
         1.2221331781        1.0891872222
         1.4505775203        1.3142693993
         1.3142693993        1.2243378999
```

H, the Hypothesis Matrix

```
    0.9444331687        2.1396606906        2.4004527379
    2.1396606906        4.8475085618        5.4383460192
    2.4004527379        5.4383460192        6.1011975631
    2.3189700315        5.2537428628        5.8940940938
    2.2097384417        5.0062732199        5.6164616708
```

```
    2.3189700315        2.2097384417
    5.2537428628        5.0062732199
    5.8940940938        5.6164616708
    5.6940206946        5.4258124278
    5.4258124278        5.1702377073
```

Multivariate Statistics and Exact F Statistics

S=1   M=1.5   N=8

Statistic	Value	F	Num DF	Den DF	Pr > F
Wilks' Lambda	0.065585	51.29	5	18	0.0001
Pillai's Trace	0.934415	51.29	5	18	0.0001
Hotelling-Lawley Trace	14.24729	51.29	5	18	0.0001
Roy's Greatest Root	14.24729	51.29	5	18	0.0001

Multivariate Test: ONLYX2

E, the Error Matrix

```
    0.3521207721        0.1323723295       -0.081582595
    0.1323723295        0.3986006993        0.2479046457
   -0.081582595         0.2479046457        1.2273557756
   -0.238505417         0.2352986084        1.2221331781
   -0.211550264         0.2252678955        1.0891872222
```

```
        -0.238505417        -0.211550264
         0.2352986084        0.2252678955
         1.2221331781        1.0891872222
         1.4505775203        1.3142693993
         1.3142693993        1.2243378999
```

H, the Hypothesis Matrix

0.0976730369	0.0792504059	-0.01314736
0.0792504059	0.0643025654	-0.010667567
-0.01314736	-0.010667567	0.0017697113
-0.020670718	-0.016771904	0.0027823991
-0.032603961	-0.026454355	0.0043886832

-0.020670718	-0.032603961
-0.016771904	-0.026454355
0.0027823991	0.0043886832
0.0043745806	0.0069000341
0.0069000341	0.0108834365

Multivariate Statistics and Exact F Statistics

S=1   M=1.5   N=8

Statistic	Value	F	Num DF	Den DF	Pr > F
Wilks' Lambda	0.654906	1.897	5	18	0.1449
Pillai's Trace	0.345094	1.897	5	18	0.1449
Hotelling-Lawley Trace	0.526938	1.897	5	18	0.1449
Roy's Greatest Root	0.526938	1.897	5	18	0.1449

A few comments may be made about the possible alternative approaches to the analysis presented here for the fish data. The actual raw data deal with the counts, namely the number, of dead fish and are multinomial in nature. The normality was obtained by applying the variance stabilizing transformation which in general may or may not stabilize the correlations between the variables. Although for this particular data set the transformation works well, there may be some extreme cases with 0% and 100% mortalities that cannot be transformed to normality. In such instances, it is advisable to use the iteratively reweighted least squares analysis which can be done using the NLIN procedure of SAS involving the nonlinear regression techniques. However, we will not pursue this analysis here.

3.5.2 Stepdown Analysis

Another linear hypothesis test, which is especially useful when there is a certain order among the response variables due to some physical interpretation of such ordering, is based on what is commonly referred to as *stepdown analysis*. To perform the analysis, let the physically meaningful ordering for consideration among the dependent variables, for a particular situation, be $y_1, y_2, \ldots, y_p$. The essential idea behind this procedure is to sequentially perform univariate tests on the univariate models associated with dependent variables, y_j conditional on (that is, fixing), $y_1, y_2, \ldots, y_{j-1}$. As a result, the hypotheses in Equation 3.11

$$H_0 : \mathbf{LB} = \mathbf{0} \text{ vs. } H_1 : \mathbf{LB} \neq \mathbf{0}$$

can be written in terms of the intersection and union of univariate subhypotheses H_{i1}, $H_{i2}, \ldots, H_{ip}$, $i = 0$ or 1 as

$$H_0 = \bigcap_{j=1}^{p} H_{0j} \text{ and } H_1 = \bigcup_{j=1}^{p} H_{1j}.$$

Thus, the null hypothesis H_0 is not rejected if and only if all $H_{0j} : \mathbf{L}\boldsymbol{\eta}_j = 0$, $j = 1, \ldots, p$, are not rejected and is rejected if at any stage a rejection occurs. The hypothesis H_{0j} is the null hypothesis on parameters $\boldsymbol{\eta}_j$ of the jth model $E(\mathbf{y}_j) = \mathbf{X}\boldsymbol{\eta}_j + \gamma_1 \mathbf{y}_1 + \cdots + \gamma_{p-1}\mathbf{y}_{p-1}$. That is, the model which corresponds to y_j, conditional on $y_1, \ldots, y_{j-1}$. By definition, if $j = 1$, it is unconditional. Further, H_{0j} can be reduced to $\mathbf{L}\boldsymbol{\beta}_j = 0$ assuming that $H_{01}, \ldots, H_{0,j-1}$ have been not rejected at previous stages. This is so since in this case $\mathbf{L}\boldsymbol{\eta}_j = 0$ implies $\mathbf{L}\boldsymbol{\beta}_j = 0$. The F statistic, F_j for testing H_{0j}, conditional on $y_1, \ldots, y_{j-1}$ follows an $F(r, n - k - j)$ distribution when H_{0j} is true. As before, $r = Rank(\mathbf{L})$. Further, $F_1, \ldots, F_k$ are all independently distributed of each other, since the conditional distributions are the same as the unconditional ones. We therefore test H_0 at the significance level α by sequentially testing $H_{01}, \ldots, H_{0p}$ using $\alpha_1, \ldots, \alpha_p$ as the levels of significance, where

$$P\left[\bigcap_{j=1}^{p} (F_j < F_{\alpha_j}(r, n - k - j)) \right] = \prod_{j=1}^{p} (1 - \alpha_j) = 1 - \alpha.$$

Thus the stepdown analysis can be easily implemented using successive univariate regression models by including at every stage all the dependent variables previously tested in the list of independent variables. Note that at any particular stage, except the first one, the matrix $\mathbf{L}$ has to be augmented by additional zero columns corresponding to the dependent variables added to the set of independent variables. The levels of significance $\alpha_1, \ldots, \alpha_p$ are to be chosen so as to satisfy $(1 - \alpha_1)(1 - \alpha_2) \cdots (1 - \alpha_p) = (1 - \alpha)$. For example, choosing α_j to be equal to, say, α^* leads to $\alpha^* = 1 - (1 - \alpha)^{1/p}$.

EXAMPLE 3   *Stepdown Analysis for Fish Data*   We again consider the fish data, given in Srivastava and Carter (1983). As shown in Example 2, the original proportions of dead fish after 8, 14, 24, 36, and 48 hours are transformed into new variables y_1 through y_5 respectively using the arcsine transformation. Since there is a definite natural ordering among the dependent variables $y_1, \ldots, y_5$ (through time), stepdown analysis is a meaningful possibility. Following the natural time ordering, we perform stepwise tests on y_j conditional on all y_i, $i < j$. If the problem is to see the significance of the effects of $x_1 = \ln(\text{DOSE})$ and $x_2 = $ average weight, on $y_1, \ldots, y_5$, then the null hypothesis H_0 of no significant effect of either x_1 or x_2 on $y_1, \ldots, y_5$, can be written as

$$H_0 = H_{01} \bigcap H_{02} \bigcap H_{03} \bigcap H_{04} \bigcap H_{05},$$

where

- H_{01}: x_1 and x_2 do not have any effect on y_1,
- H_{02}: x_1 and x_2 do not have any effect on y_2 given y_1,
- H_{03}: x_1 and x_2 do not have any effect on y_3 given y_1 and y_2,
- H_{04}: x_1 and x_2 do not have any effect on y_4 given y_1, y_2 and y_3, and
- H_{05}: x_1 and x_2 do not have any effect on y_5 given y_1, y_2, y_3 and y_4.

We test all five null hypotheses listed above in the natural time order. If the total level of significance is desired to be $\alpha = 0.05$, then for any individual null hypothesis, the significance level, assuming it to be the same for all subhypotheses, will be $\alpha^* = 1 - (1 - 0.05)^{1/5} \cong .010$. Since at the jth stage all y_i, $i < j$ are conditioned, they would appear in the right side of the model as independent variables (or *covariates*) along with x_1 and x_2.

For example, to test the hypothesis, say H_{03}, the corresponding SAS statements are

```
proc reg;
model y3 = x1 x2 y1 y2;
test x1 = 0.0, x2=0.0;
```

and the MODEL statement in the above SAS code corresponds to the univariate linear model

$$\mathbf{y}_3 = \beta_0 + \beta_1\mathbf{x}_1 + \beta_2\mathbf{x}_2 + \gamma_1\mathbf{y}_1 + \gamma_2\mathbf{y}_2 + \boldsymbol{\epsilon}_3.$$

The TEST statement tests the null hypothesis

$$H_{03} : \beta_1 = \beta_2 = 0$$

in the presence of the conditioned variables y_1 and y_2 as the other independent variables in the model. A complete SAS program, which specifically tests the statistical significance of $x_1 = ln\,(\text{DOSE})$ and of $x_2 = $ average weight using the stepdown analysis, is given in Program 3.4. Output 3.4 shows the results.

```
/* Program 3.4 */

options ls=64 ps=45 nodate nonumber;
data fish;
infile 'fish.dat';
input p1 p2 p3 p4 p5 dose wt @@;
y1=arsin(sqrt(p1));
y2=arsin(sqrt(p2));
y3=arsin(sqrt(p3));
y4=arsin(sqrt(p4));
y5=arsin(sqrt(p5));
x1=log(dose);
x2=wt;
title1 'Output 3.4 ';
title2 ' Stepdown Analysis';
/*The following program performs the Stepdown Analysis */
proc reg data=fish;
model y1=x1 x2;
fishwt: test x2=0.0;
fmodel: test x1=0.0,x2=0.0;
proc reg data=fish;
model y2=x1 x2 y1;
fishwt: test x2=0.0;
fmodel: test x1=0.0,x2=0.0;
proc reg data=fish;
model y3=x1 x2 y1 y2;
fishwt: test x2=0.0;
fmodel: test x1=0.0,x2=0.0;
proc reg data=fish;
model y4=x1 x2 y1 y2 y3;
fishwt: test x2=0.0;
fmodel: test x1=0.0,x2=0.0;
proc reg data=fish;
model y5=x1 x2 y1 y2 y3 y4;
fishwt: test x2=0.0;
fmodel: test x1=0.0,x2=0.0;
run;
```

Output 3.4

Output 3.4
Stepdown Analysis

Model: MODEL1
Dependent Variable: Y1

Analysis of Variance

Source	DF	Sum of Squares	Mean Square	F Value
Prob>F				
Model	2	0.95388	0.47694	29.798
0.0001				
Error	22	0.35212	0.01601	
C Total	24	1.30600		

Root MSE	0.12651	R-square	0.7304	
Dep Mean	0.19092	Adj R-sq	0.7059	
C.V.	66.26628			

Parameter Estimates

Variable	DF	Parameter Estimate	Standard Error	T for H0: Parameter=0
INTERCEP	1	-1.651099	0.28363283	-5.821
X1	1	0.336405	0.04379362	7.682
X2	1	-0.430791	0.17438706	-2.470

Variable	DF	Prob > |T|
INTERCEP	1	0.0001
X1	1	0.0001
X2	1	0.0217

Dependent Variable: Y1
Test: FISHWT Numerator: 0.0977 DF: 1 F value: 6.1025
              Denominator: 0.016005 DF: 22 Prob>F: 0.0217

Dependent Variable: Y1
Test: FMODEL Numerator: 0.4769   DF: 2 F value: 29.7985
              Denominator: 0.016005 DF: 22   Prob>F: 0.0001

The rejection of H_0 is attained as soon as any subhypothesis H_{0j} is rejected at the desired significance level $\alpha^*(\simeq .010)$. In view of the very small p value, which is 0.0001 (see Output 3.4), for the very first model (for y_1) we reject the null hypothesis H_0 and conclude that the fish death rates do indeed depend on either dose or weight of the fish or both.

3.6 Simultaneous Confidence Intervals

If $H_0 : \mathbf{LB} = \mathbf{0}$ is rejected, it may be of interest to provide the confidence intervals for the individual components of $\mathbf{LB}$ (or $\mathbf{B}$ if $\mathbf{L} = \mathbf{I}_{k+1}$) or the linear functions of these components. Under the assumption of the full rank of $\mathbf{X}$, a set of simultaneous confidence intervals for the linear combinations of the type $\mathbf{c'LBd}$ corresponding to the linear hypothesis $H_0 : \mathbf{LB} = \mathbf{0}$ can be constructed.

Noting that $H_0^{(\mathbf{c,d})} : \mathbf{c'LBd} = \mathbf{0}$ is true for all $\mathbf{c}$ and $\mathbf{d}$ if and only if H_0 is true, we can write H_0 as the intersection of all such $H_0^{(\mathbf{c,d})} : \mathbf{c'LBd} = \mathbf{0}$. Testing of $H_0^{(\mathbf{c,d})} : \mathbf{c'LBd} = \mathbf{0}$ can be done using the appropriate F test. Let the corresponding F statistic be $F^{(\mathbf{c,d})}$ and let F_α be the cutoff point. Then, H_0 is not rejected if and only if all $H_0^{(c,d)}$ are not rejected, that is, if and only if $\max_{\mathbf{c,d}} F^{(\mathbf{c,d})} \le F_\alpha$. In fact, the $\max_{\mathbf{c,d}} F^{(\mathbf{c,d})}$ is equal to $(n - k - 1)\lambda_{\max}$, where $\lambda_{\max}$ is the Roy's largest root test statistic, defined earlier, corresponding to H_0 indicated above. Thus $100(1 - \alpha)\%$ simultaneous confidence intervals for all linear combinations $\mathbf{c'LBd}$ are given by

$$\mathbf{c'L\hat{B}d} \pm \{\lambda_\alpha \mathbf{c'L(X'X)}^{-1}\mathbf{L'c} \cdot \mathbf{d'Ed}\}^{1/2},$$

where $\mathbf{E}$ is the error sums of squares and product matrix and λ_α, the cutoff point for $\lambda_{\max}$, is such that $P[\lambda_{\max} \le \lambda_\alpha] = 1 - \alpha$.

The tables for the cutoff points λ_α are available in Pillai (1960). Alternatively, the F statistics approximation from λ_{max} which is reported in Table 3.2 can be used. For Program 3.5, we follow the latter alternative. Thus the $(1 - \alpha)100\%$ cutoff point λ_α can be approximated by

$$\lambda_\alpha = \frac{h}{n - k - h + r - 1} F_\alpha(h, n - k - h + r - 1). \tag{3.17}$$

The calculations of the confidence intervals for a choice of $\mathbf{c}$ and $\mathbf{d}$ are illustrated in the following example using the IML procedure (see Program 3.5). Note that the matrices $\mathbf{\hat{B}}$, $(\mathbf{X'X})^{-1}$ and $\mathbf{E}$ are available as the outputs of PROC REG or PROC GLM. Selected parts of the output of Program 3.5 are shown in Output 3.5.

EXAMPLE 4   *Confidence Intervals for Cork Deposits in Cork Data*   Chapter 1 states that our interest in these data is in discovering if cork deposits are uniform in all four directions. Also, recall that an appropriate set of transformations of variables to do this would be

$$z_1 = y_1 - y_2 + y_3 - y_4, \quad z_2 = y_3 - y_4, \quad z_3 = y_1 - y_3.$$

To construct the $(1 - \alpha)100\%$ simultaneous confidence intervals on the corresponding means $\mu_1 - \mu_2 + \mu_3 - \mu_4$, $\mu_3 - \mu_4$ and $\mu_1 - \mu_3$, we write each of these means as $\mathbf{c'LBd}$ with the respective choices of $\mathbf{d}$ as the columns of

$$\begin{bmatrix} 1 & 0 & 1 \\ -1 & 0 & 0 \\ 1 & 1 & -1 \\ -1 & -1 & 0 \end{bmatrix}$$

and with $\mathbf{c} = 1, \mathbf{L} = 1$, and $\mathbf{B} = (\mu_1 \; \mu_2 \; \mu_3 \; \mu_4)$.

In Program 3.5 the matrices $\mathbf{c}$, $\mathbf{d}$, $\mathbf{L}$ and matrices $\mathbf{\hat{B}}$, $\mathbf{X'X}$ (both obtained from the output of PROC REG) and the matrix $\mathbf{E}$ (obtained from the output of PROC GLM) are explicitly specified.

In the present context,

$r_t = Rank(\mathbf{H} + \mathbf{E}) = p = 4,$
$r = Rank(\mathbf{L}) = Rank([1]) = 1,$

$n = 28,$

$k = 0,$

$s = min(r, r_t) = 1,$

$h = max(r, r_t) = 4,$

$m_1 = (|r - r_t| - 1)/2 = 1$ and

$m_2 = (n - k - r_t - 2)/2 = 11.$

As a result $\frac{n-k-h+r-1}{h}\lambda_{max} = 6\lambda_{max}$ follows an exact $F(4, 24)$ distribution (since $s=1$) $F(h, n - k - h + r - 1)$. If $\alpha = 0.05$, then using Equation 3.17 we can compute the 95% cutoff point $\lambda_{0.05}$ as $\frac{1}{6}F_{0.05}(4, 24) = 0.4627.$

```
/* Program 3.5 */

options ls=64 ps=45 nodate nonumber;
title1 'Output 3.5';
proc iml;
alpha = .05 ;
n = 28;
/*
Calculations for simultaneous confidence intervals are shown below.
   r_t =Rank (H+E)=p=4, Rank(L)=1, k=0
   df of error matrix = dferror = 27
   s = min(r, r_t) = 1, h=max(r,r_t)=4
   m1 = .5(|r-r_t| - 1) = 1
   m2 = .5(n-k-r_t-2) = 11.
   lambda  = [h/(n-k-h+r-1)]F(alpha,h, n-k-h+r-1)
*/
r_t=4;
r=1;
k=0;
s = min(r, r_t);
h=max(r,r_t);
m1 = .5*(abs(r-r_t) - 1);
m2 = .5*(n-k-r_t-2);
lambda  = (h/(n-k-h+r-1))*finv(1-alpha,h,n-k-h+r-1);
cutoff = sqrt(lambda);
/*
Cut-off point for Bonferroni intervals will be computed as follows:
  dferror = 27 ; * dferror=n-1;
  g=3.0; * g is the no. of comparisons;
  cutoff = tinv(1-(alpha/(2*g)),dferror);
  cutoff=cutoff/sqrt(dferror);
*/
xpx = {28};
e = {7840.9643 6041.3214 7787.8214 6109.3214,
     6041.3214 5938.1071 6184.6071 4627.1071,
     7787.8214 6184.6071 9450.1071 7007.6071,
     6109.3214 4627.1071 7007.6071 6102.1071};
bhat = {50.535714 46.178571 49.678571 45.178571};
l = {1} ;
c={1};
d1={1,-1,1,-1};
d2={0,0,1,-1};
d3={1,0,-1,0};
clbhatd1=c'*l*bhat*d1;
clbhatd2=c'*l*bhat*d2;
clbhatd3=c'*l*bhat*d3;
cwidth1=cutoff*sqrt((c'*l*(inv(xpx))*l'*c)*(d1'*e*d1));
```

```
cwidth2=cutoff*sqrt((c'*l*(inv(xpx))*l'*c)*(d2'*e*d2));
cwidth3=cutoff*sqrt((c'*l*(inv(xpx))*l'*c)*(d3'*e*d3));
cl11=clbhatd1-cwidth1;
cl12=clbhatd1+cwidth1;
cl21=clbhatd2-cwidth2;
cl22=clbhatd2+cwidth2;
cl31=clbhatd3-cwidth3;
cl32=clbhatd3+cwidth3;
print 'Simultaneous Confidence Intervals';
print 'For first contrast: (' cl11', '  cl12 ')';
print 'For second contrast:(' cl21', '  cl22 ')';
print 'For third contrast: (' cl31', '  cl32 ')';
run;
```

In the IML procedure code given in Program 3.5 it is necessary only to specify r_t, r, n, k and α explicitly. The rest of the parameters and the (approximate) value of λ_α are computed by the program.

Output 3.5

Output 3.5

Simultaneous Confidence Intervals

	CL11	CL12
For first contrast: (	1.2786666 ,	16.435619)

	CL21	CL22
For second contrast:(	-0.539816 ,	9.5398157)

	CL31	CL32
For third contrast: (	-4.467176 ,	6.1814617)

Output 3.5 shows that three confidence intervals are (1.2787, 16.4356), (−0.5398, 9.5398) and (−4.4672, 6.1815) respectively. It may be noted that the first of these intervals does not contain zero and therefore, it is this contrast, namely the difference between the average (or sums of) deposits in (N,S) and (E,W) direction, which caused the rejection of the hypothesis of uniform cork deposits in all four directions.

Hotelling's T^2 offers another choice of simultaneous confidence intervals. See Johnson and Wichern (1998, p. 239) for further details. These simultaneous confidence intervals have the drawback of being too wide. However, if the interest is in only a few specific linear combinations, it is possible to provide corresponding confidence intervals, which are shorter, using Bonferroni's inequality. These confidence intervals are based on the usual Student's t test for the associated univariate linear hypothesis. For instance, if $\mathbf{c}_i'\mathbf{LBd}_i$, $i = 1, \ldots, g$ are the g linear functions of interest, then $100(1-\alpha)\%$ Bonferroni's intervals are

$$\mathbf{c}_i'\mathbf{L\hat{B}d}_i \pm \frac{t_{n-k-1}(\alpha_i/2)}{\sqrt{n-k-1}}\sqrt{(\mathbf{c}_i'\mathbf{L}(\mathbf{X}'\mathbf{X})^{-1}\mathbf{L}'\mathbf{c}_i)(\mathbf{d}_i'\mathbf{Ed}_i)}, \quad i = 1, \ldots, g,$$

where $\alpha_1 + \cdots + \alpha_g = \alpha$ and $t_\nu(\delta)$ is the $100(1 - \delta)\%$ upper cutoff point from a t-distribution with ν degrees of freedom. Note that in order to compute these intervals the IML procedure as shown in Program 3.5 can be used with $\lambda_d^{1/2}$ replaced by $t_{n-k-1}(\alpha_i/2)$, for $i = 1, \ldots, g$ respectively. The corresponding statements have been included in Program 3.5 with $\alpha_i = \alpha/3$ but have been commented out. To obtain the Bonferroni intervals, replace these statements with those corresponding to simultaneous confidence intervals.

3.7   Multiple Response Surface Modeling

Response surface modeling is essentially a regression analysis problem. In situations where data are collected on a number of response (dependent) variables under the controlled levels of process or recipe (independent) variables, these responses are correlated. However, as it often happens, the levels of process or recipe variables that are optimum for one dependent variable may not be optimum for others. Therefore it is important to investigate response variables simultaneously and not individually or independently of one another, in order to also account for interrelationships. Consequently, the "best model" search for the individual response variables may not be meaningful. What is desired is a best *set* of models for these responses. A way to do this would be to simultaneously fit multivariate regression models and statistically test the significance of various terms corresponding to independent variables using multivariate methods. This can be done by the repeated use of the MTEST statement in PROC REG.

EXAMPLE 5   *Quality Improvement of Mullet Flesh*   Tseo, Deng, Cornell, Khuri, and Schmidt (1983) considered a study of quality improvement of minced mullet flesh where three process variables, namely washing temperature (TEMP), washing time (TIME), and washing ratio (WRATIO) were varied in a controlled manner in a designed experiment. The four responses, springiness (SPRNESS) in mm, TBA number (TBA), cooking loss (COOKLOSS) in % and whiteness index (WHITNESS) were observed for all 17 experiments. For the analysis, all the variables are standardized to have zero means. The standardization of independent variables was carried out to have the levels of the three variables as $+1$ and -1 in the 2^3 full factorial part of the design. To do so, we define

$$X1 = \frac{TEMP - 33}{7}, \quad X2 = \frac{TIME - 5.5}{2.7}, \quad X3 = \frac{WRATIO - 22.5}{4.5}.$$

The four response variables are standardized to have sample variances equal to 1. The standardized variables are respectively denoted by $Y1$, $Y2$, $Y3$, and $Y4$. The multiple response surfaces are fitted for $Y1$, $Y2$, $Y3$, and $Y4$ as functions of $X1$, $X2$, and $X3$.

Suppose the interest is to simultaneously fit models which contain effects only up to the second degree in $X1$, $X2$, and $X3$ and up to two variable interactions. For that, we first obtain the values for the variables $X1$, $X2$, and $X3$ as indicated above and their respective interactions defined as $X1X2 = X1 * X2$, $X1X3 = X1 * X3$, $X2X3 = X2 * X3$. Then we obtain the quadratic effects $X1SQ = X1 * X1$, $X2SQ = X2 * X2$ and $X3SQ = X3 * X3$, within the DATA step for the data set WASH. The values for variables $Y1, \ldots, Y4$ are obtained by using the STANDARD procedure where the options MEAN = 0 and STD = 1 are used to set the respective sample means at zero and respective sample standard deviations at unity for these variables which are listed in the VAR statement. Output is stored in the data set WASH2. We perform a multivariate regression analysis on WASH2 to obtain the appropriate response surfaces by performing the various significance tests.

A complete second order model would involve a total of nine terms, namely $X1$, $X2$, $X3$, $X1X2$, $X1X3$, $X2X3$, $X1SQ$, $X2SQ$, and $X3SQ$, apart from the intercept. We will confine our discussion to only three specific hypotheses

$H_0^{(1)}$ The multivariate model contains only the linear terms plus an intercept vector.

$H_0^{(2)}$ The multivariate model is quadratic without any interaction terms.

$H_0^{(3)}$ The multivariate model has only linear, two-variable interaction terms and intercept vector but no quadratic terms.

The three null hypotheses stated above can respectively be tested using the following three MTEST statements:

```
linear: mtest x1sq, x2sq, x3sq, x1x2, x1x3, x2x3/print;
nointctn: mtest x1x2, x1x3, x2x3/print;
noquad: mtest x1sq, x2sq, x3sq/print;
```

The names LINEAR, NOINTCTN and NOQUAD before the colon (:) in the respective three statements are used only for labeling purposes and are optional. The SAS code and resulting output are presented in Program 3.6 and Output 3.6.

```
/* Program 3.6 */

options ls=64 ps=45 nodate nonumber;
title1 'Output 3.6';
data wash;
input temp time wratio sprness tba cookloss whitness ;
x1 = (temp - 33)/7.0 ;
x2 = (time - 5.5)/2.7 ;
x3 = (wratio-22.5)/4.5 ;
x1sq =x1*x1;
x2sq = x2*x2;
x3sq = x3*x3;
x1x2 = x1*x2;
x1x3 = x1*x3;
x2x3 = x2*x3;
y1= sprness;
y2= tba;
y3 = cookloss;
y4= whitness ;
lines;
26.0  2.8   18.0   1.83 29.31 29.50 50.36
40.0  2.8   18.0   1.73 39.32 19.40 48.16
26.0  8.2   18.0   1.85 25.16 25.70 50.72
40.0  8.2   18.0   1.67 40.81 27.10 49.69
26.0  2.8   27.0   1.86 29.82 21.40 50.09
40.0  2.8   27.0   1.77 32.20 24.00 50.61
26.0  8.2   27.0   1.88 22.01 19.60 50.36
40.0  8.2   27.0   1.66 40.02 25.10 50.42
21.2  5.5   22.5   1.81 33.00 24.20 29.31
44.8  5.5   22.5   1.37 51.59 30.60 50.67
33.0  1.0   22.5   1.85 20.35 20.90 48.75
33.0  10.0  22.5   1.92 20.53 18.90 52.70
33.0  5.5   14.9   1.88 23.85 23.00 50.19
33.0  5.5   30.1   1.90 20.16 21.20 50.86
33.0  5.5   22.5   1.89 21.72 18.50 50.84
33.0  5.5   22.5   1.88 21.21 18.60 50.93
33.0  5.5   22.5   1.87 21.55 16.80 50.98
;
/* Source: Tseo et al. (1983).  Reprinted by permission of
the Institute of Food Technologists. */
proc standard data=wash mean=0 std=1 out=wash2 ;
var  y1 y2 y3 y4 ;
run;
proc reg data = wash2;
model y1 y2 y3 y4 = x1 x2 x3 x1sq x2sq x3sq
x1x2 x1x3 x2x3 ;
Linear: mtest x1sq, x2sq, x3sq, x1x2, x1x3,
x2x3/print;
```

```
Nointctn: mtest x1x2,x1x3,x2x3/print;
Noquad: mtest x1sq, x2sq, x3sq/print;
title2 'Quality Improvement in Mullet Flesh';
run;
proc reg data = wash2;
model y1 y2 y3 y4 = x1 x2 x3 x1sq x2sq x3sq ;
run;
```

Output 3.6

```
                          Output 3.6
                Quality Improvement in Mullet Flesh

Multivariate Test: LINEAR


                      E, the Error Matrix


     0.8657014633     0.0034272456    -0.815101433    -1.634723273
     0.0034272456     0.661373219     -0.17637356      0.6463179546
    -0.815101433     -0.17637356       1.9374527491     2.2687768511
    -1.634723273      0.6463179546     2.2687768511     6.765404456


                    H, the Hypothesis Matrix


     8.0041171357    -8.822976865    -7.893007279     6.5980348245
    -8.822976865     10.054652572     9.1165224684    -7.103378577
    -7.893007279      9.1165224684   12.622482263     -5.15027833
     6.5980348245    -7.103378577    -5.15027833       5.9966106878

              Multivariate Statistics and F Approximations

                    S=4     M=0.5     N=1

Statistic                 Value        F     Num DF  Den DF  Pr > F

Wilks' Lambda            0.001812    3.228      24   15.164  0.0107
Pillai's Trace           2.001528    1.1685     24      28   0.3436
Hotelling-Lawley Trace  88.2617      9.1939     24      10   0.0004
Roy's Greatest Root     83.82687    97.798       6       7   0.0001

 NOTE: F Statistic for Roy's Greatest Root is an upper bound.


Multivariate Test: NOINTCTN

                      E, the Error Matrix


     0.8657014633     0.0034272456    -0.815101433    -1.634723273
     0.0034272456     0.661373219     -0.17637356      0.6463179546
    -0.815101433     -0.17637356       1.9374527491     2.2687768511
    -1.634723273      0.6463179546     2.2687768511     6.765404456
```

```
                         H, the Hypothesis Matrix


     0.3238160164      -0.440143047     -0.758060209     -0.030667821
    -0.440143047        0.7018039785     0.6864395048     -0.025289745
    -0.758060209        0.6864395048     3.9583870894      0.4937367103
    -0.030667821       -0.025289745      0.4937367103      0.0844373123
```

Multivariate Statistics and F Approximations

S=3   M=0   N=1

Statistic	Value	F	Num DF	Den DF	Pr > F
Wilks' Lambda	0.061574	1.6927	12	10.875	0.1973
Pillai's Trace	1.443396	1.3909	12	18	0.2558
Hotelling-Lawley Trace	7.205689	1.6013	12	8	0.2566
Roy's Greatest Root	5.86998	8.805	4	6	0.0110

NOTE: F Statistic for Roy's Greatest Root is an upper bound.

Multivariate Test: NOQUAD

                          E, the Error Matrix

     0.8657014633   0.0034272456   -0.815101433   -1.634723273
     0.0034272456   0.661373219   -0.17637356   0.6463179546
    -0.815101433   -0.17637356   1.9374527491   2.2687768511
    -1.634723273   0.6463179546   2.2687768511   6.765404456

                         H, the Hypothesis Matrix

     7.6803011193   -8.382833817   -7.134947069   6.6287026452
    -8.382833817   9.3528485931   8.4300829636   -7.078088833
    -7.134947069   8.4300829636   8.6640951737   -5.644015041
     6.6287026452   -7.078088833   -5.644015041   5.9121733755
```

Multivariate Statistics and F Approximations

S=3        M=0        N=1

| Statistic | Value | F | Num DF | Den DF | Pr > F |
|---|---|---|---|---|---|
| Wilks' Lambda | 0.004154 | 6.2945 | 12 | 10.875 | 0.0024 |
| Pillai's Trace | 1.659428 | 1.8568 | 12 | 18 | 0.1141 |
| Hotelling-Lawley Trace | 81.05601 | 18.012 | 12 | 8 | 0.0002 |
| Roy's Greatest Root | 79.06247 | 118.59 | 4 | 6 | 0.0001 |

NOTE: F Statistic for Roy's Greatest Root is an upper bound.
```

The output resulting from the MTEST statements indicated above shows that the null hypothesis $H_0^{(1)}$, which states that the models contain only linear terms and intercepts, is probably not true. The p value corresponding to Wilks' Λ is 0.0107. Except for Pillai's trace test, all the other multivariate tests produce small p values. Thus at least some of the quadratic and/or interaction terms may be important and may need to be included in the model. Also, it may be of interest to test the hypotheses $H_0^{(2)}$ and $H_0^{(3)}$ (among many others) which exclusively test for the absence of two-variable interaction effects and the absence of quadratic effects respectively. As Output 3.6 shows, the null hypothesis $H_0^{(2)}$ is not rejected and hence, we may probably drop all the two-variable interaction terms from the model. In view of small p values corresponding to all multivariate tests except the Pillai's trace statistic, $H_0^{(3)}$ is rejected leading to the conclusion that there are at least some quadratic effects present. As a result, the equations of the four estimated response surfaces, obtained from the output corresponding to univariate analyses, (not shown), are

$$\hat{Y}1 = 0.5831 - 0.7187 * X1 - 0.0073 * X2 + 0.0667 * X3 - 0.7569 * X1SQ$$
$$+ 0.0099 * X2SQ + 0.0227 * X3SQ,$$

$$\hat{Y}2 = -0.8297 - 0.6071 * X1 - 0.0186 * X2 - 0.1314 * X3 + 0.8736 * X1SQ$$
$$+ 0.0510 * X2SQ + 0.1065 * X3SQ,$$

$$\hat{Y}3 = -1.1671 + 0.1854 * X1 - 0.0025 * X2 - 0.2660 * X3 + 0.856 * X1SQ$$
$$+ 0.2044 * X2SQ + 0.3904 * X3SQ,$$

$$\hat{Y}4 = 0.2979 - 0.4684 * X1 + 0.1212 * X2 + 0.0516 * X3 - 0.6040 * X1SQ$$
$$+ 0.1275 * X2SQ + 0.1075 * X3SQ.$$

Even the terms $X2$, $X3$, $X2SQ$, and $X3SQ$ can also be dropped (the output is not shown here), leaving the four quadratic response surfaces as functions of the variable $X1$, that is temperature only. In that sense, the four responses appear to be robust with respect to washing time and washing ratio.

3.8   General Linear Hypotheses

Sometimes our interest may be in comparing the regression coefficients from the models corresponding to different variables. For example, in a multivariate linear model with two response variables, it may be interesting to test if the intercepts, means, or some other regression coefficients are equal in the respective models corresponding to these two response variables. Such a hypothesis cannot be expressed in the form of Equation 3.11. This hypothesis can, however, be formulated as a general linear hypothesis which can be written as

$$H_0 : \mathbf{LBM} = \mathbf{0} \text{ vs. } H_1 : \mathbf{LBM} \neq \mathbf{0},$$

where r by $(k + 1)$ matrix $\mathbf{L}$ is, as earlier, of rank r and the p by s matrix $\mathbf{M}$ is of rank s. These two matrices have different roles to play and need to be chosen carefully depending on the particular hypothesis to be tested. Specifically, the premultiplied matrix $\mathbf{L}$ is used to obtain a linear function of the regression coefficients within the individual models while the postmultiplied matrix $\mathbf{M}$ does the same for the coefficients from different models but corresponding to the same set of regressors or independent variables. In other words, the matrix $\mathbf{L}$ provides a means of comparison of coefficients within models, whereas the matrix $\mathbf{M}$ offers a way for "between models" comparisons of regression coefficients. As a result, the simultaneous pre- and postmultiplication to $\mathbf{B}$ by $\mathbf{L}$ and $\mathbf{M}$ respectively provides a method for defining a general linear hypothesis involving various coefficients of $\mathbf{B}$.

For example, let $\mathbf{B} = (\beta_{ij})$, that is, except for the zeroth row of $\boldsymbol{\beta}$, the coefficient β_{ij} is the regression coefficient of the ith independent variable in the jth model (that is, the model for j^{th} dependent variable). The zeroth row of $\boldsymbol{\beta}$ contains intercepts of various models. Suppose we want to test the hypothesis that the difference between the coefficients of first and second independent variables is the same for the two univariate models involving the first two dependent variables. In other words, the null hypothesis to be tested is $\beta_{11} - \beta_{21} = \beta_{12} - \beta_{22}$, i.e., $H_0 : (\beta_{11} - \beta_{21}) - (\beta_{12} - \beta_{22}) = 0$. This equation in matrix notation is written as $\mathbf{LBM} = \mathbf{0}$ with

$$\mathbf{L} = (0 \quad 1 \quad -1 \quad 0 \quad \cdots \quad 0)$$

and

$$\mathbf{M} = \begin{bmatrix} 1 \\ -1 \\ 0 \\ . \\ . \\ . \\ 0 \end{bmatrix}.$$

These types of general linear hypotheses occur frequently in the studies of growth curves, repeated measures, and crossover designed data. Chapter 5 covers various examples of these data.

EXAMPLE 6   *Spatial Uniformity in Semiconductor Processes*   Guo and Sachs (1993) presented a case study that attempted to model and optimize the spatial uniformity of the product output characteristics at different locations in a batch of products. This example empirically models these responses using the multiple response surfaces and interprets the problem of testing the spatial uniformity as a problem of general linear hypothesis testing.

The independent variables under consideration are two flow rates denoted by $X1$ and $X2$, and the resulting dependent variables are the deposition rates at three measurement sites. We denote these by $Y1$, $Y2$, and $Y3$ respectively. We are interested in the spatial uniformity, that is, we want to achieve a uniformity between the values of $Y1$, $Y2$, and $Y3$ for the given levels of two flow rates.

This example fits the multivariate regression model for $Y1$, $Y2$, and $Y3$ in terms of $X1$ and $X2$. Assume that there is no interaction between $X1$ and $X2$ and that the effects of $X1$ and $X2$ are both linear. The individual models can be obtained by using the SAS statements

```
proc reg;
model y1 y2 y3 = x1 x2;
```

given as part of Program 3.7. Output 3.7, which is produced by Program 3.7, provides the estimates (collected from three separate univariate analyses) of regression coefficients

$$\hat{\mathbf{B}} = \begin{bmatrix} 29.8449 & 34.0764 & 41.3942 \\ 0.2940 & 0.2048 & 0.1346 \\ 0.1175 & 0.1378 & 0.0355 \end{bmatrix}$$

and hence the three models are

$$\hat{Y}1 = 29.8449 + 0.2940X1 + 0.1175X2,$$
$$\hat{Y}2 = 34.0764 + 0.2048X1 + 0.1378X2,$$
$$\hat{Y}3 = 41.3942 + 0.1346X1 + 0.0355X2.$$

In the ideal case of complete spatial uniformity, we would expect the models for $Y1$, $Y2$, and $Y3$ to be identical. We construct an appropriate null hypothesis from this interpretation of spatial uniformity.

```
/* Program 3.7 */

options ls=64 ps=45 nodate nonumber;
data semicond;
input x1 x2 y1 y2 y3;
z1=y1-y2;z2=y2-y3;
lines;
46 22 45.994 46.296 48.589
56 22 48.843 48.731 49.681
66 22 51.555 50.544 50.908
46 32 47.647 47.846 48.519
56 32 50.208 49.930 50.072
66 32 52.931 52.387 51.505
46 42 47.641 49.488 48.947
56 42 51.365 51.365 50.642
66 42 54.436 52.985 51.716
;
/* Source: Guo and Sachs (1993).  Reprinted by permission of the
   Institute of Electrical and Electronics Engineers, Inc.
   Copyright 1993 IEEE. */
proc reg data = semicond ;
model y1 y2 y3 =  x1 x2 ;
AllCoef: mtest y1-y2,y2-y3,intercept,x1,x2/print;
X1andX2: mtest y1-y2,y2-y3,x1,x2/print;
title1 ' Output 3.7 ';
title2 'Spatial Uniformity in Semiconductor Processes' ;
run;
```

Let

$$\mathbf{B} = \begin{bmatrix} \beta_{01} & \beta_{02} & \beta_{03} \\ \beta_{11} & \beta_{12} & \beta_{13} \\ \beta_{21} & \beta_{22} & \beta_{23} \end{bmatrix}.$$

Then the columns of $\mathbf{B}$ represent the regression coefficients in the models for $Y1$, $Y2$, and $Y3$ respectively. Thus the complete spatial uniformity amounts to testing the hypothesis that the three columns of $\mathbf{B}$ are identical, that is, our null hypothesis is

$$H_0: \quad \begin{aligned} \beta_{01} &= \beta_{02} = \beta_{03}, \\ \beta_{11} &= \beta_{12} = \beta_{13}, \\ \beta_{21} &= \beta_{22} = \beta_{23}; \end{aligned}$$

or,

$$H_0: \quad \begin{aligned} \beta_{01} - \beta_{02} &= 0, & \beta_{02} - \beta_{03} &= 0, \\ \beta_{11} - \beta_{12} &= 0, & \beta_{12} - \beta_{13} &= 0, \\ \beta_{21} - \beta_{22} &= 0, & \beta_{22} - \beta_{23} &= 0; \end{aligned}$$

or,

$$H_0: \begin{bmatrix} \beta_{01} & \beta_{02} & \beta_{03} \\ \beta_{11} & \beta_{12} & \beta_{13} \\ \beta_{21} & \beta_{22} & \beta_{23} \end{bmatrix} \begin{bmatrix} 1 & 0 \\ -1 & 1 \\ 0 & -1 \end{bmatrix} = \begin{bmatrix} 0 & 0 \\ 0 & 0 \\ 0 & 0 \end{bmatrix};$$

or,

$$H_0: \quad \mathbf{LBM} = \mathbf{0} \text{ with } \mathbf{L} = \mathbf{I}_3 \text{ and } \mathbf{M} = \begin{bmatrix} 1 & 0 \\ -1 & 1 \\ 0 & -1 \end{bmatrix}.$$

To test this null hypothesis, we use the MTEST statement in PROC REG. An alternative could be to use the CONTRAST and MANOVA statements in PROC GLM, but since that is applicable only for designed experiments and not otherwise, we do not make this choice here. Chapters 4 and 5 elaborate on this approach. The approach using the MTEST statement is general and applicable for all regression modeling problems even when the data are collected from designed or undesigned experiments.

If H_0 is true, then the coefficients of the three models are identical. And hence the true means (expected values) of $Y1 - Y2$ and $Y2 - Y3$ would both be zero. Therefore, in order to test H_0 we could simultaneously test the three linear hypotheses. Specifically, the null hypotheses are that the intercepts, as well as the coefficients of $X1$ and $X2$ in the models for these two variables, are all zero. This can be done using the MTEST statement after the MODEL statement. Models without intercepts can be fitted by using the NOINT option in the MODEL statement.

Consequently, to test $H_0 :$ **LBM** $= 0$ with the choice of **L** and **M** indicated above, we use the following SAS statements:

```
proc reg;
model y1 y2 y3 = x1 x2;
mtest y1-y2, y2-y3, intercept, x1, x2/print;
```

Note that the MTEST statement performs the four multivariate tests on variables $Y1 - Y2$ and $Y2 - Y3$. An alternative yet equivalent approach would have been to define $Z1 = Y1 - Y2$ and $Z2 = Y2 - Y3$ early in the DATA step after INPUT statement as shown below:

```
data semicond;
input x1 x2 y1 y2 y3;
z1=y1-y2;
z2=y2-y3;
```

and then later in REG procedure use the MODEL and MTEST statements

```
model z1 z2 = x1 x2;
mtest z1, z2, intercept, x1, x2/print;
```

In fact, we do not really need to include variables $Z1$ and $Z2$ in the MTEST statement given above. They are included by default because, if the list in the MTEST statement does not include any dependent variable, SAS automatically includes the variables being analyzed and those listed on the left side of the MODEL statement. In Program 3.7 we have, however, used the MTEST statement to avoid creating two extra variables $Z1$ and $Z2$. The PRINT option in the MTEST statement prints the corresponding **H** and **E** matrices for variables $Y1 - Y2$ and $Y2 - Y3$.

Output 3.7

```
                          Output 3.7
            Spatial Uniformity in Semiconductor Processes

Model: MODEL1
Dependent Variable: Y1


                       Analysis of Variance

                         Sum of        Mean
Source           DF      Squares       Square      F Value
Prob>F
```

Model	2	60.14535	30.07268	201.946
0.0001				
Error	6	0.89348	0.14891	
C Total	8	61.03883		

Root MSE	0.38589	R-square	0.9854
Dep Mean	50.06889	Adj R-sq	0.9805
C.V.	0.77073		

Parameter Estimates

Variable	DF	Parameter Estimate	Standard Error	T for H0: Parameter=0
INTERCEP	1	29.844889	1.02421553	29.139
X1	1	0.294000	0.01575405	18.662
X2	1	0.117500	0.01575405	7.458

Variable	DF	Prob > |T|
INTERCEP	1	0.0001
X1	1	0.0001
X2	1	0.0003

Dependent Variable: Y2

Analysis of Variance

Source	DF	Sum of Squares	Mean Square	F Value
Prob>F				
Model	2	36.54818	18.27409	251.850
0.0001				
Error	6	0.43536	0.07256	
C Total	8	36.98354		

Root MSE	0.26937	R-square	0.9882
Dep Mean	49.95244	Adj R-sq	0.9843
C.V.	0.53925		

Parameter Estimates

Variable	DF	Parameter Estimate	Standard Error	T for H0: Parameter=0
INTERCEP	1	34.076444	0.71494183	47.663
X1	1	0.204767	0.01099694	18.620
X2	1	0.137783	0.01099694	12.529

Variable	DF	Prob > |T|
INTERCEP	1	0.0001
X1	1	0.0001
X2	1	0.0001

Dependent Variable: Y3

Analysis of Variance

Source Prob>F	DF	Sum of Squares	Mean Square	F Value
Model 0.0001	2	11.61893	5.80947	183.301
Error	6	0.19016	0.03169	
C Total	8	11.80910		

Root MSE	0.17803	R-square	0.9839
Dep Mean	50.06433	Adj R-sq	0.9785
C.V.	0.35560		

Parameter Estimates

Variable	DF	Parameter Estimate	Standard Error	T for H0: Parameter=0
INTERCEP	1	41.394200	0.47250828	87.605
X1	1	0.134567	0.00726792	18.515
X2	1	0.035450	0.00726792	4.878

Variable	DF	Prob > |T|
INTERCEP	1	0.0001
X1	1	0.0001
X2	1	0.0028

Multivariate Test: ALLCOEF

E, the Error Matrix

1.9090653889	-0.761418778
-0.761418778	0.6168702222

H, the Hypothesis Matrix

5.1464346111	2.3958517778
2.3958517778	9.3527627778

Multivariate Statistics and F Approximations

S=2   M=0   N=1.5

Statistic	Value	F	Num DF	Den DF	Pr > F
Wilks' Lambda	0.008835	16.064	6	10	0.0001
Pillai's Trace	1.617641	8.4614	6	12	0.0010

```
Hotelling-Lawley Trace 41.27571    27.517      6       8  0.0001
Roy's Greatest Root    39.47972    78.959      3       6  0.0001
```

```
    NOTE: F Statistic for Roy's Greatest Root is an upper bound.
          NOTE: F Statistic for Wilks' Lambda is exact.
```

Multivariate Test: X1ANDX2

E, the Error Matrix

```
        1.9090653889        -0.761418778
        -0.761418778        0.6168702222
```

H, the Hypothesis Matrix

```
        5.0244008333        2.5131113333
        2.5131113333        9.2400906667
```

Multivariate Statistics and F Approximations

```
            S=2     M=-0.5     N=1.5
```

Statistic	Value	F	Num DF	Den DF	Pr > F
Wilks' Lambda	0.00916	23.622	4	10	0.0001
Pillai's Trace	1.605325	12.202	4	12	0.0003
Hotelling-Lawley Trace	41.0887	41.089	4	8	0.0001
Roy's Greatest Root	39.38536	118.16	2	6	0.0001

```
    NOTE: F Statistic for Roy's Greatest Root is an upper bound.
          NOTE: F Statistic for Wilks' Lambda is exact.
```

The first part of Output 3.7 presents the results of multivariate tests. The error $SS\&CP$ matrix $\mathbf{E}$ and the hypothesis $SS\&CP$ matrix $\mathbf{H}$ are given first and are then followed by four multivariate tests. All of the four multivariate tests reject the null hypothesis. For example, the value of Wilks' Λ is 0.0088, which leads to an (exact) $F(6, 10)$ statistic value of 16.064 and a p value of 0.0001. In view of this extremely small p value, there is sufficient evidence to reject H_0 and conclude the lack of spatial uniformity.

Having rejected the null hypothesis of equality of all regression coefficients including the intercepts in the three models, we may want to test if the three models differ only in their intercepts and if the respective coefficients of $X1$ and $X2$ are the same in the models for $Y1$, $Y2$ and $Y3$. Therefore, we exclude the keyword INTERCEPT in the MTEST statement. The appropriate MODEL and MTEST statements are

```
model y1 y2 y3 =x1 x2;
mtest y1-y2, y2-y3, x1, x2/print;
```

Of course, as earlier, $Y1 - Y2$ and $Y2 - Y3$ can be removed from the list in the MTEST statement if $Z1 = Y1 - Y2$ and $Z2 = Y2 - Y3$ have already been defined in the DATA step and if the variables $Z1$ and $Z2$ are analyzed in the MODEL statement. The resulting multivariate outputs would be identical. These are presented under the label, "Multivariate

Test: X1ANDX2" in the latter part of Output 3.7. As in the previous case, the null hypothesis in the present case is also rejected by all four multivariate tests leading us to believe that the deposition rates Y1, Y2, and Y3 at the three different measurement sites depend differently on the two flow rates X1 and X2.

3.9   Variance and Bias Analyses for Calibration Problems

This section presents an application of the multivariate analysis of variance techniques to test for equality of variances in several measuring devices. The approach is taken from Christensen and Blackwood (1993). Such calibration problems occur frequently in assessing the relative measurement quality of various instruments or of a particular instrument at various times.

Suppose y_{ij} $i = 1, \ldots, n$, $j = 1, \ldots, q$ are the measurements for a random sample of n items distributed around their respective true values, each measured by q instruments or at q different laboratories. If α_j is the fixed bias of the jth instrument and ϵ_{ij} are the independent random errors with zero mean and variance σ_j^2, then the problem of testing the equality of the error variances, namely

$$H_0 : \sigma_1^2 = \sigma_2^2 = \cdots = \sigma_q^2,$$

can be reduced to a MANOVA problem with the test based on the multivariate regression coefficients.

Let us define, for each of n items, $\bar{y}_i = \frac{1}{q} \sum_{j=1}^{q} y_{ij}$ as the average measurement for the ith item and $\tilde{y}_{ij} = y_{ij} - \bar{y}_i$ as the deviation of each measurement on i^{th} item from the corresponding sample mean $\bar{y}_i$. Then, Christensen and Blackwood (1993) have shown that the testing problem stated above is equivalent to testing

$$H_0 : \beta_{11} = \beta_{12} = \cdots = \beta_{1,q-1} = 0$$

in the multivariate linear model

$$\tilde{y}_{ij} = \beta_{0j} + \beta_{1j}\bar{y}_i + \epsilon_{ij},$$

$j = 1, \ldots, (q-1); i = 1, \ldots, n$. In the context of above model, the null hypothesis H_0 can be interpreted as a claim that the observed deviations from the mean do not depend on the mean itself. The above model in matrix form is written as

$$
\begin{bmatrix}
\tilde{y}_{11} & \cdots & \tilde{y}_{1,q-1} \\
\tilde{y}_{21} & \cdots & \tilde{y}_{2,q-1} \\
\cdot & & \\
\cdot & & \\
\cdot & & \\
\tilde{y}_{n1} & \cdots & \tilde{y}_{n,q-1}
\end{bmatrix}
=
\begin{bmatrix}
1 & \bar{y}_1 \\
1 & \bar{y}_2 \\
\cdot & \cdot \\
\cdot & \cdot \\
\cdot & \cdot \\
1 & \bar{y}_n
\end{bmatrix}
\begin{bmatrix}
\beta_{01} & \beta_{02} & \cdots & \beta_{0,q-1} \\
\beta_{11} & \beta_{12} & \cdots & \beta_{1,q-1}
\end{bmatrix}
+
\begin{bmatrix}
\epsilon_{11} & \cdots & \epsilon_{1\,q-1} \\
\epsilon_{21} & \cdots & \epsilon_{2\,q-1} \\
\cdot & & \\
\cdot & & \\
\cdot & & \\
\epsilon_{n1} & \cdots & \epsilon_{n\,q-1}
\end{bmatrix}
$$

or

$$\mathbf{Y} = \mathbf{X}\mathbf{B} + \mathcal{E}$$

and our null hypothesis is that the vector of the slope parameters for the $(q-1)$ variables is equal to zero. This is one of the standard null hypotheses that can be easily tested using PROC GLM or PROC REG.

In the above discussion, $\tilde{y}_{iq}, i = 1, \ldots, n$ have not been included in the model. In general, any other set, $\tilde{y}_{ij}, i = 1, \ldots, n$ could have been dropped instead. Although the data thus created would differ, the resulting test statistics and hence the conclusions are invariant of any such choice.

EXAMPLE 7 *Equality of Variances in Calibration of Thermocouples* Christensen and Blackwood (1993) described a study in which 64 measurements on a high temperature furnace were taken by each of five thermocouples that had been bound together and inserted in the furnace. The objective is to test if all five thermocouples have the same precision. In other words, in this example we want to test the hypothesis of the equality of the variances for the five thermocouples.

The data and the corresponding SAS code for analysis are presented in Program 3.8. From the five temperature variables TC1 through TC5 corresponding to these thermocouples and their average TCBAR = $(TC1 + TC2 + \cdots + TC5)/5$ the variables Y1TILDA through Y5TILDA are defined by taking the differences from their mean TCBAR. For these data, since $q = 5$, we only need to take $q - 1 = 4$ of these five variables as the response variables to fit a multivariate regression model with TCBAR as the independent variable. We can choose any four of these five as the response variable. As mentioned previously, the values of the test statistics and the corresponding P values are unaffected by any such choice.

Testing the hypothesis of equality of variances of TC1 through TC5 is equivalent to testing the hypothesis that the slope parameters corresponding to the independent variable TCBAR are all zero in the corresponding multivariate linear model. This hypothesis is tested by using the SAS code

```
proc glm;
model y2tilda y3tilda y4tilda y5tilda = tcbar/nouni;
manova h = tcbar/printe printh;
```

or alternatively by using

```
proc reg;
model y2tilda y3tilda y4tilda y5tilda = tcbar;
EqualVar: mtest tcbar/print;
```

The latter choice is used while executing the Program 3.8, and the output is presented in Output 3.8. All the multivariate tests are exact in this case and are also equivalent. Corresponding to the observed value of $F(4, 59)$ as 3.6449, the p value is equal to .0101 leading to the rejection of the null hypothesis of equality of the five variances. The next step may be to determine the equality of various variances in subgroups. The slope coefficients for regressing Y2TILDA through Y5TILDA on TCBAR respectively, obtained from the corresponding outputs of univariate analysis, (not shown), are 0.1288, 0.0242, −0.0697, and −0.0808. Because the five response variables sum to zero in this model the sum of all five coefficients (corresponding to Y1TILDA through Y5TILDA) equals zero. Thus by difference, the slope coefficient for regressing Y1TILDA on TCBAR is −0.0025. Further, the smaller regression coefficients correspond to smaller variances. Hence the five variances can be ordered as

$$\sigma_5^2, \sigma_4^2, \ \sigma_1^2, \ \sigma_3^2, \ \sigma_2^2.$$

Using the Student-Newman-Keuls test (Kuehl, 1994), Christensen and Blackwood (1993) summarized the grouping as given below.

$$\left[\sigma_5^2, \sigma_4^2, \ \left(\sigma_1^2\right), \ \sigma_3^2, \ \sigma_2^2\right).$$

See Christensen and Blackwood (1993) for further details.

```
/* Program 3.8 */

options ls=64 ps=45 nodate nonumber;
title1 'Output 3.8';
title2 'Testing the Precisions of Five Thermocouples';
data calib ;
infile 'thermoco.dat' obs=64;
input  tc1  tc2  tc3  tc4  tc5  @@ ;
tcbar = (tc1+tc2+tc3+tc4+tc5)/5 ;
y1tilda = tc1 - tcbar ;
y2tilda = tc2 - tcbar ;
y3tilda = tc3 - tcbar ;
y4tilda = tc4 - tcbar ;
y5tilda = tc5 - tcbar ;

/*
proc glm data = calib;
model  y2tilda y3tilda y4tilda y5tilda = tcbar /nouni;
manova h = tcbar/printe printh;
*/
proc reg data = calib ;
model  y2tilda y3tilda y4tilda y5tilda = tcbar;
EqualVar:mtest tcbar/print;
Bias_Var:mtest intercept, tcbar/print;
run;
```

Output 3.8

```
                         Output 3.8
           Testing the Precisions of Five Thermocouples


Multivariate Test: EQUALVAR


                    E, the Error Matrix


    0.0475701116    -0.011049524    -0.009544794    -0.013674773
    -0.011049524     0.0129013449    0.0015628105    -0.003486742
    -0.009544794     0.0015628105    0.0075360191    -0.0005665
    -0.013674773    -0.003486742    -0.0005665       0.0173898899



                  H, the Hypothesis Matrix


    0.0036643259     0.0006892739   -0.001983269    -0.002297977
    0.0006892739     0.0001296551   -0.000373061    -0.000432258
    -0.001983269    -0.000373061     0.0010734184    0.00124375
    -0.002297977    -0.000432258     0.00124375      0.0014411101

          Multivariate Statistics and Exact F Statistics

                    S=1     M=1     N=28.5

    Statistic              Value      F     Num DF  Den DF  Pr > F


    Wilks' Lambda          0.801854  3.6449      4      59  0.0101
    Pillai's Trace         0.198146  3.6449      4      59  0.0101
```

```
Hotelling-Lawley Trace  0.247109    3.6449        4       59  0.0101
Roy's Greatest Root     0.247109    3.6449        4       59  0.0101
```

Multivariate Test: BIAS_VAR

E, the Error Matrix

```
 0.0475701116   -0.011049524   -0.009544794   -0.013674773
-0.011049524     0.0129013449   0.0015628105  -0.003486742
-0.009544794     0.0015628105   0.0075360191  -0.0005665
-0.013674773    -0.003486742   -0.0005665      0.0173898899
```

H, the Hypothesis Matrix

```
 295.33648078  -263.4663775    67.96997654   168.39596614
-263.4663775    235.03968978  -60.63831058  -150.2289007
  67.96997654   -60.63831058   15.645075922   38.758738355
 168.39596614  -150.2289007    38.758738355   96.021841751
```

Multivariate Statistics and F Approximations

S=2   M=0.5   N=28.5

Statistic	Value	F	Num DF	Den DF	Pr > F
Wilks' Lambda	0.000018	3484.1	8	118	0.0001
Pillai's Trace	1.148626	20.237	8	120	0.0001
Hotelling-Lawley Trace	47904.34	347306	8	116	0.0001
Roy's Greatest Root	47904.16	718562	4	60	0.0001

NOTE: F Statistic for Roy's Greatest Root is an upper bound.
NOTE: F Statistic for Wilks' Lambda is exact.

Another problem of interest in calibration is to test for the equality of both the device variances σ_i^2 and biases α_i^2, $i = 1, \ldots, q$. Christensen and Blackwood (1993) showed that with $\tilde{y}_{ij} = y_{ij} - \bar{y}_i$, $j = 1, \ldots, q - 1$ as defined earlier, the null hypothesis of equality of variances and equality of biases is equivalent to testing

$$H_0: \quad \begin{aligned} \beta_{01} &= \beta_{02} = \cdots = \beta_{0,q-1} = 0, \\ \beta_{11} &= \beta_{12} = \cdots = \beta_{1,q-1} = 0. \end{aligned}$$

in the linear model $\tilde{y}_{ij} = \beta_{0j} + \beta_{1j}\bar{y}_i + \epsilon_{ij}$, $j = 1, \ldots, q - 1$; $i = 1, \ldots, n$. Thus the hypothesis is on slope coefficients as well as the intercepts. To test this null hypothesis using SAS, we can, as earlier, use the MTEST statement in PROC REG. The SAS code for this is given at the end of Program 3.8. All four multivariate tests lead to very small p values, for example, the Wilks' Λ results in the observed value of exact F(8, 118) as 3484.124, leading to a p value of 0.0001. Consequently, we conclude that for at least two thermocouples either the variances, or the biases, or both of the temperatures are unequal. For further discussion of analysis involving biases and variances, see Christensen and Blackwood (1993).

3.10   Regression Diagnostics

As in univariate regression, to assess various assumptions on the multivariate regression model and the validity of the model, diagnostics should be an integral part of the analysis. Such diagnostics tools also include data scrutiny to notice any striking or unusual features present in the data. Certain quantities such as the matrices of predicted values, $\hat{\mathbf{Y}} = \mathbf{X}\hat{\mathbf{B}} = \mathbf{X}(\mathbf{X}'\mathbf{X})^{-1}\mathbf{X}'\mathbf{Y} = \mathbf{P}\mathbf{Y}$, of residuals, $\hat{\mathcal{E}} = \mathbf{Y} - \hat{\mathbf{Y}} = (\mathbf{I} - \mathbf{P})\mathbf{Y}$, and of estimated error variances and covariances, $\hat{\mathbf{\Sigma}} = \mathbf{E}/(n - k - 1) = \mathbf{Y}'(\mathbf{I} - \mathbf{P})\mathbf{Y}/(n - k - 1)$, where $\mathbf{P} = \mathbf{X}(\mathbf{X}'\mathbf{X})^{-1}\mathbf{X}' = (p_{ij})$ is the projection matrix also known as the *hat matrix*, play a key role in such analyses. We will present certain aspects of these analyses in this section.

3.10.1   Assessing the Multivariate Normality of Error

One of the basic assumptions needed to perform the statistical tests under Equation 3.2 is that the rows of $\mathcal{E}$ are independent and the multivariate normally distributed. This assumption can be checked graphically using a $Q - Q$ plot or analytically by applying formal statistical tests, as presented in Chapter 1, on the residuals.

Let $\epsilon_1', \ldots, \epsilon_n'$ be the rows of $\mathcal{E}$ and $\hat{\epsilon}_1', \ldots, \hat{\epsilon}_n'$ be that of $\hat{\mathcal{E}}$. If the multivariate normality assumption holds, the squared distances $d_i^2 = \hat{\epsilon}_i' \hat{\mathbf{\Sigma}}^{-1} \hat{\epsilon}_i/(1 - p_{ii})$, $i = 1, \ldots, n$ will be distributed approximately as chi-squares each with p degrees of freedom. Although these are not theoretically independent, for a model well fit the correlations will be weak. Hence a $Q - Q$ plot of d_i^2 against the χ_p^2 quantiles will indicate a multivariate normality assumption being acceptable if the points on the plot fall around a 45° angle line passing through the origin. Formal tests were used in Chapter 1 for testing the multivariate normality of data. The same tests can be adopted in the present context after some modifications for the residual vectors $\hat{\epsilon}_1', \ldots, \hat{\epsilon}_n'$.

EXAMPLE 8   ***Multivariate Normality Test***   We use the data of Dr. William D. Rohwer reported in Timm (1975, p. 281, 345). Interest is in predicting the performance of a school child on three standardized tests, namely, Peabody Picture Vocabulary Test (y_1), Raven Progressive Matrices Test (y_2), and Student Achievement Test (y_3) given the sums of the numbers of correct items out of 20 (on two exposures) to five types of paired-associated (PA) tasks. The five PA tasks are: named (x_1), still (x_2), named still (x_3), named action (x_4), and sentence still (x_5). The data correspond to 32 randomly selected school children in an upper-class, white residential school.

We fit the multivariate linear regression model with three responses y_1, y_2, y_3 and five independent variables $x_1, \ldots, x_5$ as

$$\mathbf{Y}_{32 \times 3} = \mathbf{X}_{32 \times 6}\mathbf{B}_{6 \times 3} + \mathcal{E}_{32 \times 3},$$

In the SAS Program 3.9, the OUTPUT statement

```
output out = b, r=e1 e2 e3 p=yh1 yh2 yh3 h=p_ii;
```

is used to store the residuals (option R = E1 E2 E3), predicted values (option P = YH1 YH2 YH3), and the diagonal elements of the hat matrix (option H=P_II) in a SAS data set named B using the option OUT=B. The residuals are then used to calculate d_i^2. Selected parts of the output are shown in Ouput 3.9.

```
/* Program 3.9 */

options ls=64 ps=45 nodate nonumber;
title1 'Output 3.9';
title2 'Multivariate Regression Diagnostics';
data rohwer;
infile 'rohwer.dat' firstobs=3;
input ind y1 y2 y3 x1-x5;
run;

/* Store all the residuals, predicted values and the diagonal
elements (p_ii) of the projection matrix, (X(X'X)^-1X'), in a
SAS data set (we call it B in the following) */
proc reg data=rohwer;
model y1-y3=x1-x5/noprint;
output out=b r=e1 e2 e3 p=yh1 yh2 yh3 h=p_ii;

/* Test for multivariate normality using the skewness and kurtosis
measures of the residuals.  Since the sample means of the residuals
are zeros, Program 1.2 essentially works. */
proc iml;
use b;
read all var {e1 e2 e3} into y;
n = nrow(y) ;
p = ncol(y) ;
dfchi = p*(p+1)*(p+2)/6;
q = i(n) - (1/n)*j(n,n,1);
s = (1/(n))*y'*q*y; /* Use the ML estimate of Sigma */
s_inv = inv(s);
g_matrix = q*y*s_inv*y'*q;
beta1hat = ( sum(g_matrix#g_matrix#g_matrix) )/(n*n);
beta2hat =trace( g_matrix#g_matrix )/n;
kappa1 = n*beta1hat/6;
kappa2 = (beta2hat - p*(p+2) ) /sqrt(8*p*(p+2)/n);
pvalskew = 1 - probchi(kappa1,dfchi);
pvalkurt = 2*( 1 - probnorm(abs(kappa2)) ) ;
print beta1hat ;
print kappa1 ;
print pvalskew;
print beta2hat ;
print kappa2 ;
print pvalkurt;

/* Q-Q plot for checking the multivariate normality using
Mahalanobis distance of the residuals;*/
data b;
set b;
totn=32.0;  /* totn is the number of observations */
k=5.0;  /* k is the no. of indep. variables */
p=3.0; /* p is the number of dep. variables */
proc princomp data=b cov std out=c noprint;
var e1-e3;
data sqd;
set c;
student=_n_;
dsq=uss(of prin1-prin3);
dsq=dsq*(totn-k-1)/(totn-1);
* Divide the distances by (1-p_ii);
dsq=dsq/(1-p_ii);
```

```
data qqp;
set sqd;
proc sort;
by dsq;
data qqp;
set qqp;
stdnt_rs=student;
chisq=cinv(((_n_-.5)/ totn),p);
proc print data=qqp;
var stdnt_rs dsq chisq;

* goptions for the gplot;
filename gsasfile "prog39a.graph";
goptions reset=all gaccess=gsasfile autofeed dev=pslmono;
goptions horigin=1in vorigin=2in;
goptions hsize=6in vsize=8in;

title1 h=1.5 'Q-Q Plot for Assessing Normality';
title2 j=l 'Output 3.9';
title3 'Multivariate Regression Diagnostics';
proc gplot data=qqp;
plot dsq*chisq='star';
label dsq = 'Mahalanobis D Square'
            chisq= 'Chi-Square Quantile';
run;

* Q-Q plot for detection of outliers using Robust distance;
*              (Section 3.10.3);
data qqprd;
set sqd;
rdsq=((totn-k-2)*dsq/(totn-k-1))/(1-(dsq/(totn-k-1)));
proc sort;
by rdsq;
data qqprd;
set qqprd;
stdnt_rd=student;
chisq=cinv(((_n_-.5)/ totn),p);
proc print data=qqprd;
var stdnt_rd rdsq chisq;
filename gsasfile "prog39b.graph";
title1 h=1.5 'Q-Q Plot of Robust Squared Distances';
title2 j=l 'Output 3.9';
title3 'Multivariate Regression Diagnostics';
proc gplot;
plot rdsq*chisq='star';
label rdsq = 'Robust Mahalanobis D Square'
            chisq= 'Chi-Square Quantile';
run;

* Influence Measures for Multivariate Regression;
*              (Section 3.10.4);
* Diagonal Elements of Hat (or Projection) Matrix;
data hat;
set sqd;
proc sort;
by p_ii;
data hat;
set hat;
stdnt_h=student;
```

```
keep stdnt_h p_ii;
run;

* Cook's type of distance for detection of influential obs.;
data cookd;
set sqd;
csq=dsq*p_ii/((1-p_ii)*(k+1));
proc sort;
by csq;
data cookd;
set cookd;
stdnt_co=student;
keep stdnt_co csq;
run;

* Welsch-Kuh type statistic for detection of influential obs.;
data wks;
set qqprd;
wksq=rdsq*p_ii/(1-p_ii);
proc sort;
by wksq;
data wks;
set wks;
stdnt_wk=student;
keep stdnt_wk wksq;
run;

* Covariance Ratio for detection of influential obs.;
data cvr;
set qqprd;
covr=((totn-k-2)/(totn-k-2+rdsq))**(k+1)/(1-p_ii)**p;
covr=covr*((totn-k-1)/(totn-k-2))**((k+1)*p);
proc sort;
by covr;
data cvr;
set cvr;
stdnt_cv=student;
keep stdnt_cv covr;
run;

* Display of Influence Measures;
data display;
merge hat cookd wks cvr;
title2 'Multivariate Regression Diagnostics';
title3 'Influence Measures';
proc print data=display noobs;
var stdnt_h p_ii stdnt_co csq stdnt_wk wksq stdnt_cv covr;
run;
```

The $Q-Q$ plot of ordered d_i^2 (denoted by DSQ in Program 3.9) against the quantiles of χ_3^2 (denoted by CHISQ) presented in Output 3.9 seems to indicate no violation of multivariate normality assumption.

Using a modification of Program 1.2 (code provided in Program 3.9) Mardia's skewness and kurtosis tests (see Chapter 1) are performed on the residuals. The p value for the test based on skewness (denoted by PVALSKEW in Program 3.9) is 0.8044 and that for the test based on kurtosis (denoted by PVALKURT) is 0.4192. The corresponding output has been eliminated. In view of these large p values, we see no evidence of any violation of multivariate normality assumption. One can also perform the test due to Mudholkar,

Output 3.9

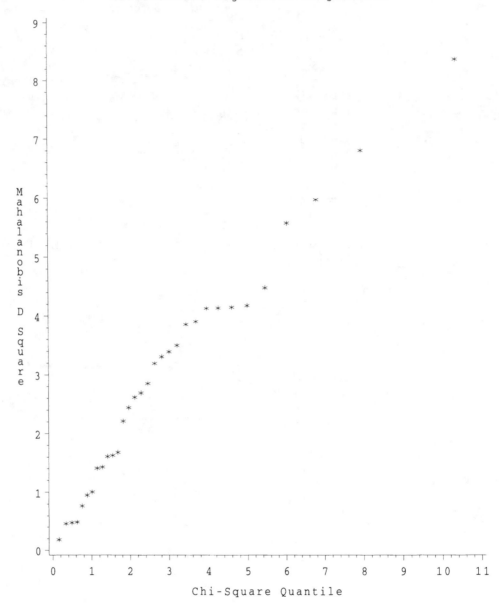

Output 3.9

Q–Q Plot for Assessing Normality

Multivariate Regression Diagnostics

McDermott, and Srivastava (1992). For the present dataset, the conclusion obtained is the same.

3.10.2   Assessing Dispersion Homogeneity

One of the important components of residual analysis in univariate regression is the residual plots. A plot of (studentized) residual values drawn against the predicted values may indicate various possible violations of the assumptions of the regression model. For example, a plot which funnels out (or funnels in) is considered to be an indication of heterogeneity of error variance, in that the error variance may depend on certain functions of the predicted values or the independent variables.

For multivariate regression, a similar residual plot may be suggested based on the following motivation. Suppose in the univariate situation the absolute values of the (studentized) residuals (instead of the residuals) are plotted against the absolute values of the corresponding predicted values. Then various shapes and features in this plot can be utilized to determine the violations of different aspects of the assumptions in the model. Since in this plot the absolute values of the studentized residuals are being used, a plot in the original residuals which was funneling out, for example, will now have only the upper half of the funnel in this plot. This fact can be utilized to suggest a residual plot in (d_i^ϵ, d_i^y), $i = 1, \ldots, n$, where

$$d_i^\epsilon = \sqrt{\frac{\hat{\boldsymbol{\epsilon}}_i' \hat{\boldsymbol{\Sigma}}^{-1} \hat{\boldsymbol{\epsilon}}_i}{1 - p_{ii}}} \quad \text{and} \quad d_i^y = \sqrt{\frac{\hat{\mathbf{y}}_i' \hat{\boldsymbol{\Sigma}}^{-1} \hat{\mathbf{y}}_i}{p_{ii}}},$$

and where $\hat{\mathbf{y}}_i'$ is the ith row of $\hat{\mathbf{Y}}$, $i = 1, \ldots, n$. We note that the quantities d_i^ϵ and d_i^y respectively are the Mahalanobis distances of the vectors $\hat{\boldsymbol{\epsilon}}_i$ and $\hat{\mathbf{y}}_i$ from the origin. We will call such a plot based on these two distances a *D-D plot*.

EXAMPLE 9   *D-D Plot, Air Pollution Data*   Daily measurements on independent variables, wind speed (x_1) and amount of solar radiation (x_2), and dependent variables indicating the amounts of $CO(y_1)$, $NO(y_2)$, $NO_2(y_3)$, O_3 (y_4) and $HC(y_5)$ in the atmosphere were recorded at noon in the Los Angeles area on 42 different days. One of the problems is to predict the air-pollutants measured in terms of y_1 through y_5 given the two predictors x_1 and x_2. We fit a multivariate regression model by taking only y_3 and y_4 as dependent variables and x_1 and x_2 as the independent variables. The logarithmic transformation was applied on all four variables. The D-D plot to assess the homogeneity of dispersion is considered here. Thus, we plot $(d_i^\epsilon$ *vs.* $d_i^y)$, which are computed using the formulas given above. The data set as well as the SAS code are given in Program 3.10. The data are courtesy of G. C. Tiao. In the Program 3.10 the distances d_i^y and d_i^ϵ are denoted respectively by DSQ1 and DSQ2. The corresponding D-D plot is presented in Output 3.10.

```
/* Program 3.10 */

options ls=64 ps=45 nodate nonumber;
title1 'Output 3.10';
title2 'Multivariate Regression: D-D Plot';
* The data set containing independent and dependent variables.;
data pollut;
infile 'airpol.dat' firstobs=6;
input x1 x2 y1-y5;
* Make transformations if necessary;
data pollut;
set pollut;
x1=log(x1);
```

```
                    x2=log(x2);
                    y1=log(y1);
                    y2=log(y2);
                    y3=log(y3);
                    y4=log(y4);
                    y5=log(y5);
                    proc reg data=pollut;
                    model y3 y4=x1 x2/noprint;
                    output out=b r=e1 e2 p=yh1 yh2 h=p_ii;

                    /* Univariate residual plots;
                    proc reg data=pollut;
                    model y3 y4=x1 x2 /noprint;
                    plot student.*p.;
                    title2 'Univariate Residual Plots';
                    run;
                    */


                    * The Mahalanobis distances of residuals and predicted values;
                    proc iml;
                    use b;
                    read all var {e1 e2} into ehat;
                    read all var {yh1 yh2} into yhat;
                    read all var {p_ii} into h;
                    n = nrow(ehat);
                    p = ncol(ehat);
                    k = 2.0; *The no. of independent variables in the model;
                    sig=t(ehat)*ehat;
                    sig=sig/(n-k-1); /* Estimated Sigma */
                    dsqe=j(n,1,0);
                    dsqyh=j(n,1,0);
                    do i=1 to n;
                    dsqe[i,1]=ehat[i,]*inv(sig)*t(ehat[i,])/(1-h[i]);
                    dsqyh[i,1]=yhat[i,]*inv(sig)*t(yhat[i,])/h[i];
                    end;
                    dsqe=sqrt(dsqe);
                    dsqyh=sqrt(dsqyh);
                    dsqpair=dsqyh||dsqe;
                    varnames={dsq1 dsq2};
                    create ndat1 from dsqpair (|colname=varnames|);
                    append from dsqpair;
                    close ndat1;
                    * D-D Plot;
                    data ndat1;
                    set ndat1;
                    filename gsasfile "prog310.graph";
                    goptions reset=all gaccess=gsasfile autofeed dev=pslmono;
                    goptions horigin=1in vorigin=2in;
                    goptions hsize=6in vsize=8in;
                    title1 h=1.5 'D-D Plot';
                    title2 j=l 'Output 3.10';
                    title3 'Residual Analysis: Multivariate Regression';
                    proc gplot data=ndat1;
                    plot dsq2*dsq1='star';
                    label dsq1='Distance of Predicted Value'
                                dsq2='Distance of Residual';
                    run;
```

Output 3.10

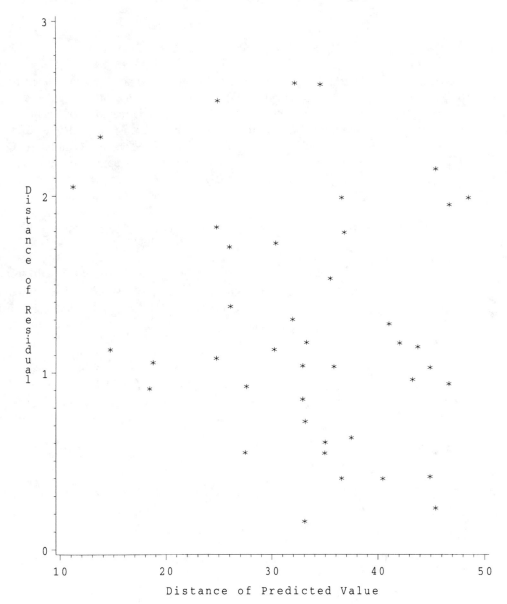

D – D Plot

Output 3.10
Residual Analysis: Multivariate Regression

The plot clearly seems to show an increasing trend indicating that there may be some heterogeneity of error dispersion. A look at individual univariate studentized plots (not shown here) also supports this conclusion.

3.10.3   Outliers

Formal or graphical tests for multivariate normality may sometimes fail due to the presence of outliers in the data. Observations which are unusual in the sense that they violate certain model assumptions are outliers. A simple method of detection of outliers in a multivariate normal distribution case has been discussed in Chapter 2. It involves plotting robust Mahalanobis distances against the corresponding quantiles (Q-Q plot) of an appropriate chi-square distribution. This method can be applied on residuals to detect any outliers in multivariate regression analysis set up.

Analogous to the discussion in Chapter 2, we define

$$D_i^2 = \frac{1}{1 - p_{ii}} \hat{\epsilon}_i' \hat{\Sigma}_{(i)} \hat{\epsilon}_i,$$

where $\hat{\Sigma}_{(i)} = \hat{\mathcal{E}}_{(i)}' \hat{\mathcal{E}}_{(i)} / (n - k - 2)$, and $\hat{\mathcal{E}}_{(i)}$ is the $n - 1$ by p residual matrix obtained by fitting the multivariate regression model $\mathbf{Y} = \mathbf{XB} + \mathcal{E}$ without the ith observation. As in Chapter 2, a relationship between the squared Mahalanobis distance d_i^2 and its robust version, D_i^2 exists and is given by

$$D_i^2 = \frac{(n - k - 2)d_i^2}{(n - k - 1)(1 - \frac{d_i^2}{n-k-1})},$$

which is also a Hotelling T^2 statistic. In the absence of any outliers the quantity

$$F = \frac{D_i^2}{n - k - 2} \frac{n - k - 1 - p}{p},$$

follows an $F_{p,n-k-1-p}$ distribution. However, for large samples (such that $n - k - 1 - p$ is large), the distribution of D_i^2 is approximately χ_p^2 and this fact will be utilized to identify outliers in a Q-Q plot.

EXAMPLE 8   ***Detection of Outliers, Rohwer's Data (continued)***   SAS code given in Program 3.9 also computes d_i^2 (DSQ) and D_i^2 (RDSQ) and then prints them in an increasing order of magnitude for this data set. The D_i^2 values along with the corresponding chi-square quantiles (CHISQ) are listed in Output 3.9 along with a Q-Q plot of these values. The largest D_i^2 is 11.87 corresponding to the 25th observation (listed under the variable STDNT_RD in the output). Since this is not very large compared to the corresponding χ_3^2 quantile we conclude that there are no outliers in this data set. The corresponding Q-Q plot also supports this conclusion.

Output 3.9
continued

Output 3.9
Multivariate Regression Diagnostics
Q-Q Plot of Robust Squared Distances

OBS	STDNT_RD	RDSQ	CHISQ
1	15	0.1772	0.1559
2	10	0.4521	0.3360
3	6	0.4698	0.4864
4	4	0.4809	0.6253
5	2	0.7605	0.7585
6	18	0.9489	0.8888
7	22	1.0047	1.0181
8	26	1.4352	1.1475
9	16	1.4545	1.2780
10	11	1.6509	1.4103
11	28	1.6719	1.5452
12	24	1.7299	1.6834
13	13	2.3245	1.8256
14	19	2.5938	1.9725
15	20	2.7996	2.1250
16	3	2.8882	2.2838
17	30	3.0811	2.4501
18	7	3.5062	2.6250
19	1	3.6494	2.8099
20	12	3.7542	3.0065
21	27	3.8898	3.2169
22	5	4.3593	3.4438
23	23	4.4198	3.6906
24	8	4.7224	3.9617
25	17	4.7321	4.2636
26	29	4.7502	4.6049
27	32	4.7893	4.9989
28	9	5.2072	5.4670
29	21	6.8286	6.0464
30	31	7.4628	6.8124
31	14	8.8648	7.9586
32	25	11.8711	10.3762

If so desired, formal statistical tests to detect outliers in the context of multivariate regression can be used. One such test has been derived by Naik (1989). Instead of $\hat{\epsilon}_i$ he utilized the uncorrelated best linear unbiased scalar (BLUS) residuals (Theil, 1971). The resulting test statistic is Mardia's kurtosis measure of BLUS residuals. We will not discuss this approach here. However, SAS code using PROC IML can easily be developed for this method.

3.10.4   Influential Observations

Influential observations are those unusual observations which upon dropping from the analysis yield results which are drastically different from the results that were otherwise obtained. They are often contrasted from outliers by the fact that the presence of outliers in the data does not affect the results to the extent the influential observations do. One of the important developments in the recent univariate regression analysis literature has been the introduction of the concept of influential observations and statistical methods for de-

Output 3.9
continued

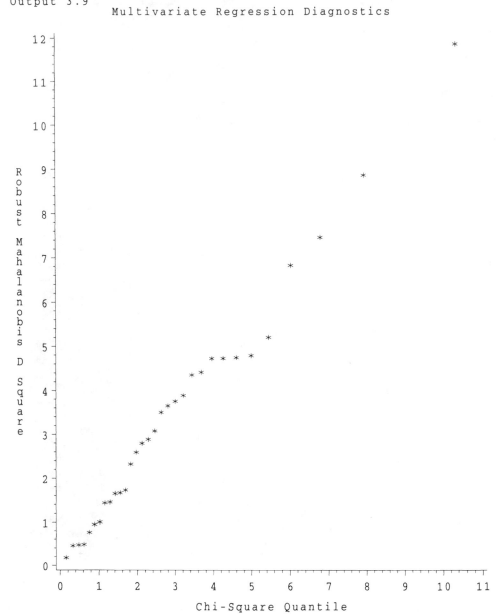

tection of these observations. See Cook and Weisberg (1982). Here we discuss some of the approaches considered in Hossain and Naik (1989). It must be emphasized that these methods are not substitutes for one another and must be used in conjunction with each other.

The Hat Matrix Approach   The projection or hat matrix is defined as $\mathbf{P} = \mathbf{X}(\mathbf{X}'\mathbf{X})^{-1}\mathbf{X}'$. Let p_{ii} be the ith diagonal element of $\mathbf{P}$. It is known that $0 \leq p_{ii} \leq 1$ and $tr\,\mathbf{P} = \sum_{i=1}^{n} p_{ii} = k + 1$. Hence the average of p_{ii}, $i = 1, \ldots, n$, is $\bar{p} = \frac{tr(\mathbf{P})}{n} = \frac{k+1}{n}$. Belsley, Kuh, and Welsch (1980) defined the observation i corresponding to which $p_{ii} > 2\bar{p}$ as a leverage point. A leverage point is a potential influential observation. The influence in this case is entirely due to one or more independent variables.

An observation may be influential entirely or in part due to one or more dependent variables. Such observations can be determined by using the residuals. For example, the distances d_i^2 or the robust distances D_i^2 computed from the residuals can be used for this purpose. Thus an observation detected as an outlier using D_i^2 has the potential to be an influential observation.

Cook Type Distance   Cook (1977) introduced a distance measure to detect the influential observations in univariate regression. For the multivariate regression we can define Cook type distance (Hossain and Naik, 1989) for the ith observation as

$$C_i = \left(\frac{1}{k+1}\right)\left(\frac{p_{ii}}{1 - p_{ii}}\right)d_i^2, \quad i = 1, \ldots, n.$$

The observations with a large value of C_i are considered as potentially influential. Since $\frac{d_i^2}{n-k-1} \sim \text{Beta}\left(\frac{p}{2}, \frac{n-k-1-p}{2}\right)$, to obtain a cutoff point for assessing the largeness of C_i we may use

$$\left(\frac{1}{k+1}\right)\left(\frac{p_{ii}}{1 - p_{ii}}\right)(n - k - 1)\text{Beta}\left(1 - \alpha, \frac{p}{2}, \frac{n-k-1-p}{2}\right)$$

where Beta $(1 - \alpha, \frac{p}{2}, \frac{n-k-1-p}{2})$, is the upper α probability cutoff value of a Beta distribution with the parameters $\frac{p}{2}$ and $\frac{n-k-1-p}{2}$. To have the convenience of the same cutoff point we substitute $\bar{p}$ for p_{ii}. Thus the approximate cutoff point for C_i is simply $C_\alpha = $ Beta $(1 - \alpha, \frac{p}{2}, \frac{n-k-1-p}{2})$.

Welsch-Kuh Type Statistic   Belsley, Kuh and Welsch (1980) introduced several contenders for Cook distance in the context of univariate regression. One such statistic attributed originally to Welsch and Kuh (1977) that has been generalized to multivariate regression (Hossain and Naik, 1989) is

$$WK_i = \frac{p_{ii}}{1 - p_{ii}}D_i^2.$$

The observations corresponding to the large WK_i are considered as potentially influential. As indicated earlier, D_i^2 is same as Hotelling's T^2 statistic. Therefore, an α probability cutoff point for the WK_i statistic should be $\frac{p_{ii}}{1 - p_{ii}} p \frac{(n-k-2)}{(n-k-1-p)} F_{1-\alpha, p, n-k-1-p}$, where $F_{1-\alpha, p, n-k-1-p}$ is the upper α probability cutoff point of the F distribution with p and $n - k - 1 - p$ degrees of freedom. As in the case of Cook type distance we replace p_{ii} by $\bar{p}$ resulting in an approximate cutoff point for WK_i given by

$$WK_\alpha = \frac{k+1}{n-k-1}\frac{p(n-k-2)}{n-k-1-p}F_{1-\alpha, p, n-k-1-p}.$$

Covariance Ratio   The influence of the ith observation on the variance covariance matrix of $\hat{\mathbf{B}}$, the matrix of estimated regression coefficients, can be measured by the covariance ratio defined in terms of sample generalized variances with and without considering the ith observation,

$$R_i = \left(\frac{1}{1 - p_{ii}} \right)^p \left\{ \frac{|\hat{\Sigma}_{(i)}|}{|\hat{\Sigma}|} \right\}^{k+1}.$$

The observations corresponding to both low and high values of covariance ratio are considered as potentially influential. In order to find the cutoff points for R_i we use the fact that

$$\frac{|\hat{\Sigma}_{(i)}|}{|\hat{\Sigma}|} = \left(\frac{n - k - 2}{n - k - 1} \right)^{-p} \left(1 + \frac{D_i^2}{n - k - 2} \right)^{-1}$$

and that (Rao, 1973, p. 555)

$$\left(1 + \frac{D_i^2}{n - k - 2} \right)^{-1} \sim \text{Beta} \left(\frac{n - k - 1 - p}{2}, \frac{p}{2} \right).$$

Using these, and replacing p_{ii} by $\bar{p}$ as done earlier an approximate lower cutoff point for R_i is

$$R_L = \left(\frac{1}{1 - \bar{p}} \right)^p \left(\frac{n - k - 1}{n - k - 2} \right)^{p(k+1)} \left\{ \text{Beta} \left(\frac{\alpha}{2}, \frac{n - k - 1 - p}{2}, \frac{p}{2} \right) \right\}^{k+1}.$$

For an upper cutoff point, $\frac{\alpha}{2}$ is replaced by $1 - \frac{\alpha}{2}$. But p_{ii} this time will be replaced by $2\bar{p}$ (Belsley et al., 1980) resulting in the approximate upper cutoff point

$$R_U = \left(\frac{1}{1 - 2\bar{p}} \right)^p \left(\frac{n - k - 1}{n - k - 2} \right)^{p(k+1)} \left\{ \text{Beta} \left(1 - \frac{\alpha}{2}, \frac{n - k - 1 - p}{2}, \frac{p}{2} \right) \right\}^{k+1}.$$

EXAMPLE 8   *Identification of Influential Observations, Rohwer's Data (continued)*   Rohwer's data are considered for the illustration of all four methods described above. Program 3.9 includes the necessary SAS code. Output 3.9 shows the results. The output lists all the diagonal elements $p_{11}, \ldots, p_{nn}$ of the hat matrix $\mathbf{P}$ (under the variable P_II) in the increasing order of magnitude. The largest two of these are $p_{55} = 0.5682$ and $p_{10,10} = 0.4516$ corresponding to the 5th and 10th observations (see under STDNT_H) respectively. Both of these values are greater than $2\bar{p} = 2\frac{k+1}{n} = 0.3750$ and hence are deemed as the leverage points or potentially influential.

Output 3.9
continued

Output 3.9
Multivariate Regression Diagnostics
Influence Measures

S T D N T _ H	P _ I I	S T D N T _ C C O	C C S Q	S T D N T _ W K	W K S Q	S T D N T _ C C V	C O V R
23	0.04455	4	0.00645	4	0.03795	25	0.3287
7	0.04531	6	0.01458	6	0.08572	14	0.4920
9	0.05131	15	0.01519	15	0.08826	31	0.5594

4	0.07314	7	0.02530	7	0.16640	21	0.7485
32	0.07321	23	0.03036	23	0.20610	9	0.7624
20	0.07881	28	0.03422	28	0.21065	23	0.8746
30	0.08655	2	0.03576	2	0.21258	32	0.8891
31	0.08922	20	0.03733	20	0.23950	7	1.0593
13	0.10375	22	0.04025	22	0.24151	17	1.2668
28	0.11190	9	0.04040	26	0.25993	8	1.2851
14	0.12650	26	0.04261	13	0.26907	30	1.3235
21	0.14024	13	0.04267	11	0.28094	20	1.3707
3	0.14173	30	0.04505	9	0.28165	12	1.5332
11	0.14543	11	0.04568	30	0.29195	1	1.5475
26	0.15334	32	0.05503	18	0.33957	13	1.6506
6	0.15432	18	0.05671	10	0.37234	3	1.6629
25	0.15713	10	0.06339	32	0.37831	28	1.9612
1	0.16701	24	0.07294	24	0.44995	29	2.1183
12	0.17050	3	0.07411	3	0.47696	11	2.2117
17	0.17321	31	0.09758	16	0.72235	4	2.2694
8	0.17661	1	0.11067	31	0.73104	26	2.3879
22	0.19380	12	0.11629	1	0.73167	24	2.7132
24	0.20642	16	0.11832	12	0.77168	6	2.9955
2	0.21845	17	0.14448	17	0.99133	22	3.0522
18	0.26354	8	0.14768	8	1.01291	19	3.2436
19	0.29836	21	0.15164	19	1.10294	27	3.3582
29	0.30427	14	0.16427	21	1.11383	2	3.5453
16	0.33183	19	0.17321	14	1.28379	18	4.0558
15	0.33247	25	0.26008	29	2.07746	16	4.8372
27	0.36726	29	0.30260	25	2.21298	15	6.5280
10	0.45161	27	0.33866	27	2.25782	5	9.5932
5	0.56821	5	0.84672	5	5.73670	10	11.0314

The program also computes the Cook type distances (denoted by CSQ in the program). With the level of significance $\alpha = 0.05$, the corresponding cutoff point is $C_\alpha = C_{0.05} =$ Beta $(0.05, 1.5, 11.5) = 0.2831$. This is calculated using the SAS function BETAINV. Using this, observations 5, 27, and 29 are identified as potential influential observations (see under the variable STDNT_CO in Output 3.9). At the same level of significance the cutoff point for WK_i (WKSQ) is $WK_\alpha = 2.2801$. Only the 5th observation (see under STDNT_WK) with $WK_5 = 5.7367$ qualifies as a potential influential observation under this criterion.

Finally, at $\alpha = 0.05$ and for the covariance ratio criterion the lower and upper cutoff points are computed respectively as $C_L = 0.3463$ and $C_U = 7.8549$. Upon computation, we find (see the output under the variables COVAR and STDNT_CV) that $R_{25} = 0.3282$, $R_{10} = 11.0314$, and $R_5 = 9.5932$ are three values which do not fall between R_L and R_U. Hence these three are identified as potential influential observations. These results are summarized in Table 3.4.

TABLE 3.4 Influential Observation Identification

Method	Influential Obs.
Hat Matrix	5, 10
Cook Type	5, 27, 29
Welsch Kuh Type	5
Covariance Ratio	5, 10, 25

A few comments may be made after examining the above table. All the measures declare the 5th observation as influential. The influence is perhaps mainly due to the independent variables since it is a leverage point. An examination of the data set indicates that this observation has the largest $X1$ value of 20. This is almost twice as large as the $X1$ values for the other observations except that for the 10th observation. Observation 5 also has the largest value for variable $X3$ (=21). Observation 10 has fairly large values for many independent variables. Observation 25 has large values for the variables $Y1$ and $Y2$ but small value for $Y3$. It is identified only by the covariance ratio. Observations 27 and 29 were identified only by Cook type distance although no striking features in these two observations are immediately apparent.

3.11   Concluding Remarks

In closing this chapter, we emphasize that the inference in multivariate regression can be sensitive to the assumption of multivariate normality as well as the presence of unusual observations. We have briefly described some of the approaches to these problems here. It is a good practice to apply these checks before performing the formal multivariate analysis described earlier in this chapter. Some of the test statistics in a normal MANOVA are more sensitive to nonnormality than others. In general, if the data are not strictly normal, Pillai's test is the most robust (in terms of preserving the power of the test) of the MANOVA tests and therefore is especially recommended for such situations.

Multivariate Analysis of Experimental Data

4.1 Introduction 117
4.2 Balanced and Unbalanced Data 120
4.3 One-Way Classification 123
4.4 Two-Way Classification 129
4.5 Blocking 137
4.6 Fractional Factorial Experiments 139
4.7 Analysis of Covariance 145
4.8 Concluding Remarks 149

4.1 Introduction

In the previous chapter, we considered the multivariate linear regression model

$$\mathbf{Y} = \mathbf{XB} + \mathcal{E}. \tag{4.1}$$

In the model, the

- n by p matrix $\mathbf{Y}$ contains the random observations on p dependent variables,
- $k + 1$ by p matrix $\mathbf{B}$ is the matrix of unknown parameters,
- n by p matrix $\mathcal{E}$ is the matrix of random errors such that each row of $\mathcal{E}$ is a p variate normal vector with mean vector zero and variance-covariance matrix $\mathbf{\Sigma}$. The matrix $\mathbf{\Sigma}$ is assumed to be a p by p positive definite matrix.
- n by $k + 1$ matrix $\mathbf{X}$ was assumed to be of full rank, that is, Rank $(\mathbf{X}) = k + 1$.

There are, however, situations especially those involving the analysis of classical experimental designs where the assumption $Rank\,(\mathbf{X}) = k + 1$ cannot be made. This in turn requires certain suitable modifications in the estimation and testing procedures. In fact, following the same sequence of development as in the previous chapter, a generalized theory has been developed, which contains the results of the previous chapter as the special "full rank" case.

Let us assume that $Rank\,(\mathbf{X}) = r < k + 1$. It was pointed out in Chapter 3 that in this case, the solution to the normal equations in Equation 3.3 is not unique. If $(\mathbf{X}'\mathbf{X})^{-}$ is a generalized inverse of $\mathbf{X}'\mathbf{X}$, then correspondingly a least square solution is given by

$$\hat{\mathbf{B}}^{(g)} = (\mathbf{X}'\mathbf{X})^{-}\mathbf{X}'\mathbf{Y},$$

which will depend on the particular choice of the generalized inverse. As a result, the matrix $\mathbf{B}$ is not (uniquely) estimable. The following example illustrates this case.

EXAMPLE 1   *Checking Estimability, Jackson's Laboratories Comparison Data*   Jackson (1991) considered a situation where samples were tested in three different laboratories using two different methods. Each of the laboratories received four samples and each of the samples was divided into subsamples to be tested by these two methods. As a result, the observations on the subsamples arising out of the same sample are correlated leading to the data as the four bivariate vector observations per laboratory. The data are shown in Table 4.1.

TABLE 4.1 Laboratory Data

Laboratory	Method 1	Method 2
1	10.1	10.5
	9.3	9.5
	9.7	10.0
	10.9	11.4
2	10.0	9.8
	9.5	9.7
	9.7	9.8
	10.8	10.7
3	11.3	10.1
	10.7	9.8
	10.8	10.1
	10.5	9.6

Let $\mathbf{y}_{ij}$ be the 2 by 1 vector of observation on the jth sample sent to the ith laboratory, $j = 1, \ldots, 4; i = 1, 2, 3$. If we assume that $\mathbf{y}_{ij}$ has a bivariate distribution with a structured mean vector $\begin{bmatrix} \mu_1 + \tau_{i1} \\ \mu_2 + \tau_{i2} \end{bmatrix}$ and the variance covariance matrix $\boldsymbol{\Sigma}$, then we can write our model as

$$(y_{ij1}, y_{ij2}) = (\mu_1 + \tau_{i1}, \mu_2 + \tau_{i2}) + (\epsilon_{ij1}, \epsilon_{ij2}),$$

$$i = 1, 2, 3, \quad j = 1, \ldots, 4.$$

Stacking these equations one below the other for $j = 1, \ldots, 4$ and (then for) $i = 1, 2, 3$, leads to

$$
\begin{bmatrix}
\mathbf{y}'_{11} \\
\vdots \\
\mathbf{y}'_{14} \\
\cdots \\
\mathbf{y}'_{21} \\
\vdots \\
\mathbf{y}'_{24} \\
\cdots \\
\mathbf{y}'_{31} \\
\vdots \\
\mathbf{y}'_{34}
\end{bmatrix}_{12 \times 2}
=
\begin{bmatrix}
1 & 1 & 0 & 0 \\
\vdots & \vdots & \vdots & \vdots \\
1 & 1 & 0 & 0 \\
\cdots & \cdots & \cdots & \cdots \\
1 & 0 & 1 & 0 \\
\vdots & \vdots & \vdots & \vdots \\
1 & 0 & 1 & 0 \\
\cdots & \cdots & \cdots & \cdots \\
1 & 0 & 0 & 1 \\
\vdots & \vdots & \vdots & \vdots \\
1 & 0 & 0 & 1
\end{bmatrix}_{12 \times 4}
\begin{bmatrix}
\mu_1 & \mu_2 \\
\tau_{11} & \tau_{12} \\
\tau_{21} & \tau_{22} \\
\tau_{31} & \tau_{32}
\end{bmatrix}_{4 \times 2}
+
\begin{bmatrix}
\boldsymbol{\epsilon}'_{11} \\
\vdots \\
\boldsymbol{\epsilon}'_{14} \\
\cdots \\
\boldsymbol{\epsilon}'_{21} \\
\vdots \\
\boldsymbol{\epsilon}'_{24} \\
\cdots \\
\boldsymbol{\epsilon}'_{31} \\
\vdots \\
\boldsymbol{\epsilon}'_{34}
\end{bmatrix}_{12 \times 2}
,
$$

where $\boldsymbol{\epsilon}_{ij} = \begin{pmatrix} \epsilon_{ij1} \\ \epsilon_{ij2} \end{pmatrix}$ represents the sample-to-sample variation. The model represented by the set of equations given above is in the form of Equation 4.1, with

$$
\mathbf{X} = \begin{bmatrix}
 & \mathbf{1}_4 & \mathbf{0} & \mathbf{0} \\
\mathbf{1}_{12} & \mathbf{0} & \mathbf{1}_4 & \mathbf{0} \\
 & \mathbf{0} & \mathbf{0} & \mathbf{1}_4
\end{bmatrix},
$$

where $\mathbf{1}_q$ represents a q by 1 vector of unit elements. Since all the elements of $\mathbf{X}$ are either zero or one, with zero representing the absence and one representing the presence of the particular parameter in the individual equation, this can be considered as a situation where the regression is performed on the dummy variables. It is easy to see that since the last three columns of the matrix $\mathbf{X}$ above add to the first column, the first column is linearly dependent on the last three columns. A similar statement can be made about the linear dependence of any other columns on the remaining three. As a result, the matrix $\mathbf{X}$ is not of full column rank. In fact, $Rank\ (\mathbf{X}) = 3$, as the last three columns of $\mathbf{X}$ form a linearly independent set of vectors. Now as $Rank\ (\mathbf{X}'\mathbf{X}) = Rank\ (\mathbf{X}) = 3$, the 4 by 4 matrix $\mathbf{X}'\mathbf{X}$ is singular, thereby not admitting the inverse $(\mathbf{X}'\mathbf{X})^{-1}$. Therefore the least squares system of linear equations corresponding to Equation 4.1,

$$\mathbf{X}'\mathbf{X}\mathbf{B} = \mathbf{X}'\mathbf{Y}$$

does not admit a unique solution $\hat{\mathbf{B}}$. As a result, for $k = 1, 2,\ (\mu_k, \tau_{1k}, \tau_{2k}, \tau_{3k})$ *cannot* be uniquely estimated.

If we want to estimate the mean measurement for each of the two methods, then the quantities of interest are $\nu_{ik} = \mu_k + \tau_{ik},\ i = 1, 2, 3,\ k = 1, 2$. We may also be interested in comparing the laboratories, that is, in estimating the differences between the true means for the three laboratories, namely $\nu_{1k} - \nu_{2k} = \tau_{1k} - \tau_{2k},\ \nu_{1k} - \nu_{3k} = \tau_{1k} - \tau_{3k}$ and $\nu_{2k} - \nu_{3k} = \tau_{2k} - \tau_{3k},\ k = 1, 2$. Even though $(\mu_k, \tau_{1k}, \tau_{2k}, \tau_{3k})$ cannot be uniquely estimated, the unique estimates of these differences are available, regardless of what generalized inverse is used to obtain the solution $(\hat{\mu}_k, \hat{\tau}_{1k}, \hat{\tau}_{2k}, \hat{\tau}_{3k}),\ k = 1, 2$ of $\mathbf{X}'\mathbf{X}\mathbf{B} = \mathbf{X}'\mathbf{Y}$. Thus, even though the matrix $\mathbf{B}$ is not estimable, certain linear functions of $\mathbf{B}$ are still estimable. Specifically, as mentioned in Chapter 3, a nonrandom linear function $\mathbf{c}'\mathbf{B}$, where $\mathbf{c} \neq \mathbf{0}$ is *estimable* if and only if

$$(\mathbf{X}'\mathbf{X})(\mathbf{X}'\mathbf{X})^{-}\mathbf{c} = \mathbf{c}. \tag{4.2}$$

Quite appropriately, a linear hypothesis is called *testable* if it involves only the estimable functions of $\mathbf{B}$.

For the first laboratory, the vector of mean measurements for each of the two methods $(\nu_{11},\ \nu_{12})$ is given by

$$(\nu_{11},\ \nu_{12}) = (\mu_1 + \tau_{11},\ \mu_2 + \tau_{12}) = (1\ 1\ 0\ 0) \begin{bmatrix} \mu_1 & \mu_2 \\ \tau_{11} & \tau_{12} \\ \tau_{21} & \tau_{22} \\ \tau_{31} & \tau_{32} \end{bmatrix} = \mathbf{c}'\mathbf{B},$$

with $\mathbf{c}' = (1\ 1\ 0\ 0)$. The choices of respective $\mathbf{c}'$ vectors for the other two laboratory means are obtained in the same way. These are $(1\ 0\ 1\ 0)$ and $(1\ 0\ 0\ 1)$. Similarly, for the differences between the laboratory means the three choices of $\mathbf{c}$ are

$$\mathbf{c}' = (0\ 1\ -1\ 0),\quad (0\ 1\ 0\ -1),\quad \text{and}\quad (0\ 0\ 1\ -1).$$

It can be theoretically shown (Searle, 1971) that all the above choices of $\mathbf{c}$ satisfy Equation 4.2 and hence all the laboratory means and their pairwise differences are estimable. It is equivalent to saying that all of the above choices of $\mathbf{c}'$ can be expressed as the linear function of the rows of $\mathbf{X}$. That this is true in our example is easily verified by the visual examination of the rows of our matrix $\mathbf{X}$. The actual rank of the matrix $\mathbf{X}$ would depend on the particular design and the particular statistical model. For a one-way classification model with k groups, the rank of $\mathbf{X}_{n \times (k+1)}$ is k. This deficiency in rank of $\mathbf{X}$ affects the tests for the statistical significance in many ways. First of all, such tests can be performed only for the testable linear hypotheses. That given, all the univariate and multivariate tests can still be adopted after making a simple yet important modification. When the hypothesis is linear, the quantity $r = Rank\ (\mathbf{X})$ replaces $(k + 1)$ in most formulas of Chapter 3.

4.2 Balanced and Unbalanced Data

When the design is balanced in the sense that each group has the same number of measurements or certain orthogonality conditions are met (Searle, 1971), the analysis is relatively much simpler with respect to the computations as well as interpretations. In this case for a given response variable, the (univariate) ANOVA partitioning (Searle, 1971) of the corrected total sums of squares into various sources of variation specified by the model is unique. This simplicity is unfortunately lost as soon as the underlying design becomes unbalanced. The partitioning of the corrected total sums of squares is no longer unique in that it depends on the model and the various submodels of it as specified by the order in which various sums of squares are extracted. For example, suppose we have an unbalanced (univariate) two-way classification design with interaction, for which the statistical model is

$$y_{ijk} = \mu + \alpha_i + \beta_j + (\alpha\beta)_{ij} + \epsilon_{ijk}$$

$$i = 1, \ldots, a; \ j = 1, \ldots, b; \ k = 1, \ldots, n_{ij}.$$

Denoting the main effects as A and B and the interaction effect as AB and following the notation of Searle (1971), the corrected model sum of squares $R(A, B, AB|\mu)$ can be partitioned in the following two alternative ways:

$$R(A, B, AB|\mu) = R(A|\mu) + R(B|\mu, A) + R(AB|\mu, A, B) \tag{4.3}$$

or

$$R(A, B, AB|\mu) = R(B|\mu) + R(A|\mu, B) + R(AB|\mu, A, B), \tag{4.4}$$

where $R(A|\mu)$ is the sum of squares due to A after correcting for μ, and $R(B|\mu, A)$ is the sum of squares due to B after correcting for μ and the variable A (i.e., after discounting the effect of A). Other quantities are similarly defined. Unless the design is balanced, $R(B|\mu, A) \neq R(B|\mu)$ and $R(A|\mu, B) \neq R(A|\mu)$. The complexity increases further for the higher order unbalanced designs. As a result, SAS computes the four types of sums of squares, commonly referred to as Type I through Type IV sums of squares. A brief summary of these four sums of squares, adopted from Littell, Freund and Spector (1991) follows.

- The Type I sums of squares represent a partitioning of the model sum of squares into component sums of squares due to each variable or interaction as it is added sequentially to the model in the order prescribed by the MODEL statement (Littell, Freund, and Spector, 1991, p. 20). They are often referred to as sequential sums of squares. In view of their dependence on the order prescribed the corresponding partitioning of the model sum of squares is not unique. For example, for a three-way classification model with all possible interactions in variables A, B, and C the MODEL statement

```
model y = a b c a*b a*c b*c a*b*c/ss1 ss2 ss3 ss4;
```

  results in the Type I sum of squares (generated by the use of option SS1) for, say, A*C as the one which is adjusted for all the previous terms in the model: A, B, C, and A*B.

- The Type II sums of squares for a particular variable represent the increase in the model sum of squares. This increase is due to adding the particular variable or interaction to a model that already contains all the other variables *and* interactions in the MODEL statement which do not *notationally* contain the particular variable or interaction (Littell, Freund, and Spector, 1991, p. 21). For example, for the MODEL statement given above, the Type II sums of squares for A*C represents the increase in the model sum of squares by adding A*C while A, B, C, A*B, and B*C have already been included in the model.

The three-factor interaction is not included in this because the notational symbol A*B*C contains the symbol A*C. Type II sums of squares do not depend on the order in which the variables and interactions are listed in the MODEL statement. In general, Type II sums of squares for various variables and interactions do not add up to the model sum of squares. Type II sums of squares are commonly called partial sums of squares.

- The Type III sums of squares are also a kind of partial sums of squares (Littell, Freund, and Spector, 1991, p. 156). They differ from Type II sums of squares in that a particular sum of squares represents increase in the model sum of squares due to adding the particular variable or interaction to a model that contains all the other variables and interactions listed in the MODEL statement. For example, for the MODEL statement given above the Type III sums of squares for A*C represent the increase in the model sum of squares by adding A*C while all the remaining terms in the right-hand side of the MODEL statement, A, B, C, A*B, B*C, and A*B*C, have already been included in the model. As in the case of Type II, Type III sums of squares also do not depend on the order in which the variables and interactions are listed in the MODEL statement. Further, in general Type III sums of squares for various variables and interactions do not add up to the model sum of squares.

- In case there are empty cells, the Type IV sums of squares are recommended. Unfortunately, they can be discussed only in the general framework of estimable functions and their constructions. For cross-classified unbalanced data, these are not unique when there are empty cells in that they depend on the way the data may have been arranged. When there are no empty cells, Type IV sums of squares are identical to Type III sums of squares (Littell, Freund, and Spector, 1991, p. 156).

For details, see the *SAS/STAT User's Guide, Version 6, Fourth Edition*, Littell, Freund and Spector (1991), and Milliken and Johnson (1991).

For multivariate analysis purposes, when the data are multivariate in nature we analogously define the sums of squares and crossproducts (SS&CP) matrices rather than just the sums of squares. The partitioning that is essentially similar to ANOVA partitioning, called in the literature MANOVA partitioning (M for multivariate), can be done for the corrected total SS&CP matrix. As is true in the univariate case, we will encounter problems related to the nonuniqueness of this partitioning for the unbalanced data. Needless to say, the interpretations similar to those mentioned in the references given above can be assigned to various types of analyses to help in choosing the appropriate MANOVA partitioning and/or analysis.

Based on Milliken and Johnson (1991) we make the following recommendations for two-way classification models. For most higher order models, a straightforward modification of these recommendations will be applicable, in most situations.

- Type III SS&CP matrices are appropriate when the interest is in comparing the effects of the experimental variables. The corresponding null hypotheses are equivalent to the hypotheses tested in the balanced classifications. Specifically, for a multivariate version of the two-way classification model stated earlier, the hypotheses being tested are

  i. $\boldsymbol{\alpha}_i + b^{-1}\sum_{j=1}^{b}(\boldsymbol{\alpha\beta})_{ij}$ are all equal, $i = 1, \ldots, a$.

  ii. $\boldsymbol{\beta}_j + a^{-1}\sum_{i=1}^{a}(\boldsymbol{\alpha\beta})_{ij}$ are all equal, $j = 1, \ldots, b$.

  iii. $(\boldsymbol{\alpha\beta})_{i'j'} - b^{-1}\sum_{j=1}^{b}(\boldsymbol{\alpha\beta})_{i'j} - a^{-1}\sum_{i=1}^{a}(\boldsymbol{\alpha\beta})_{ij'} + (ab)^{-1}\sum_{i=1}^{a}\sum_{j=1}^{b}(\boldsymbol{\alpha\beta})_{ij} = \mathbf{0}$ for all $i' = 1, \ldots, a; \; j' = 1, \ldots, b$.

- For model-building purposes such as in response surface modeling, where we want to predict the responses, Type I and/or Type II SS&CP is desirable. Usually, since the terms are to be added sequentially in the process of model building, Type I analysis may be more appropriate.

- In survey designs and observational studies such as in sociology, where the data are collected "passively," rather than "actively" generated under a designed experiment, the number of observations per cell will be approximately proportional to the actual relative frequencies of these cells in the reference population. As a result, the weighted averages with observed cell sizes as weights may be of interest in the course of analyzing the data. In this case, it is advisable to attempt and carefully interpret the two possible sequential analyses using Type I SS&CP leading to the partitioning given by Equations 4.3 and 4.4. Of course, the three-way or other higher order cross-classified designs would require several sequential analyses.

Remember that underlying any test statistic or significant effect as shown in any computer output, there is a specific statement in the null hypothesis which is being tested. In the case of designed experiments, the cell sizes are determined by the experimenter or by certain circumstances which are beyond the control of the experimenter. The effects, significant or not, are the characteristics of the reference population and in no way should be a function of the design parameters such as the cell sizes. It makes no intuitive sense that a null hypothesis would involve these parameters of the particular design. It is, therefore, very important that any appropriate null hypothesis is *a priori* identified before declaring an effect significant or nonsignificant. This is preferable to retroactively identifying what the hypothesis is, corresponding to a significant or nonsignificant p value associated with a particular test statistic. In fact, in the case of highly unbalanced designs, the SS&CP matrices for the *notationally same* effects (in the computer output) under Type I, II, or III analyses may correspond to very different null hypotheses. Not surprisingly, one often obtains mutually conflicting conclusions from these analyses. Of course, the best solution to this problem is to construct a design which is as balanced as possible.

The issues related to which of the three sums of squares is appropriate have been the subject of considerable discussion for the past several decades. See Goodnight (1976) and Searle (1987). These issues do not seem to have subsided or been adequately settled or clarified as evident from the recent contributions to this topic. See Dallal (1992), De Long (1994), Goldstein (1994) and Searle (1994). It is thus inevitable not to find a consensus on various modes of analyses considered in the specific examples in this book. Wherever possible, we attempt to intuitively justify the type of analysis chosen, while at the same time deliberately avoiding the complex notational and mathematical representations of the underlying hypotheses.

Type IV analysis is appropriate in the case of missing observations when, for certain cells, the cell frequency is zero. In this case, none of the Type I, II, or III analyses may be entirely satisfactory, and may be difficult to interpret. The Type IV hypotheses are constructed to have balance in the cell mean coefficients in such cases. As a result, meaningful interpretations can be assigned to the underlying hypotheses being tested by these SS&CP matrices. For designs with missing observations, PROC GLM automatically generates certain Type IV hypotheses which can be identified by examining the list of estimable functions generated by SAS under the given design. Unfortunately, the resulting hypotheses being tested may themselves depend on the numbering of the variables. Consequently, the very same set of treatments in the same data set, if renumbered or reordered, may result in a different set of Type IV SS&CP matrices. See Milliken and Johnson (1991) for an especially readable discussion in which the authors devote an entire chapter to these and other related issues, of course in a univariate setting.

The preceding discussion about the unbalanced designs pertains only to the cases where there are an unequal number of observations per cell or where balancedness conditions (Searle, 1971) on the cell sizes are not satisfied. Imbalance may also occur in cases when, for some observations or experiments, the data are available only on some of the response variables and not available on the others. This situation, although quite common in practice, cannot be handled in the standard multivariate analysis of variance setup. As a result, for any multivariate analysis procedure, observations with missing values for one or more response variables are automatically deleted by SAS before any analysis. This is not necessarily an ideal choice because missingness of observations may not have a random pattern and there may be an underlying selection mechanism due to which the observations were missing. In this case, due to selection bias in data, ignoring the missing values may severely affect the analysis and consequently may result in misleading conclusions.

4.3   One-Way Classification

In the previous section, we considered an interlaboratory study where four bivariate observations corresponding to two different methods were made in three different laboratories. The purpose of the study was to compare the three laboratories and decide if these laboratories provide, on the average, the same bivariate measurements. The three groups or classes of interest were three laboratories, which define a categorical variable (or factor) with three levels represented by these laboratories. In general, a one-way classification model can be defined for a variable with a levels or groups. If we denote by y_{ij}, the p by 1 vector of responses on the jth unit of the ith group, then we can write

$$y_{ij} = \mu + \tau_i + \epsilon_{ij}, \quad j = 1, \ldots, n_i; \ i = 1, \ldots, a,$$

where ϵ_{ij} is the p by 1 random vector corresponding to error, and is assumed to have a zero vector as the mean and the variance-covariance matrix Σ. The surplus or slack effect of the ith group is represented by the p by 1 vector τ_i and the p by 1 vector μ is the overall mean. The n_i is the number of observations in the ith group. If $n_1 = \cdots = n_a$, then the design is balanced. A usual assumption, though not crucial but only convenient, is to take $\sum_{i=1}^{a} n_i \tau_i = \mathbf{0}$.

This assumption implies that the weighted sum of treatment effects is zero, thereby making μ the overall average across all treatment groups. In fact, any other linear restriction on $\tau_1, \ldots, \tau_a$ can be used instead so long as it provides an additional equation which is linearly independent of the system of normal equations given in Equation 3.3. The purpose of making such an assumption is to devise a convenient method to find an appropriate generalized inverse of $\mathbf{X'X}$, where $\mathbf{X}$ is the corresponding design matrix when the above model is represented as a multivariate linear model given in Equation 3.1. In fact, PROC GLM makes the alternative assumption of $\tau_a = \mathbf{0}$ rather than the traditional assumption of $\sum_{i=1}^{a} n_i \tau_i = \mathbf{0}$ adopted by various multivariate analysis and experimental design books. This assumption amounts to setting the effect of last treatment to zero, thereby making μ the mean of the last group. A model with this assumption is often referred to as the *reference cell model*.

As mentioned earlier, since the choice of the generalized inverse is immaterial when estimating an estimable linear function or performing a testable linear hypothesis, what linear restriction is placed on τ_i does not affect the subsequent analysis in any way. Since for a one-way classification model $Rank\ (\mathbf{X}_{n \times (a+1)}) = a$, the rank of $\mathbf{X}$ is short only by one. As a result, only one linear restriction on τ_i is needed. For the higher order classifications, the number of linearly independent restrictions needed is equal to the rank deficiency of $\mathbf{X}$.

To test the multivariate null hypothesis of no differences in the group means, that is, $H_0 : \tau_1 = \tau_2 = \cdots = \tau_a$, it is possible to use any of the four multivariate tests described in Chapter 3, after making the appropriate modifications in the degrees of freedom. Specifically, the quantity $(a + 1)$ (which was $(k + 1)$ in the notation of Chapter 3), which was the rank of $\mathbf{X}$ in the full rank model of Chapter 3, would be replaced by a, the actual rank of the matrix $\mathbf{X}$.

EXAMPLE 1   *Hypothesis Testing, Laboratories Comparison Data (continued)*   We return to the Jackson (1991) data as presented in Table 4.1. The objective of simultaneously comparing the three laboratories translates to the bivariate null hypothesis,

$$H_0 : \quad \tau_1 = \tau_2 = \tau_3$$

against the alternative

$$H_1: \quad \text{At least two } \tau_i \text{ are different from each other.}$$

The null hypothesis is testable, as seen earlier, and the four different multivariate tests, namely Wilks' Λ, Pillai's trace, Hotelling-Lawley's trace, and Roy's maximum root test, are available to test H_0. Further, the design is balanced with no missing values and hence the four types of analyses are equivalent, all resulting in identical SS&CP matrices (in fact, for one-way classification, this is true even for unbalanced data). The SAS code to do this analysis is presented in Program 4.1. The program produces Output 4.1.

```
/* Program  4.1 */

options ls = 64 ps=45 nodate nonumber;
data jack;
input lab method1 method2;
lines;
1 10.1 10.5
1 9.3 9.5
1 9.7 10.0
1 10.9 11.4
2 10.0 9.8
2 9.5 9.7
2 9.7 9.8
2 10.8 10.7
3 11.3 10.1
3 10.7 9.8
3 10.8 10.1
3 10.5 9.6
;
/* Source: Jackson (1991, p. 301). Principal Components.  Copyright
   1991 John Wiley & Sons, Inc.  Reprinted by permission of
   John Wiley & Sons, Inc. */

Title1 'Output 4.1' ;
title2 'Balanced One-Way MANOVA';
proc glm data = jack;
class lab;
model method1 method2 = lab/nouni;
manova h = lab/printe printh ;
run;
/* proc glm data = jack ;
class lab ;
model method1 method2 = lab/nouni;
contrast 'Test: lab eff.' lab 1 -1  0,
                          lab 1  0  -1;
manova/printe printh;
run; */
```

The independent variable which defines the classification is denoted by LAB and the two methods specified as METHOD1 and METHOD2 are the dependent variables. We perform the analysis using the GLM procedure. The MANOVA statement performs multivariate analysis. It is important that the variable LAB is specified in the CLASS statement. This enables SAS to create the appropriate **X** matrix. We could have used any other numeric or nonnumeric coding for the values taken by the class variable LAB, since classification variables can be either character or numeric.

To test the null hypothesis, it suffices to indicate the variable LAB as H=LAB in the MANOVA statement. The PRINTE and PRINTH options enable us to print the SS&CP

matrices corresponding to the error and the null hypothesis H_0. We could have also specified the type of SS&CP matrices to be used in the analysis in the MODEL statement but since the four types of analyses are identical in this case, it is not necessary to specify one type over another. As a result, SAS uses the default, Type III analysis.

Output 4.1

```
                              Output 4.1
                      Balanced One-Way MANOVA


                      E = Error SS&CP Matrix


                            METHOD1              METHOD2


          METHOD1           2.7275               2.63
          METHOD2           2.63                 2.81



          H = Type III SS&CP Matrix for LAB


                            METHOD1              METHOD2


          METHOD1           1.815                -0.605
          METHOD2           -0.605               0.4466666667



          Manova Test Criteria and F Approximations for
              the Hypothesis of no Overall LAB Effect
      H = Type III SS&CP Matrix for LAB    E = Error SS&CP Matrix


                      S=2     M=-0.5     N=3


Statistic                    Value        F    Num DF  Den DF  Pr > F


Wilks' Lambda                0.069895    11.13      4      16   0.0002
Pillai's Trace               0.971691     4.2522    4      18   0.0135
Hotelling-Lawley Trace      12.71214     22.246     4      14   0.0001
Roy's Greatest Root         12.66516     56.993     2       9   0.0001


    NOTE: F Statistic for Roy's Greatest Root is an upper bound.
        NOTE: F Statistic for Wilks' Lambda is exact.
```

For the present data set, the number of data points $n = 12$ and the number of dependent variables $p = 2$. The null hypothesis can be written as

$$H_0 : \begin{bmatrix} 0 & 1 & -1 & 0 \\ 0 & 1 & 0 & -1 \end{bmatrix} \begin{bmatrix} \mu' \\ \tau_1' \\ \tau_2' \\ \tau_3' \end{bmatrix} = 0 \quad \text{or} \quad \mathbf{LB} = \mathbf{0}$$

and since the left-most matrix in H_0, that is $\mathbf{L}$, has rank 2, the value of $r = Rank\,(\mathbf{L}) = 2$ (see Table 3.2). The four test statistics corresponding to the null hypothesis are shown in Output 4.1. Recall that according to Table 3.3, the transformation of Wilks' Λ to F statistic is exact, since $p = 2$ here. As a result,

$$F = \left(\frac{n-k-1}{r}\right)\left(\frac{1-\sqrt{\Lambda}}{\sqrt{\Lambda}}\right) = \left(\frac{12-3-1}{2}\right)\left(\frac{1-\sqrt{\Lambda}}{\sqrt{\Lambda}}\right) = \frac{8}{2}\left(\frac{1-\sqrt{\Lambda}}{\sqrt{\Lambda}}\right)$$

follows an F (4, 16) distribution. Corresponding to the observed value of F = 11.1299 with df (4,16), the p value is 0.0002. Consequently, we conclude that there is sufficient evidence against H_0 and that there is a significant difference between the laboratories. We reach essentially the same conclusions under the other three test criteria. The output also presents the corresponding SS&CP matrices for error and the hypothesis as results of PRINTE and PRINTH options in the the MANOVA statement. These were respectively denoted by $\mathbf{E}$ and $\mathbf{H}$ in the previous chapter. It may also be noted that as an alternative to H=LAB, one could also use the following CONTRAST statement,

```
contrast 'Test: lab eff.'   lab 1 -1 0,
                            lab 1 0 -1;
```

It is so since the above statement specifies the hypothesis $\tau_1 = \tau_2$ and $\tau_2 = \tau_3$ together which are then equivalent to our H_0 stated earlier. For completeness we have included this code in Program 4.1 but have commented it out to suppress the corresponding output.

EXAMPLE 2   ***An Unbalanced One-Way Classification, Diabetic Patients Study Data***   Crowder and Hand (1990, p. 8) provided this example of unbalanced data. Two groups of subjects, an eight-member normal control group and a six-member group of diabetic patients without complications, were to be compared as part of a medical experiment. The subjects performed a small physical task, and the measurements were recorded on each of the subjects during various subsequent time points. The data in Table 4.2 are these measurements after one minute, five minutes, and ten minutes after performing the task. The question of interest concerns differences between the two groups. In other words, we want to investigate if the two groups differ from each other in their abilities to perform the specified physical task.

This one-way classification data has GROUP as the CLASS variable. On each of the 14 subjects, a trivariate vector of data representing the three measurements at one, five, and ten minutes after performing the physical task, is available. If the respective population mean vectors for the two groups on these measurements are represented as $\boldsymbol{\mu}^{(1)} = \boldsymbol{\mu} + \boldsymbol{\tau}_1$ and $\boldsymbol{\mu}^{(2)} = \boldsymbol{\mu} + \boldsymbol{\tau}_2$, then the matrix $\mathbf{B}$ of regression coefficients can be written as

$$\mathbf{B}_{3\times 3} = \begin{bmatrix} \boldsymbol{\mu}' \\ \boldsymbol{\tau}'_1 \\ \boldsymbol{\tau}'_2 \end{bmatrix}.$$

TABLE 4.2 Effect of a Physical Task on Hospital Patients

		Time		
	Subject	1	5	10
Group 1	1	7.6	8.7	7.0
	2	10.1	8.9	8.6
	3	11.2	9.5	9.4
	4	10.8	11.5	11.4
	5	3.9	4.1	3.7
	6	6.7	7.3	6.6
	7	2.2	2.5	2.4
	8	2.1	2.0	2.0
Group 2	9	8.5	5.6	8.4
	10	7.5	5.0	9.5
	11	12.9	13.6	15.3
	12	8.8	7.9	7.3
	13	5.5	6.4	6.4
	14	3.2	3.4	3.2

To test the equality of the treatment effects (that is, the two groups' abilities to complete the specified physical task) between the two groups for all the three time points, the null hypothesis is

$$H_0: \quad \begin{bmatrix} 0 & 1-1 \end{bmatrix} \begin{bmatrix} \mu_1 & \mu_2 & \mu_3 \\ \tau_{11} & \tau_{12} & \tau_{13} \\ \tau_{21} & \tau_{22} & \tau_{23} \end{bmatrix} = \mathbf{0}$$

that is,

$$H_0: \mathbf{LB} = \mathbf{0},$$

which can be tested as in Example 1. However, in the present context a more realistic hypothesis may be to test that the amount of change in measurements from one minute to five minutes is equal for the two groups and that the change between the five minutes and ten minutes is equal for the two groups. These can be represented as

$$H_0: \begin{bmatrix} 0 & 1 & -1 \end{bmatrix} \begin{bmatrix} \mu_1 & \mu_2 & \mu_3 \\ \tau_{11} & \tau_{12} & \tau_{13} \\ \tau_{21} & \tau_{22} & \tau_{23} \end{bmatrix} \begin{bmatrix} 1 & 0 \\ -1 & 1 \\ 0 & -1 \end{bmatrix} = \mathbf{0}$$

or

$$H_0: \mathbf{LBM} = \mathbf{0} \tag{4.5}$$

with $\mathbf{L} = (0 \; 1 \; -1)$, and $\mathbf{M} = \begin{bmatrix} 1 & 0 \\ -1 & 1 \\ 0 & -1 \end{bmatrix}$.

The above representation deserves some further explanation. Let us first premultiply $\mathbf{B}$ to $\mathbf{M}$, resulting in

$$\mathbf{BM} = \begin{bmatrix} \mu_1 - \mu_2 & \mu_2 - \mu_3 \\ \tau_{11} - \tau_{12} & \tau_{12} - \tau_{13} \\ \tau_{21} - \tau_{22} & \tau_{22} - \tau_{23} \end{bmatrix}.$$

The entries in the first row of the above matrix represent the successive differences in the intercept or the overall mean for the three time points. The second row represents the successive treatment differences for Group 1 and the third row represents the same for Group 2. Since we want to compare these differences for the two groups, this is accomplished by premultiplying $\mathbf{BM}$ by $\mathbf{L} = (0 \; 1 \; -1)$ and equating the product to zero. This results in the simplification of $H_0: \mathbf{LBM} = \mathbf{0}$ to

$$H_0: \begin{bmatrix} (\tau_{11} - \tau_{12}) - (\tau_{21} - \tau_{22}) \\ (\tau_{12} - \tau_{13}) - (\tau_{22} - \tau_{23}) \end{bmatrix}' = \begin{bmatrix} 0 \\ 0 \end{bmatrix}'.$$

The choices of either $\mathbf{L}$ or $\mathbf{M}$ indicated here are not unique. For example, $\mathbf{L} = (0 \; -1 \; 1)$ and $\mathbf{M} = \begin{bmatrix} 1 & 1 \\ -1 & 0 \\ 0 & -1 \end{bmatrix}$ are the other equally legitimate choices for $\mathbf{L}$ and $\mathbf{M}$. The tests for the hypotheses of the type $H_0: \mathbf{LBM} = \mathbf{0}$ were described in Chapter 3. In SAS, this objective is attained by specifying the $\mathbf{M}$ matrix in the MANOVA statement. SAS automatically identifies the corresponding $\mathbf{L}$ matrix from the specification H = GROUP.

The $\mathbf{M}$ matrix can be specified using one of the two different yet equivalent ways. We can either explicitly specify all the entries of $\mathbf{M}$ in the M= specification of the MANOVA statement as

```
m = (1 -1   0,
     0  1 -1);
```

or ask SAS to create it so as to correspond to the measurement differences of interest. The latter is achieved by using the algebraic statements which, in the present context, are

```
m = min1 - min5,
    min5 - min10;
```

where MIN1, MIN5, and MIN10 were the names assigned in Program 4.2 to the measure-ments at 1, 5, and 10 minutes after performing the physical task. It may also be pointed out that when using the former choice, the assignment is column after column separated by commas. Similarly, when the respective columns are written in different lines of the program, the matrix in the SAS code may visually resemble $\mathbf{M}'$ and not $\mathbf{M}$. In Program 4.2, we have used the latter alternative. See Output 4.2 for the results.

```
/* Program 4.2 */

options ls=64 ps=45 nonumber nodate;
data phytask ;
input group min1 min5 min10 ;
lines ;
1 7.6  8.7  7.0
1 10.1 8.9  8.6
1 11.2 9.5  9.4
1 10.8 11.5 11.4
1 3.9  4.1  3.7
1 6.7  7.3  6.6
1 2.2  2.5  2.4
1 2.1  2.0  2.0
2 8.5  5.6  8.4
2 7.5  5.0  9.5
2 12.9 13.6 15.3
2 8.8  7.9  7.3
2 5.5  6.4  6.4
2 3.2  3.4  3.2
;
/* Source: Crowder and Hand (1990, p. 8). */

title1 'Output 4.2';
title2 'Unbalanced One-Way MANOVA';
proc glm data = phytask;
class group;
model min1 min5 min10 = group/nouni;
manova h = group m = min1-min5,
                     min5-min10/printe printh ;
  manova h = intercept m = min1-min5,
                     min5-min10/printe printh ;
run;
```

Output 4.2

```
               Output 4.2
          Unbalanced One-Way MANOVA


      Manova Test Criteria and Exact F Statistics for
           the Hypothesis of no Overall GROUP Effect
      on the variables defined by the M Matrix Transformation
    H = Type III SS&CP Matrix for GROUP    E = Error SS&CP Matrix


                  S=1     M=0     N=4.5

Statistic              Value       F    Num DF  Den DF  Pr > F

Wilks' Lambda          0.65534  2.8926       2      11  0.0979
```

```
Pillai's Trace              0.34466   2.8926    2    11   0.0979
Hotelling-Lawley Trace 0.525925   2.8926    2    11   0.0979
Roy's Greatest Root    0.525925   2.8926    2    11   0.0979


              Manova Test Criteria and Exact F Statistics for
               the Hypothesis of no Overall INTERCEPT Effect
           on the variables defined by the M Matrix Transformation
      H = Type III SS&CP Matrix for INTERCEPT    E = Error SS&CP Matrix


                       S=1     M=0     N=4.5

      Statistic                Value      F    Num DF  Den DF  Pr > F


      Wilks' Lambda          0.872037  0.8071    2      11    0.4709
      Pillai's Trace         0.127963  0.8071    2      11    0.4709
      Hotelling-Lawley Trace 0.146741  0.8071    2      11    0.4709
      Roy's Greatest Root    0.146741  0.8071    2      11    0.4709
```

Suppose, in addition, that we are also interested in testing the null hypothesis that the changes with respect to time in the levels of overall means (intercepts) are zero. This amounts to testing $H_0 : \mu_1 = \mu_2 = \mu_3$ or

$$H_0 : \begin{bmatrix} \mu_1 - \mu_2 \\ \mu_2 - \mu_3 \end{bmatrix} = \mathbf{0}. \tag{4.6}$$

With the choice of $\mathbf{M}$ the same as earlier and $\mathbf{L} = (1\ 0\ 0)$, this hypothesis also reduces to the form $H_0 : \mathbf{LBM} = \mathbf{0}$. As earlier, the corresponding $\mathbf{M}$ will be specified through the M= specification of the MANOVA statement. However, the choice of $\mathbf{L}$ in this case is specified by indicating H = INTERCEPT.

The null hypotheses in Equations 4.5 and 4.6 are tested using Program 4.2. Output 4.2 presents portions of the resulting output. We use the default Type III analysis since all the four types of analyses are identical in this case.

In both the cases, since $\mathbf{L}$ is a nonzero row vector, it is of rank 1. Consequently, all four multivariate test criteria lead to an exact and identical F test statistic. For the hypothesis in Equation 4.5, the p value corresponding to the test statistic is 0.0979, which indicates that there is some evidence, though it is not very strong, against the null hypothesis. However, with respect to the null hypothesis in Equation 4.6, there is not enough evidence to reject H_0 (p value = 0.4709) and hence we conclude that levels of overall mean are the same for the three periods.

Certain other ways of analyzing repeated measures data are discussed in Chapter 5.

4.4   Two-Way Classification

In one-way classification models, the interest is in comparing the treatment effects which correspond to a single variable. When there are two variables, say A and B, various treatments are obtained by combining the various levels of variable A with those of variable B. If A is at a different levels and B is at b levels, then assuming that all possible levels of A can be attempted with those of B, the experiment consists of ab treatment combinations. In such a case, we say that A and B are *crossed* with each other, and the design is often referred to as a two-way classification.

If each ab treatment is tried an equal number of times, then the resulting design is balanced. Such designs usually lead to a simpler analysis in that the corrected total SS&CP

matrix has a unique partitioning: the SS&CP matrices corresponding to variables A and B, the interaction AB, and the error. In fact, this same uniqueness of partitioning can be ensured if r_{ij}, the number of observations in the $(i, j)^{th}$ cell, that is, corresponding to the ith level of variable A and the jth level of variable B, $i = 1, \ldots, a$, $j = 1, \ldots, b$ is such that

$$r_{ij} = \frac{r_{i.} r_{.j}}{r_{..}},$$

where

$$r_{i.} = \sum_{j=1}^{b} r_{ij}, \quad r_{.j} = \sum_{i=1}^{a} r_{ij}, \quad \text{and} \quad r_{..} = \sum_{i=1}^{a} r_{i.} = \sum_{j=1}^{b} r_{.j}.$$

EXAMPLE 3   *A Balanced Two-Way Classification, Weight Loss in Mice*   Morrison (1976, p. 190) presented a two-way classification study to compare the loss in weights of male and female mice under three different drugs. Four mice of each sex were randomly assigned to each of the three drugs and weight losses were measured at the end of the first and second weeks. The resulting bivariate data correspond to a balanced two-way classification with interaction:

$$y_{ijk} = \mu + \alpha_i + \beta_j + (\alpha\beta)_{ij} + \epsilon_{ijk}$$

$$i = 1, 2, \quad j = 1, 2, 3, \quad k = 1, 2, 3, 4,$$

where suffix i indicates the particular level of variable SEX, and j indicates the particular level of variable DRUG.

The purpose in this example is to test the significance of the effects of SEX, the effects of DRUG, and their interaction. Program 4.3 provides the needed SAS code. The output is presented as Output 4.3.

```
/* Program 4.3 */

options ls=64 ps=45 nodate nonumber;
data wtloss ;
input sex $ drug $  week1 week2 ;
lines;
male    a  5  6
male    a  5  4
male    a  9  9
male    a  7  6
male    b  7  6
male    b  7  7
male    b  9 12
male    b  6  8
male    c 21 15
male    c 14 11
male    c 17 12
male    c 12 10
female a  7 10
female a  6  6
female a  9  7
female a  8 10
female b 10 13
female b  8  7
female b  7  6
female b  6  9
female c 16 12
female c 14  9
```

```
female c 14  8
female c 10  5
;
/* Source: Morrison (1976, p. 190). Multivariate Statistical
   Methods, McGraw-Hill, Inc.  Reproduced with permission
   of  McGraw-Hill, Inc. */

proc glm data = wtloss ;
class sex drug ;
model week1 week2= sex|drug/nouni ;
*model week1 week2= sex drug sex*drug/nouni ;
manova h = sex drug sex*drug/printe printh ;
title1 'Output 4.3';
title2 'Balanced Two-Way MANOVA';
run;
```

The MANOVA statement in the program asks for the individual multivariate testing for the variables SEX, DRUG, and the interaction SEX*DRUG. The PRINTE and PRINTH options are used to print the resulting SS&CP matrices corresponding to the error and the particular hypotheses in each of the three tests.

We discuss the test for the interaction first. Note that the same SS&CP matrix $\mathbf{E}$ for error will be used for the tests for main effects as well, unless the model is modified. Here

$$\mathbf{E} = \begin{bmatrix} 94.5 & 76.5 \\ 76.5 & 114.0 \end{bmatrix}.$$

The SS&CP matrix corresponding to the hypothesis of no interaction is

$$\mathbf{H}_{int} = \begin{bmatrix} 14.3333 & 21.3333 \\ 21.3333 & 32.3333 \end{bmatrix}.$$

As a result, Wilks' Λ is computed as $\Lambda_{int} = \frac{|\mathbf{E}|}{|\mathbf{E}+\mathbf{H}_{int}|} = 0.7744$. From Table 3.3, $F_{int} = \frac{17}{2}\{\frac{1-\sqrt{0.7744}}{\sqrt{0.7744}}\} = 1.1593$ is the observed value from an F(4, 34) distribution. As the corresponding p value $= 0.3459$ is quite large, we do not reject the hypothesis of no interaction between the variables SEX and DRUG. Other test statistics were calculated using the corresponding formulas, and their respective p values also support this conclusion. As a result, it may be assumed that an additive model for SEX and DRUG is valid.

In the absence of possible interaction, we may want to perform tests of significance on the main variables. A similar calculation provides for variable SEX: $\Lambda_{sex} = 0.9925$, $F_{sex} = 0.0639$ with $df = (2, 17)$ (p value $= 0.9383$) and for variable DRUG: $\Lambda_{drug} = 0.1686$, $F_{drug} = 12.1991$ with $df = (4, 34)$ (p value $= 0.0001$). As a result, we conclude that the variable DRUG has a significant effect on weight loss, but the sex of the rats does not play any important role; that is, rats of either sex lost weight in the same way.

Output 4.3

```
                          Output 4.3
                   Balanced Two-Way MANOVA


          Manova Test Criteria and Exact F Statistics for
                the Hypothesis of no Overall SEX Effect
       H = Type III SS&CP Matrix for SEX    E = Error SS&CP Matrix


                     S=1      M=0      N=7.5


Statistic                 Value        F    Num DF  Den DF  Pr > F


Wilks' Lambda           0.992537   0.0639        2      17  0.9383
```

Pillai's Trace	0.007463	0.0639	2	17	0.9383
Hotelling-Lawley Trace	0.007519	0.0639	2	17	0.9383
Roy's Greatest Root	0.007519	0.0639	2	17	0.9383

Manova Test Criteria and F Approximations for
the Hypothesis of no Overall DRUG Effect
H = Type III SS&CP Matrix for DRUG   E = Error SS&CP Matrix

S=2   M=-0.5   N=7.5

Statistic	Value	F	Num DF	Den DF	Pr > F
Wilks' Lambda	0.16863	12.199	4	34	0.0001
Pillai's Trace	0.880378	7.0769	4	36	0.0003
Hotelling-Lawley Trace	4.639537	18.558	4	32	0.0001
Roy's Greatest Root	4.576027	41.184	2	18	0.0001

NOTE: F Statistic for Roy's Greatest Root is an upper bound.
NOTE: F Statistic for Wilks' Lambda is exact.

Manova Test Criteria and F Approximations for
the Hypothesis of no Overall SEX*DRUG Effect
H = Type III SS&CP Matrix for SEX*DRUG   E = Error SS&CP Matrix

S=2   M=-0.5   N=7.5

Statistic	Value	F	Num DF	Den DF	Pr > F
Wilks' Lambda	0.774362	1.1593	4	34	0.3459
Pillai's Trace	0.226949	1.152	4	36	0.3481
Hotelling-Lawley Trace	0.289692	1.1588	4	32	0.3473
Roy's Greatest Root	0.283723	2.5535	2	18	0.1056

NOTE: F Statistic for Roy's Greatest Root is an upper bound.
NOTE: F Statistic for Wilks' Lambda is exact.

When the design is not balanced, the partitioning of corrected total SS&CP is not unique. This issue has already been addressed in Section 4.2. As mentioned earlier, for most comparison purposes, when the variables are being treated as categorical, and the purpose is merely to compare or identify treatment effects, Type III analysis can be adopted. However, for model-building purposes, where we are implicitly performing a selection of variables analyses to obtain an appropriate model, the analysis using the sequential sums of squares and crossproducts (Type I) may be appropriate.

The following example with unbalanced data provides an illustration.

EXAMPLE 4   ***Optimization of Uniformity and Selectivity in Etching Process***   In manufacturing in the integrated circuit industry, the process must etch layers uniformly across the wafers. This study investigated whether pressure (PRESS) and power (POWER) are the two main explanatory variables which considerably affect various response variables. The list of response variables includes the uniformity of the etching of the two layers (UNIF1 and UNIF2) and the selectivity (SELECT), which is defined as the ratio of the etch rates of the two layers. The problems in this example are first to see the effects of PRESS and POWER on the response variables indicated above and second to fit a model for the three response variables in terms of the variables PRESS and POWER and possibly their interaction. Observations at each of two levels of POWER (240 and 290 watts) and three levels of PRESS (90, 110 and 130 torr) were made with a total of 15 observations leading to an unbalanced design. Three different response variables were observed in each of these 15 experiments.

The data are shown as part of Program 4.4. For each of the fifteen experiments, as the three responses have resulted from the same experiment and are taken on the same wafer, they will be correlated and hence the three separate univariate analyses may give misleading results. It seems logical to perform a multivariate analysis of variance to draw meaningful results. In addition, since we also want to find a suitable choice of the levels of POWER and PRESS, to optimize the three responses, we also want to find suitable regression models for the uniformity of the two etches and the selectivity. Here a Type I analysis seems more appropriate than Type III; the three analyses are not going to be identical since the design under consideration is not balanced.

It is possible that the variable PRESS may be more influential than the variable POWER. This possibility indicates that, in the right-hand side of the MODEL statement, PRESS should precede POWER and their interaction PRESS*POWER. To obtain a Type I analysis, we use the following sequential SS&CP partitioning. See Equations 4.3 and 4.4 for notations.

$$R\,(PRESS,\ POWER,\ PRESS*POWER\,|\,INTERCEPT)$$

$$=\ R\,(PRESS\,|\,INTERCEPT) + R\,(POWER\,|\,INTERCEPT,\ PRESS)$$

$$+\ R\,(PRESS*POWER\,|\,INTERCEPT,\ PRESS,\ POWER).$$

In Program 4.4, we ask for all the three types of analyses, namely Types I, II, and III. The appropriate SS&CP matrices computed by the statement:

```
manova h = press power press*power/printe printh;
```

for various hypotheses are indicated below. We have used the same notations as in Equations 4.3 and 4.4, with the understanding that the corresponding values are SS&CP matrices and not just the sums of squares.

Type I: R (PRESS | INTERCEPT),
     R (POWER | INTERCEPT, PRESS),
     R (PRESS*POWER | INTERCEPT, PRESS, POWER),
     Error SS&CP.

Type II: R (PRESS | INTERCEPT, POWER),
     R (POWER | INTERCEPT, PRESS),
     R (PRESS*POWER | INTERCEPT, PRESS, POWER),
     Error SS&CP.

Type III: R (PRESS | INTERCEPT, POWER, PRESS*POWER),
     R (POWER | INTERCEPT, PRESS, PRESS*POWER),
     R (PRESS*POWER | INTERCEPT, PRESS, POWER),
     Error SS&CP.

```
/* Program 4.4 */

options ls=64 ps=45 nodate nonumber;
data etch;
input press power etch1 etch2 unif1 unif2 ;
select = etch1/etch2 ;
x1 = (press-265)/25 ;
```

```
x2 = (power-110)/20 ;
x2sq = x2*x2 ;
lines ;
240   90   793 300 13.2 25.1
240   90   830 372 15.1 24.6
240   90   843 389 14.2 25.7
240  110  1075 400 15.8 25.9
240  110  1102 410 14.9 25.1
240  130  1060 397 15.3 24.9
240  130  1049 427 14.7 23.8
290   90   973 350  7.4 18.3
290   90   998 373  8.3 17.7
290  110   940 365  8.0 16.9
290  110   935 365  7.1 17.2
290  110   953 342  8.9 17.4
290  110   928 340  7.3 16.6
290  130  1020 402  8.6 16.3
290  130  1034 409  7.5 15.5
;
title1 'Output 4.4';
title2 'Unbalanced Two-Way Classification: MANOVA' ;
title3 'Effects of Factors on Etch Uniformity and Selectivity';
proc glm data = etch ;
class press power ;
model select unif1 unif2= press power press*power/ss1 nouni ;
manova h = press power press*power/printe printh ;

proc glm data = etch ;
class press power ;
model select unif1 unif2=press power press*power/ss2 nouni ;
manova h = press power press*power/printe printh ;

proc glm data = etch ;
class press power ;
model select unif1 unif2=press power press*power/ss3 nouni ;
manova h = press power press*power/printe printh ;

proc glm data = etch ;
model select unif1 unif2  = x1 x2 x2sq /ss1 nouni;
manova h = x1 x2 x2sq /printe printh ;

/* proc glm data = etch ;
model select unif1 unif2  = x1 x2 x2sq /ss2 nouni;
manova h = x1 x2 x2sq /printe printh ;

proc glm data = etch ;
model select unif1 unif2  = x1 x2 x2sq /ss3 nouni;
manova h = x1 x2 x2sq /printe printh; */
```

A part of the results of Program 4.4 appears in Output 4.4. Since PRESS may play a more important role than POWER, and hence if possible should be included in the model when fitting a response surface, we examine the Type I analysis next. We first consider the interaction term for testing. It is so because a significant interaction causes the main effect tests to be meaningless. After discounting for the effects of PRESS and POWER, we find that the interaction PRESS*POWER is not significant. The p value for Wilks' Λ is 0.3605. The other multivariate tests also provide comparable p values and similar conclusions. We therefore look sequentially at the statistical significance of the main effect PRESS and then POWER after discounting for PRESS.

The analysis shows that the variable PRESS is highly significant with the corresponding calculated value of $F(3, 7)=264.6643$ under all four criteria. These tests are all equivalent for PRESS but not for POWER and PRESS*POWER. This is so since the rank of the corresponding **L** matrix is 1 for PRESS whereas it is 2 for the other two hypotheses. The corresponding p values are all very small. For POWER, significance is observed but only marginally. For example, corresponding to Wilks' test the p value is 0.0477. Note that the test for the variable POWER was performed after discounting for the effect of PRESS. As a result, we conclude that the variable POWER also has some effect on the response variables. It may be pointed out that these conclusions are further supported by Type II as well as Type III analyses outputs, which are not included here.

Output 4.4

```
                          Output 4.4
              Unbalanced Two-Way Classification: MANOVA
           Effects of Factors on Etch Uniformity and Selectivity


               Manova Test Criteria and Exact F Statistics for
                  the Hypothesis of no Overall PRESS Effect
           H = Type I SS&CP Matrix for PRESS   E = Error SS&CP Matrix


                       S=1     M=0.5    N=2.5
```

Statistic	Value	F	Num DF	Den DF	Pr > F
Wilks' Lambda	0.008739	264.66	3	7	0.0001
Pillai's Trace	0.991261	264.66	3	7	0.0001
Hotelling-Lawley Trace	113.4276	264.66	3	7	0.0001
Roy's Greatest Root	113.4276	264.66	3	7	0.0001

```
               Manova Test Criteria and F Approximations for
                  the Hypothesis of no Overall POWER Effect
           H = Type I SS&CP Matrix for POWER   E = Error SS&CP Matrix


                       S=2     M=0     N=2.5
```

Statistic	Value	F	Num DF	Den DF	Pr > F
Wilks' Lambda	0.199491	2.8908	6	14	0.0477
Pillai's Trace	1.014505	2.7452	6	16	0.0498
Hotelling-Lawley Trace	2.940054	2.9401	6	12	0.0529
Roy's Greatest Root	2.513229	6.7019	3	8	0.0142

```
        NOTE: F Statistic for Roy's Greatest Root is an upper bound.
              NOTE: F Statistic for Wilks' Lambda is exact.



               Manova Test Criteria and F Approximations for
               the Hypothesis of no Overall PRESS*POWER Effect
        H = Type I SS&CP Matrix for PRESS*POWER   E = Error SS&CP Matrix


                       S=2     M=0     N=2.5
```

Statistic	Value	F	Num DF	Den DF	Pr > F
Wilks' Lambda	0.435362	1.203	6	14	0.3605
Pillai's Trace	0.582269	1.0952	6	16	0.4068

```
Hotelling-Lawley Trace 1.256443    1.2564        6       12   0.3457
Roy's Greatest Root     1.223339    3.2622        3        8   0.0805
```

NOTE: F Statistic for Roy's Greatest Root is an upper bound.
NOTE: F Statistic for Wilks' Lambda is exact.

Manova Test Criteria and Exact F Statistics for
the Hypothesis of no Overall X1 Effect
H = Type I SS&CP Matrix for X1   E = Error SS&CP Matrix

S=1   M=0.5   N=3.5

Statistic	Value	F	Num DF	Den DF	Pr > F
Wilks' Lambda	0.010972	270.42	3	9	0.0001
Pillai's Trace	0.989028	270.42	3	9	0.0001
Hotelling-Lawley Trace	90.1392	270.42	3	9	0.0001
Roy's Greatest Root	90.1392	270.42	3	9	0.0001

Manova Test Criteria and Exact F Statistics for
the Hypothesis of no Overall X2 Effect
H = Type I SS&CP Matrix for X2   E = Error SS&CP Matrix

S=1   M=0.5   N=3.5

Statistic	Value	F	Num DF	Den DF	Pr > F
Wilks' Lambda	0.294806	7.1762	3	9	0.0092
Pillai's Trace	0.705194	7.1762	3	9	0.0092
Hotelling-Lawley Trace	2.392064	7.1762	3	9	0.0092
Roy's Greatest Root	2.392064	7.1762	3	9	0.0092

Manova Test Criteria and Exact F Statistics for
the Hypothesis of no Overall X2SQ Effect
H = Type I SS&CP Matrix for X2SQ   E = Error SS&CP Matrix

S=1   M=0.5   N=3.5

Statistic	Value	F	Num DF	Den DF	Pr > F
Wilks' Lambda	0.830186	0.6136	3	9	0.6231
Pillai's Trace	0.169814	0.6136	3	9	0.6231
Hotelling-Lawley Trace	0.20455	0.6136	3	9	0.6231
Roy's Greatest Root	0.20455	0.6136	3	9	0.6231

Since our present goal is to predict the optimum combination of PRESS and POWER, both of which are the continuous variables, we obtain the appropriate response surface models by examining the importance of various terms in the model. We thus transform PRESS and POWER as $x_1 = (PRESS - 265)/25$ and $x_2 = (POWER - 110)/20$. Since PRESS is at two levels and POWER at three, only the following terms can be included in the model:

$$X1 = x_1, \ X2 = x_2, \ X2SQ = x_2^2, \ X1X2 = x_1x_2, \text{ and } X1X2SQ = x_1x_2^2.$$

However, since we have already found the interaction PRESS*POWER to be nonsignificant, we may not want to include the terms $x_1 x_2$ and $x_1 x_2^2$ in the model. As a result, for the three response variables we simultaneously fit three models with x_1, x_2, and x_2^2 as the independent variables. As earlier, we rely primarily on the Type I analysis, with the sequence of the terms as specified. Since in regression modeling the quadratic effect x_2^2 is treated as a new variable and not as a function of x_2, the Type I and Type II SS&CP matrices and the corresponding test statistics for x_2^2 are identical.

Output 4.4 shows the result of testing various hypotheses under Type I analysis. All four multivariate tests are equivalent in this case. In view of the small p value($= 0.0001$) for the tests on x_1 (PRESS), it is found to be statistically significant. Even after discounting the effect of PRESS, the variable POWER represented by x_2 is also significant with p value $= 0.0092$. The quadratic effect of POWER, however, is not significant as evident from the large p value($= 0.6231$) for x_2^2. Note that a significant effect obtained through a multivariate analysis does not necessarily imply that the effect is significant for each response variable. The conclusions drawn from a univariate analysis for an individual response variable may or may not be in complete agreement with those derived by performing a multivariate analysis of all the response variables collectively. This is true also in the present example where it can be verified in univariate significance testing that none of the three variables x_1, x_2, and x_2^2 appear to have any significant effect on selectivity.

4.5   Blocking

In order to remove the additional variability in the data due to other external sources, blocking is often desired. Likewise, if the external variability is present due to two independent sources or is present in two orthogonal directions, two-way blocking using the Latin square design is often used. If the data are available for all the cells in the Latin square, the orthogonality of the two blocking variables and the treatment is automatically accomplished. Hence the Type I, II, III, and IV analyses are identical. A problem, however, occurs if the data are not available for certain cells or the blocks are of unequal size. The question in that case is, which analysis is appropriate? As the treatments are to be compared only after eliminating the effects due to blocking variables, the Type I analysis is clearly the appropriate choice with the treatment variables listed after the blocking variables (in an appropriate sequence) in the MODEL statement of PROC GLM. As an illustration, see the following example, where data are collected under a Latin square design setup for all the cells except two.

EXAMPLE 5   **Experiments in Blocks, Comparison of Corn Varieties**   We consider a part of the data from Srivastava and Carter (1983, p. 107), where a certain area of land was used for testing four varieties of corn represented by four levels of the variable VARIETY. Due to the slope of the land, differences from north to south (NS) and from east to west (EW) were possible. As a result, the experiment was conducted using a Latin square design with the corresponding layout given in Table 4.3.

Note that for our analysis the experiments corresponding to $A_3 B_2$ and $A_4 B_4$ are not included in Table 4.3, so the design is unbalanced. For each experiment, two characteristics, namely the height of the plant (HEIGHT) and the yield (YIELD), were measured. Thus, the additive bivariate model ($p = 2$) containing the variable VARIETY as well as the two blocking variables NS and EW will be fitted for the response variables HEIGHT and YIELD.

As indicated earlier, sequential MANOVA partitioning (Type I) is used to adjust the treatment SS&CP matrix for the block effects. The corresponding MODEL statement is

```
model height yield = ew ns variety/ss1;
```

TABLE 4.3 Data on Corn Yield and Plant Height

| | | | *East* | - | *West* | |
			A1	A2	A3	A4
		Variety	C2	C3	C4	C1
	B1	Height	65	68	67	68
		Yield	24	21	26	27
		Variety	C3	C2	–	C4
	B2	Height	66	63	–	67
		Yield	20	23	–	24
North—South		Variety	C1	C4	C2	C3
	B3	Height	65	67	64	63
		Yield	24	25	19	20
		Variety	C4	C1	C3	–
	B4	Height	65	64	64	–
		Yield	26	25	25	–

An alternative is to use the SS3 option instead of SS1 to get all the SS&CP matrices adjusted for all the remaining terms in the right side of the MODEL statement. However, if in the MODEL statement given above, VARIETY was specified before NS and EW, then SS3 and not SS1 would be the correct option since under the SS1 option, the SS&CP matrix for variety would be adjusted only for the intercept.

Therefore, since the MODEL statement in Program 4.5 lists VARIETY first, we have chosen the Type III analysis and specified the SS3 option. The multivariate tests based on Type III analysis (that is, after adjusting for the two directions in the Latin squares, in this case) are presented as Output 4.5. Based on any of the four multivariate tests, there does not appear to be significant difference between the four varieties; the p value for Wilks' Λ is 0.2638. If there were indeed a significant difference, it would also have been interesting to perform the pairwise comparisons of the four varieties. This can be done using the CONTRAST statement. For instance, if we wanted to compare the varieties C_2 and C_4, then the corresponding SAS statement, which should follow the MODEL statement but must precede the MANOVA statement, is

```
contrast  'c2 vs. c4' variety 0 1 0 -1;
```

For the optional phrase $c2$ versus $c4$ enclosed by single quotation marks (' '), we could have used any other alternative identifier appropriately indicating the type of contrast.

```
/* Program 4.5 */

options ls=64 ps=45 nodate nonumber;
data corn1;
input ew  $ ns $ variety  $ height yield;
lines ;
a1 b1 c2 65 24
a1 b2 c3 66 20
a1 b3 c1 65 24
a1 b4 c4 65 26
a2 b1 c3 68 21
a2 b2 c2 63 23
a2 b3 c4 67 25
a2 b4 c1 64 25
```

```
a3 b1 c4 67 26
a3 b3 c2 64 19
a3 b4 c3 64 25
a4 b1 c1 68 27
a4 b2 c4 67 24
a4 b3 c3 63 20
;
/* Source: Srivastava and Carter (1983, p. 109). */

title1 'Output 4.5';
title2 "Latin Square Design: Corn Yield and Plant Height";
proc glm data = corn1;
class ew ns variety;
model height yield = variety ew ns/ss3 nouni;
manova h =variety/printe printh;
run;
```

Output 4.5

```
                           Output 4.5
              Latin Square Design: Corn Yield and Plant Height




                 Manova Test Criteria and F Approximations for
                    the Hypothesis of no Overall VARIETY Effect
            H = Type III SS&CP Matrix for VARIETY   E = Error SS&CP Matrix


                      S=2      M=0     N=0.5

   Statistic               Value        F    Num DF  Den DF  Pr > F

   Wilks' Lambda          0.135447   1.7172      6       6   0.2638
   Pillai's Trace         1.166323   1.8653      6       8   0.2033
   Hotelling-Lawley Trace 4.154999   1.385       6       4   0.3925
   Roy's Greatest Root    3.522509   4.6967      3       4   0.0846

      NOTE: F Statistic for Roy's Greatest Root is an upper bound.
           NOTE: F Statistic for Wilks' Lambda is exact.
```

Since the pairwise comparisons are not meaningful in the present context of nonsignificant effect on VARIETY, we do not pursue this analysis further for this example.

4.6 Fractional Factorial Experiments

As the number of variables in an experiment increases, so does the total number of all possible combinations or treatments obtained by combining various levels of these variables. As a result, the number of experiments needed to obtain data on the corresponding full factorial design may soon become overwhelming. One way to reduce the total number of experiments is to carefully choose the combinations of variable levels, so that the information on all main effects and certain important lower order interactions can still be extracted from these experiments. This, of course, requires assuming the absence of certain interactions, for example. Also it is necessary to choose the design carefully to ensure that various important interactions which are expected to be significant can still be estimated. When the effects of two variables or interactions on a response variable cannot be distinguished from one another they are said to be *confounded*. In general, it is best to avoid confounding important variables or interactions. In fact, the confounding scheme of the design plays an important role in the choice of fractions of factorial experiments. These are often termed *fractional factorial experiments*.

There is a vast amount of literature available on the construction and analysis of the fractional factorial experiments. On the topic of analysis, John (1971) and Montgomery (1991) provide excellent discussions at different mathematical levels, the latter being more accessible to nonmathematical audiences. However, only univariate analysis is considered in both of these references. As for construction, SAS/QC software provides a number of design generation choices including fractional factorial designs through PROC FACTEX. Also available are certain SAS macros to generate more advanced designs in the same reference. See *SAS/QC Software, Usage and Reference*, for details.

The multivariate nature of the data in response variables does not play any role in the choice of fractional factorial designs or on the confounding scheme. In fact, these issues relate only to the independent variables and are usually decided prior to the experimentation. However, a multivariate analysis of various response variables may be more appropriate due to possible correlations between various response variables. The standard multivariate tests can be performed as described earlier. For various main effects and interactions use either PROC GLM or the REG procedure. The latter is appropriate when all the independent variables are continuous, or when the variables are categorical, if they are only at two levels. Since fractional factorial designs are balanced designs, the ANOVA procedure is also applicable, when we are only interested in comparison and the significance or nonsignificance of various variables and interactions and not in modeling.

It should also be noted that multivariate analysis may sometimes be restrictive in that it tests that a given variable or interaction is significant or nonsignificant collectively for all the response variables. As a result, it is possible to miss the very strong effect of a particular variable or interaction on one response variable simply because it did not significantly affect the other response variables. It is therefore advisable that various univariate analyses accompany the multivariate results to ensure that the two results do not drastically contradict each other.

EXAMPLE 6   *A 2^{8-3} Fractional Factorial Experiment, Modeling of a Chemical Process*   Daniel and Riblett (1954) described a chemical experiment with eight variables, A to H, in 32 runs. Two response variables under consideration were catalyst activity (ACTIVT) and selectivity (SELECTVT). The experiment was run as a one-eighth fraction of a full factorial (that is, a 2^{8-3} fractional factorial design) with all the variables at two levels: $+1$ and -1. The design allowed the estimation and testing of all the main effects and the following two variable interactions: AB, AC, AD, AE, AF, AG, AH, BG, BH, CE, CF, CG, CH, DG, DH, EG, EH, FG, FH, and GH. All other interactions were assumed to be nonexistent.

The purpose of the analysis was to first identify important main variables and interactions using the statistical tests, then estimate their effects and finally obtain a bivariate regression model for the catalyst activity and selectivity.

The data corresponding to this study are presented in Program 4.6. This example also illustrates the use of PROC REG as an occasionally more efficient alternative to PROC GLM. PROC REG is especially helpful here since for two-level fractional factorial experiments, the computation of various effects can be easily achieved through the estimation of certain regression coefficients as we shall see next.

For two variables, say A and B, each at two levels, denoted by $+1$ and -1, interaction can be represented by their ordinary algebraic product $A*B$. We can correspondingly create the values of the new independent variables representing various interactions by simply multiplying the columns of the values (coded as ± 1) of appropriate main variables. In Program 4.6, variables AB, AC, etc., are defined using this rule only. Having done that, we fit the models for the variables ACTIVT and SELECTVT using all the main variables and interactions indicated earlier, as the independent variables. Since these variables are being treated as continuous and not categorical, we can use PROC REG. All the main variables and interactions are defined as continuous variables which take values $+1$ and -1 for this data set. The estimated regression coefficients are saved in the output file EST1. Next the MODEL statement is specified for the two dependent variables ACTIVT and SELECTVT. Since the underlying design is a fractional factorial and hence balanced, all three types of analyses are identical. We have thus used the default choice of Type III SS. Also, since the effects are orthogonal (fractional factorial designs are always orthogonal), dropping a particular interaction or main variable term from the model would not alter the estimated regression coefficients of other terms in the reduced model.

In order to test the joint effect of a set of independent variables or an individual independent variable on the response variables, we use the MTEST statement. We first test if the bivariate model with only main variables as independent variables is adequate. To do so we include the list of all the independent variables (separated by commas) corresponding to the two variable interactions in the MTEST statement. The label ONLYMAIN is used to indicate that under the null hypothesis stated by MTEST, the reduced model contains only the main effects. The hypothesis in the MTEST statement can also be specified as an equation or a set of equations. For example, an alternative way to write the MTEST statement is

```
onlymain: mtest ab=0, ac=0, ad=0, ae=0, af=0, ag=0,
         ah=0, bg=0, bh=0, ce=0, cf=0, cg=0, ch=0,
         dg=0, dh=0, eg=0, eh=0, fg=0, fh=0, gh=0;
```

The remaining MTEST statements in Program 4.6 are used to individually test the hypotheses on the particular interactions. The four multivariate tests for these individual tests are equivalent as well as exact. See Table 3.3 for details.

```
/* Program 4.6 */

options ls = 64 ps=45 nodate nonumber;
data actselct ;
infile 'chemist.dat' obs=32;
input a b c d e f g h activt selectvt;
ab = a*b;
ac = a*c;
ad = a*d;
ae = a*e;
af = a*f;
ag = a*g;
ah = a*h;
bg = b*g;
bh = b*h;
ce = c*e;
cf = c*f;
cg = c*g;
ch = c*h;
dg = d*g;
dh = d*h;
eg = e*g;
eh = e*h;
fg = f*g;
fh = f*h;
gh = g*h;
title1 'Output 4.6';
title2 'Fractional Factorial Experiment: Modeling a
                              Chemical Process';
```

```
proc reg outest = est1  data = actselct ;
model activt selectvt = a b c d e f g h ab ac ad ae af
ag ah bg bh ce cf cg ch dg dh eg eh fg fh gh ;
onlymain: mtest ab, ac, ad, ae, af, ag, ah, bg, bh,
ce, cf, cg, ch, dg, dh, eg, eh, fg, fh, gh ;
ag_eq_0: mtest ag;
/*
ab_eq_0: mtest ab ;
ac_eq_0: mtest ac ;
ad_eq_0: mtest ad ;
ae_eq_0: mtest ae ;
af_eq_0: mtest af ;
ah_eq_0: mtest ah ;
bg_eq_0: mtest bg ;
bh_eq_0: mtest bh ;
ce_eq_0: mtest ce ;
cf_eq_0: mtest cf ;
cg_eq_0: mtest cg ;
ch_eq_0: mtest ch ;
dg_eq_0: mtest dg ;
dh_eq_0: mtest dh ;
eg_eq_0: mtest eg ;
eh_eq_0: mtest eh ;
fg_eq_0: mtest fg ;
fh_eq_0: mtest fh ;
gh_eq_0: mtest gh ;
*/
data effects;
set est1 ;
eff_a=2*a;
eff_b=2*b;
eff_c=2*c;
eff_d=2*d;
eff_e=2*e;
eff_f=2*f;
eff_g=2*g;
eff_h=2*h;
eff_ab=2*ab;
eff_ac=2*ac;
eff_ad=2*ad;
eff_ae=2*ae;
eff_af=2*af;
eff_ag=2*ag;
eff_ah=2*ah ;
eff_bg=2*bg ;
eff_bh=2*bh ;
eff_ce=2*ce ;
eff_cf=2*cf ;
eff_cg=2*cg ;
eff_ch=2*ch ;
eff_dg=2*dg ;
eff_dh=2*dh ;
eff_eg=2*eg ;
eff_eh=2*eh ;
eff_fg=2*fg ;
eff_fh=2*fh ;
eff_gh=2*gh ;
proc print data = effects ;
var _depvar_ eff_a eff_b eff_c eff_d eff_e eff_f
```

```
eff_g eff_h ;
title2 'Effects for Main Factors';
title3 'Coefficients are Half of the Effect of the Contrasts';
proc reg data = actselct ;
model activt = a b c d e f g h ;
model selectvt = activt a b c d e f g h ;
run;
```

Selected parts of the output are presented as Output 4.6. The null hypothesis (all the two-variable interactions are zero) is not rejected by all four multivariate tests. For instance, the p value corresponding to Wilks' Λ is 0.8258. Similarly, all the individual null hypotheses for the interactions are also not rejected at a 5% level of significance. In the output, we show the values of various (equivalent) test statistics and the corresponding p values (=0.7824) for $H_0 : AG = 0$ only.

Output 4.6

```
                              Output 4.6
        Fractional Factorial Experiment: Modeling a Chemical Process


Multivariate Test: ONLYMAIN

            Multivariate Statistics and F Approximations

                     S=2      M=8.5     N=0

Statistic                   Value       F    Num DF  Den DF  Pr > F

Wilks' Lambda             0.020639   0.5961     40       4   0.8258
Pillai's Trace            1.662822   0.7397     40       6   0.7425
Hotelling-Lawley Trace   14.33703    0.3584     40       2   0.9266
Roy's Greatest Root      11.44314    1.7165     20       3   0.3667

    NOTE: F Statistic for Roy's Greatest Root is an upper bound.
         NOTE: F Statistic for Wilks' Lambda is exact.



Multivariate Test: AG_EQ_0

            Multivariate Statistics and Exact F Statistics

                     S=1      M=0      N=0

Statistic                   Value       F    Num DF  Den DF  Pr > F

Wilks' Lambda             0.78244    0.2781      2       2   0.7824
Pillai's Trace            0.21756    0.2781      2       2   0.7824
Hotelling-Lawley Trace   0.278053   0.2781      2       2   0.7824
Roy's Greatest Root      0.278053   0.2781      2       2   0.7824


                     Effects for Main Factors
           Coefficients are Half of the Effect of the Contrasts


     OBS    _DEPVAR_     EFF_A      EFF_B      EFF_C      EFF_D

      1     ACTIVT      0.03437    0.11563   -0.03437    0.024375
      2     SELECTVT    0.71250    0.55000    0.85000   -0.062500
```

OBS	EFF_E	EFF_F	EFF_G	EFF_H
1	-0.17062	-0.02937	0.18813	0.08688
2	-1.00000	-1.11250	-0.53750	-0.48750

We also want to compute the effects for various main variables. Since all the variables are at two levels and the design is orthogonal, the computation of effects is especially simple. In this case, the effect of a particular variable (or interaction, if it exists) is nothing but twice the estimated value of the corresponding regression coefficient. Since all the regression coefficients of the two models corresponding to two response variables, namely ACTIVT and SELECTVT, have been output in a data set EST1, we define a new data set EFFECTS, where effects for various main variables and the interactions are computed. For example, the regression coefficient of the variable A in the model for ACTIVT is 0.01719; correspondingly, the effect of A is $2 \times (0.01719) = 0.03438$ as shown in the output.

An alternative way to compute the values of various effects would have been to use PROC GLM, treat all the main variables as classification variables, and use the ESTIMATE statement. For example, the corresponding statement for the effect of A is

```
estimate 'factor a'  a  1  -1;
```

It would, however, require that the variable A be declared as a classification variable, thereby not permitting direct computation of the regression coefficients.

It is also possible to identify significant effects using the stepdown analysis discussed in Chapter 3. Roy, Gnanadesikan, and Srivastava (1971) point out that in the context of the present data set, it is known that the response variable ACTIVT is observable with greater precision than the response variable SELECTVT. As a result, we decided that ACTIVT is more important than SELECTVT and there is a natural ordering between the two variables. Thus the overall bivariate null hypothesis of no treatment effect can be tested by first considering the hypothesis of no treatment effect on the response ACTIVT marginally. Then we can consider the hypothesis of no treatment effect on the response variable SELECTVT conditional on ACTIVT (that is, on SELECTVT after adjusting for the covariate ACTIVT). Since any effect is merely a multiple of corresponding regression coefficients for this design, we can directly apply a stepdown analysis on these regression coefficients only.

For illustration, we consider only the main factor model by assuming that all the interactions are negligible and hence our regression model contains only the main variables. In the stepdown analysis (see Chapter 3), the overall null hypothesis is not rejected if and only if the corresponding hypotheses at all the stages are not rejected. We will therefore reject the overall null hypothesis if it is rejected at either the first stage (that is, in the model for ACTIVT) or the second stage (that is, in the model for SELECTVT, with ACTIVT as a covariate) or at both the stages. As a result, an occurrence of a small p value at either of the two stages leads to the rejection of the overall null hypothesis. Of course, the individual levels of significance, α_1 and α_2 for the two hypotheses, should be appropriately decided. See Section 3.5.2.

The corresponding two models are fitted by using the last three lines of Program 4.6. As shown in Output 4.6, in the first model (for the variable ACTIVT), the effects E and G have small p values and hence are statistically significant. In the model for SELECTVT adjusted for the covariate ACTIVT, we observe that A, C, E, and F are statistically significant at a 5% level of significance. Consequently, we conclude that all the effects except B, D, and H are significant.

In the above example, we have seen how univariate analysis of covariance can be used in the analysis to adjust one response variable for the other and to test a multivariate hypothesis using the stepdown approach. However, keep in mind that situations exist in which genuine covariates are available as the independent variables. In such situations any comparison of various variables needs to be done after adjusting for the effects of these covariates. We discuss this issue in the next section.

4.7   Analysis of Covariance

When we want to compare various treatments, but the responses are affected by not only the particular treatments but also by certain other variables termed covariates or concomitant variables, we need to modify the analysis to account for these covariates and eliminate their effects. In other words, to make a fair comparison of various treatments, the data on the response variables need to be made comparable by first adjusting for the covariates. These situations commonly occur in social, biological, medical, physical, and other sciences.

For analyzing these data we utilize the following model

$$\mathbf{Y} = \mathbf{XB} + \mathbf{Z\Gamma} + \mathcal{E}, \tag{4.7}$$

where the matrices $\mathbf{Y}$ and $\mathcal{E}$ are defined as before. The term $\mathbf{XB}$ represents the design part of the model with a rank of the n by $k + 1$ matrix $\mathbf{X}$ equal to r. The n by q matrix $\mathbf{Z}$ is the matrix of data on the covariates with $Rank(\mathbf{Z}) = q$, and the q by p matrix $\mathbf{\Gamma}$ is the matrix of unknown parameters representing the regression of $\mathbf{Y}$ on $\mathbf{Z}$. Hence the term $\mathbf{Z\Gamma}$ in the model in Equation 4.7 represents the covariate part of the model. First we want to test the significance of some or all covariates in $\mathbf{Z}$ by testing the corresponding rows of $\mathbf{\Gamma}$ to be zero. Second we want to test the linear hypotheses about $\mathbf{B}$, after adjusting for the effects of the variables $\mathbf{Z}$, to answer the usual questions discussed in the earlier sections. We rewrite Equation 4.7 in the standard linear model form as

$$\mathbf{Y} = (\mathbf{X} \ \mathbf{Z}) \begin{bmatrix} \mathbf{B} \\ \mathbf{\Gamma} \end{bmatrix} + \mathcal{E} = \mathbf{W\Phi} + \mathcal{E}.$$

Then using the usual least squares principle and assuming that the $Rank(\mathbf{Z}) < n - Rank(\mathbf{X})$, the least square solutions for $\mathbf{\Gamma}$ and $\mathbf{B}$ respectively are

$$\hat{\mathbf{\Gamma}} = (\mathbf{Z'QZ})^- \mathbf{Z'QY}$$

and

$$\hat{\mathbf{B}} = (\mathbf{X'X})^- \mathbf{X'}(\mathbf{Y} - \mathbf{Z}\hat{\mathbf{\Gamma}}),$$

where $\mathbf{Q} = \mathbf{I} - \mathbf{X}(\mathbf{X'X})^- \mathbf{X'}$.

Now for the first test $H_0^{(a)} : \mathbf{\Gamma} = \mathbf{0}$, that is, covariates have no effect on the response variables, we use the matrices

$$\mathbf{H} = \mathbf{Y'QZ}(\mathbf{Z'QZ})^- \mathbf{Z'QY} = \hat{\mathbf{\Gamma}}' \mathbf{Z'QY}$$

and

$$\mathbf{E} = \mathbf{Y'QY} - \hat{\mathbf{\Gamma}}' \mathbf{Z'QY}.$$

When $H_0^{(a)}$ is true, then assuming $n > q + r$, $\mathbf{H}$ and $\mathbf{E}$ are independently distributed as $W_p(q, \ \Sigma)$ and $W_p(n - q - r, \ \Sigma)$ respectively. Using these matrices the usual multivariate tests can be used to test $H_0^{(a)}$. Next, for the second test $H_0^{(b)} : \mathbf{LB} = \mathbf{0}$, the same $\mathbf{E}$ matrix is used and the matrix $\mathbf{H}$ is determined using the model

$$\mathbf{Y} = \mathbf{W\Phi} + \mathcal{E}.$$

Since $H_0^{(b)} : \mathbf{LB} = \mathbf{0}$ can be written as $\mathbf{L}_1\Phi = \mathbf{0}$ with $\mathbf{L}_1 = (\mathbf{L} : \mathbf{0})$, the $\mathbf{H}$ matrix for $H_0^{(b)}$ is same as that for $\mathbf{L}_1\Phi = \mathbf{0}$. We use PROC GLM to test these hypotheses, as is illustrated in the next example.

EXAMPLE 7   *Comparisons in the Presence of Covariates, A Flammability Study*   Consider a situation where the interest is in comparing the effects of various types of foams and fabrics used in carpets on carpet flammability. The experiment was designed to determine the most heat-resistant foam and fabric after determining if there were any significant differences between various types of foams and fabrics. The problem appears to fit in the multivariate two-way classification setup. Three types of foams, namely *A*, *B*, and *C*, and three types of fabric materials denoted by *X*, *Y*, and *Z* were used, leading to nine possible compositions for the carpets. Two specimens of equal size (by volume) were taken and separately subjected to flame under identical temperature, pressure, and space. The heat releases at 5, 10, and 15 minutes (HR5, HR10, and HR15) were observed in each experiment.

One important issue, however, needs to be addressed. Although the specimens are all supposedly of the same volume, the amount of heat release relates more to the weight of the specimens than to the volume. Due to different densities for various types of foams and fabrics, the equality of volumes does not necessarily imply the equality of weights of all these specimens. As a result, for a fair comparison, the values of heat releases need to be adjusted for the differing weights of the various specimens. The weight to response relationship does not depend on any other factor. Thus, all effects and contrast tests discussed below are performed at the overall average weight.

These fictitious data inspired by an actual experiment are presented as part of Program 4.7. A two-way classification model with interaction in the classification variables FOAM and FABRIC is fitted for the response variables, HR5, HR10, and HR15. The weight of the specimen (WT) is taken as the covariate.

```
/* Program 4.7 */

options ls=64 ps=45 nodate nonumber;
title1 'Output 4.7';
title2 'Analysis of Covariance';

data heat ;
input foam $ fabric $ hr5 hr10 hr15 wt;
lines ;
foam_a fabric_x 9.2 18.3 20.4 10.3
foam_a fabric_x 9.5 17.8 21.1 10.1
foam_a fabric_y 10.2 15.9 18.9 10.5
foam_a fabric_y 9.9 16.4  19.2  9.7
foam_a fabric_z 7.1 12.8 16.7 9.8
foam_a fabric_z 7.3 12.6 16.9 9.9
foam_b fabric_x 8.2 12.3 15.9 9.5
foam_b fabric_x 8.0 13.4 15.4 9.3
foam_b fabric_y 9.4 17.7 21.4 11.0
foam_b fabric_y 9.9 16.9 21.6 10.8
foam_b fabric_z 8.8 14.7 20.1 9.3
foam_b fabric_z 8.1 14.1 17.4 7.7
foam_c fabric_x 7.7 12.5 17.3 10.0
foam_c fabric_x 7.4 13.3 18.1 10.5
foam_c fabric_y 8.7 13.9 18.4 9.8
foam_c fabric_y 8.8 13.5 19.1 9.8
foam_c fabric_z 7.7 14.4 18.7 8.5
foam_c fabric_z 7.8 15.2 18.1 9.0
;
```

```
proc glm data = heat;
class foam fabric ;
model hr5 hr10 hr15=wt foam fabric foam*fabric/ss1 nouni;
contrast '(a,z) vs. (c,x)'
intercept 0 foam 1 0 -1 fabric -1 0 1
                            foam*fabric  0 0 1 0 0 0 -1 0 0 ;
contrast 'foam a vs b ' foam 1 -1 0 ;
manova h = foam fabric foam*fabric/ printe printh ;
run;

/*
proc glm data = heat ;
class foam fabric ;
model hr5 hr10 hr15=wt foam fabric foam*fabric/ss3 nouni;
lsmeans foam fabric foam*fabric ;
contrast 'Foam a vs b ' foam 1 -1 0 ;
contrast '(a,z) vs. (c,x)'
intercept 0 foam 1 0 -1 fabric -1 0 1
                    foam*fabric  0 0 1 0 0 0 -1 0 0 ;
manova h = foam fabric foam*fabric/ printe printh ;
run;
*/
```

Even though the design appears to be balanced in the variables FOAM and FABRIC, the balancedness is lost due to the presence of the covariate WT as it changes from specimen to specimen. The various types of SS&CP matrices are therefore not identical and a careful analysis of the data is needed.

Since the effects and the SS&CP matrices of the variable FOAM, FABRIC, and FOAM*FABRIC are all to be adjusted for WT, a sequential partitioning of the total SS&CP matrix is appropriate with the WT variable listed first in the corresponding MODEL statement. The partitioning results in all the subsequent SS&CP matrices adjusted at least for this covariate. As far as the other two variables are concerned, there does not appear to be any reason to prefer one over the other. If we want a Type I analysis, we should examine the output resulting from two possible orders in the MODEL statement, namely

```
model hr5 hr10 hr15 = wt foam fabric foam*fabric/ss1;
```

and

```
model hr5 hr10 hr15 = wt fabric foam foam*fabric/ss1;
```

hoping for consistency in the conclusions. We have, however, chosen to limit our output for the first of these statements.

An examination of Output 4.7 reveals that the interaction FOAM*FABRIC is highly significant under all of the four test criteria. In view of this, it makes sense to conduct various pairwise comparisons for the nine treatments to decide which treatments are similar and which are not. This unfortunately requires as many as 36 pairwise comparisons; in general it is not advisable to perform too many tests since in a large number of pairwise tests, some are likely to appear to be significant just by chance. Based on the least square cell means computed by the LSMEANS statement, it appears that the treatments (A, Z) and (C, X) are comparable with relatively low values for heat release at various time points. The output from the LSMEANS statement is not shown to save space. Suppose we want to see if these two preferred treatments are significantly different from each other. Such a comparison can be made using the CONTRAST statement. Note that the CONTRAST statement should always appear before a MANOVA statement.

Output 4.7

```
                            Output 4.7
                      Analysis of Covariance


              Manova Test Criteria and F Approximations for
               the Hypothesis of no Overall FOAM*FABRIC Effect
         H = Type I SS&CP Matrix for FOAM*FABRIC   E = Error SS&CP Matrix


                     S=3     M=0    N=2

Statistic                Value        F     Num DF  Den DF  Pr > F


Wilks' Lambda           0.001717   13.602      12   16.166  0.0001
Pillai's Trace          2.071646    4.4631     12      24   0.0009
Hotelling-Lawley Trace  110.095    42.815      12      14   0.0001
Roy's Greatest Root     107.144   214.29        4       8   0.0001


      NOTE: F Statistic for Roy's Greatest Root is an upper bound.



              Manova Test Criteria and Exact F Statistics for
               the Hypothesis of no Overall (a,z) vs. (c,x) Effect
                 H = Contrast SS&CP Matrix for (a,z) vs. (c,x)
                         E = Error SS&CP Matrix


                     S=1     M=0.5    N=2

Statistic                Value        F     Num DF  Den DF  Pr > F


Wilks' Lambda           0.76977    0.5982       3       6   0.6392
Pillai's Trace          0.23023    0.5982       3       6   0.6392
Hotelling-Lawley Trace  0.299089   0.5982       3       6   0.6392
Roy's Greatest Root     0.299089   0.5982       3       6   0.6392
```

In order to identify an appropriate CONTRAST statement, it is helpful to write down the two-way classification model (the covariate term in the model can be ignored for this purpose) for the 1 by 3 response vector HR

$$HR_{ijk} = INTERCEPT + FOAM_i + FABRIC_j$$

$$+ (FOAM * FABRIC)_{ij} + ERROR_{ijk},$$

where $i = A, B, C$, $j = X, Y, Z$ and $k = 1, 2$.

Our interest is in the contrast $E(HR_{AZk} - HR_{CXk})$, where E indicates the expected value. Dropping the replication suffix $'k'$ for convenience, this can be written as

$$(INTERCEPT - INTERCEPT)$$

$$+ (FOAM_A - FOAM_C) + (FABRIC_Z - FABRIC_X)$$

$$+ ((FOAM_A * FABRIC_Z) - (FOAM_C * FABRIC_X))$$

$$= 0 \times INTERCEPT + (1\ 0\ -1) \begin{bmatrix} FOAM_A \\ FOAM_B \\ FOAM_C \end{bmatrix} + (-1\ 0\ 1) \begin{bmatrix} FABRIC_X \\ FABRIC_Y \\ FABRIC_Z \end{bmatrix}$$

$$+ (0\,0\,1\,:\,0\,0\,0\,:\,-1\,0\,0)\begin{bmatrix} FOAM_A * FABRIC_X \\ FOAM_A * FABRIC_Y \\ FOAM_A * FABRIC_Z \\ FOAM_B * FABRIC_X \\ FOAM_B * FABRIC_Y \\ FOAM_B * FABRIC_Z \\ FOAM_C * FABRIC_X \\ FOAM_C * FABRIC_Y \\ FOAM_C * FABRIC_Z \end{bmatrix}.$$

The above representation indicates that in order to get the contrast between treatments (A, Z) and (C, X),

- the coefficient for intercept is zero,
- the vector of coefficients for the vector of foams $(A, B, C)'$ is $(1\,0\,-1)'$,
- the vector of fabrics $(X, Y, Z)'$ is $(-1\,0\,1)'$,
- the vector of coefficients for the 9 by 1 vector of interactions

$$((A * X),\ (A * Y),\ (A * Z),\ (B * X),\ (B * Y),\ (B * Z),$$

$$(C * X),\ (C * Y),\ (C * Z))'$$

is obtained by respectively putting 1 and -1 at the places corresponding to $(A * Z)$ and $(C * X)$ and zeros elsewhere as follows.

All of this is specified in the CONTRAST statement as

```
contrast 'label' intercept 0 foam 1 0 -1 fabric -1 0 1
                 foam*fabric 0 0 1 0 0 0 -1 0 0;
```

The name (A, Z) versus (C, X) enclosed within single quotation marks (' ') in Program 4.7 is used as the label. A label is required in the CONTRAST statement.

For the desired contrast, it is possible to use any of the four multivariate tests. In the present case, since the rank of underlying $\mathbf{L}$ matrix is one (there is only a single contrast) all four tests are identical and exact. The corresponding observed value of the F(3, 6) test statistic is 0.5982 leading to a p value of 0.6392. Hence the null hypothesis of no overall difference between (A, Z) and (C, X) treatments *cannot* be rejected.

Although it is not quite relevant in the present context (because of highly significant interaction), if the interest were to compare the effect of Foam A with that of Foam B, the CONTRAST statement in simplified form could be written as

```
contrast 'label' foam 1 -1 0;
```

It is so, since in this case the coefficients of INTERCEPT, the vector of FABRIC, and the vector of FOAM*FABRIC all have zero coefficients and hence need not be explicitly specified in the CONTRAST statement.

Note that in the data presented here, the heat releases at various time points are the repeated measures on the same specimen. Further analysis may be possible using the repeated measures techniques. We address these techniques in Chapters 5 and 6.

4.8 Concluding Remarks

It must be remembered that all the analyses presented in this chapter assume the equality of the variance-covariance matrices of the rows of matrix $\mathcal{E}$. If in a multiway classification, this assumption of the homogeneity of the variance-covariance matrix is not satisfied, the analysis presented here may not be appropriate. We therefore strongly recommend that some appropriate tests for the homogeneity of the variance-covariance matrices be applied to the data prior to performing any multivariate analysis of variance. The equality of variance-covariance matrices can be tested using the DISCRIM procedure. See *SAS/STAT User's Guide, Version 6, Fourth Edition, Volume 1*, for details.

Analysis of Repeated Measures Data

5.1 Introduction   151
5.2 Single Population   152
5.3 *k* Populations   176
5.4 Factorial Designs   195
5.5 Analysis in the Presence of Covariates   207
5.6 The Growth Curve Models   219
5.7 Crossover Designs   236
5.8 Concluding Remarks   246

5.1 Introduction

In many experiments, several treatments are applied to the same experimental unit at different time points, or only a single treatment is applied to a subject but the measurements on the same characteristic or set of characteristics are taken on more than one occasion. The data collected under these or similar kinds of experimental setups are often referred to as *repeated measures data* and require extra care in their analyses.

A common reason for taking repeated measures on the same subject in many biological, medical, psychological, and sociological experiments is the fact that there is usually more variability in the measurements between the subjects than within a given subject. Thus, to control the variability, subjects are taken as the blocks. As a result, treatments applied to the same subject provide a more comparable set of measurements than several parallel groups subjected to different treatments. The analysis, however, is complicated by the fact that the measurements taken on the same subject will most likely be correlated. The obvious lack of ability to randomize in such situations often prevents one from using the standard block design related experimental design methodology. Therefore it is necessary to incorporate this special feature of the data in the modeling and analysis.

Within the domain of repeated measures, there are certain subtle differences in the analyses, depending on the design and the data collection scheme. For example, a situation in which three different drugs are all tried on a group of 30 patients at different time periods (and possibly in different sequences) is different in design and analysis from the one in which each of the three drugs is given to a different group of ten patients who are all observed over a certain period of time. These features are very important in choosing an appropriate model, in deciding the appropriate hypotheses to be tested, and in constructing the corresponding statistical tests.

Repeated measures designs also arise naturally in many other research or industrial contexts. For example, an auto maker may be interested in the number of problems various models of cars may have over time. In order to study this, he may decide to follow up on a specific group of cars in each model for a given length of time. Similarly a soft drink manufacturer may want to compare her drink with those from some of her competitors and to do so she may decide to conduct a double-blind taste test on a group of potential consumers. A psychologist may be interested in comparing the performance of students at various schools and may administer a battery of several tests to sample groups of students from these schools. The common aspect in all these problems is that the data are multivariate in nature: on each subject we have a vector of repeated measurements which are correlated within themselves but are independent for different subjects.

This chapter considers various experimental situations where repeated measures data may arise, and concentrates on the analysis of such data. Of course, the particular approach to these analyses depends on the particular data collection scheme and therefore, various sections have been arranged by the designs under which data are collected. While most of the chapter emphasizes analyses under various designs, at the end of this chapter we also provide certain methods to generate certain relatively complex crossover designs which are extensively used in repeated measures studies.

5.2   Single Population

5.2.1   Profile Analysis

EXAMPLE 1   ***Profile of Memory Data***   Srivastava and Carter (1983, p. 201) presented an example where a group of ten subjects was given a memory test three times. The purpose of the study was to test if there were any differences in the test scores for the three trials. In other words, if $\mu' = (\mu_1, \mu_2, \mu_3)$ is the vector of true mean scores at three occasions, then we wish to test $H_0 : \mu_1 = \mu_2 = \mu_3$. A graphical representation of the elements of μ (i.e., graph of μ_i versus i) is called a *profile of the vector* μ. Using this terminology, our null hypothesis represents the hypothesis of a horizontal profile. The sample profile, that is, a profile for the sample mean vector $\bar{\mathbf{Y}}$ is given in Output 5.1 generated by Program 5.1.

```
/* Program  5.1 */

options ls = 64 ps=45 nodate nonumber;
data memory ;
input y1 y2 y3 ;
lines ;
19 18 18
15 14 14
13 14 15
12 11 12
16 15 15
17 18 19
12 11 11
13 15 12
19 20 20
18 18 17
;
/* Source: Srivastava and Carter (1983, p. 201). */
filename gsasfile "prog51.graph";
goptions gaccess=gsasfile dev=pslmono;
goptions horigin=1in vorigin=2in;
goptions hsize=5in vsize=7in;
title1 h=1.5 'Profile for Memory Data';
title2 j=l 'Output 5.1' ;
proc summary data=memory;
var y1 y2 y3;
output out=new mean=my1-my3;
data plot;
```

```
set new;
array my{3} my1-my3;
do test =1 to 3;
Response=my(test);
output;
end;
drop my1-my3;
proc gplot data = plot;
plot response*test /vaxis=axis1 haxis=axis2 vminor=3
legend=legend1 ;
axis1  order =(14 to 16) label =(a=90 h=1.2 'Response');
axis2 offset=(2) label=(h=1.2 'Test');
symbol1 v=+ i = join;
legend1 across = 3;
run;
```

Output 5.1

Profile for Memory Data

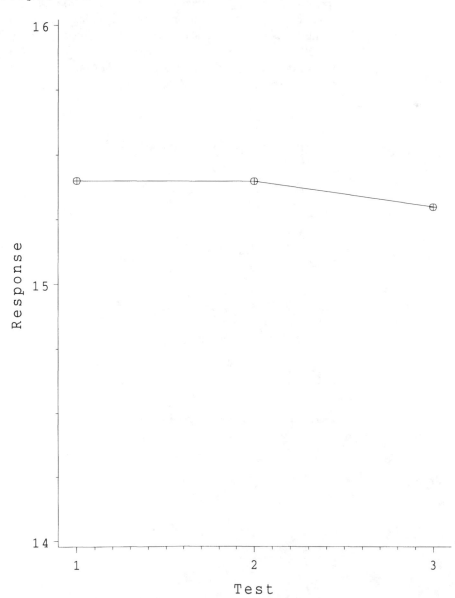

The data can be thought of as collected under a randomized complete block design in which subjects are the blocks and the three trials are the treatments. Since the group of subjects is a random sample, the block effect is assumed to be random. If we can assume equal correlation between the three trials then the above data can be analyzed as the univariate randomized complete block design. This, however, may be a questionable assumption and its validity would need to be examined before any such analysis. We will come back to this analysis later. If no such assumption is made, it may be more appropriate to assume a general correlation structure for the three trials and analyze the data using multivariate techniques. The null hypothesis under consideration can be written as $\mathbf{C}\mu = \mathbf{0}$, where

$$\mathbf{C} = \begin{bmatrix} 1 & 0 & -1 \\ 0 & 1 & -1 \end{bmatrix}.$$

Since for the underlying multivariate linear model given in Equation 3.1, viz., $\mathbf{Y} = \mathbf{XB} + \mathcal{E}$, $\mathbf{B} = (\mu_1, \mu_2, \mu_3) = \mu'$, the above hypothesis can be written as $H_0 : \mathbf{BM} = \mathbf{0}$ with $\mathbf{M} = \mathbf{C}'$ and can be tested using the multivariate approach presented in Chapters 3 and 4. Note that the multivariate linear model has only intercepts and no independent variables on the right-hand side. These kinds of models have already been examined in the previous chapters. To analyze these data, we use the first PROC GLM statement given in Program 5.2. Selected parts of the corresponding output appear in Output 5.2. Note that the null hypothesis stated above will not be rejected (p values for all four tests are equal to 0.9639). In this case all four multivariate tests are exact and equivalent. This analysis can also be done using a REPEATED statement (to be discussed later) as shown in the latter part of Program 5.2.

```
/* Program 5.2 */

options ls = 64 ps=45 nodate nonumber;
title1 'Output 5.2';
title2  'Analysis of Memory data';
data memory ;
input y1 y2 y3 @@;
lines ;
19 18 18 15 14 14 13 14 15
12 11 12 16 15 15 17 18 19
12 11 11 13 15 12 19 20 20 18 18 17
;
proc glm data = memory ;
model y1 y2 y3 = /nouni ;
manova h=intercept m=(1 0 -1,
     0 1 -1)/printe printh;
run;
proc glm data = memory ;
model y1 y2 y3 = /nouni ;
repeated test 3 profile/printe printm; run;
```

Output 5.2

```
                    Output 5.2
               Analysis of Memory data


             General Linear Models Procedure
             Multivariate Analysis of Variance


          M Matrix Describing Transformed Variables


                    Y1              Y2              Y3

       MVAR1         1               0              -1
       MVAR2         0               1              -1
```

E = Error SS&CP Matrix

	MVAR1	MVAR2
MVAR1	14.9	7.9
MVAR2	7.9	12.9

Manova Test Criteria and Exact F Statistics for
the Hypothesis of no Overall INTERCEPT Effect
on the variables defined by the M Matrix Transformation
H = Type III SS&CP Matrix for INTERCEPT   E = Error SS&CP Matrix

S=1   M=0   N=3

Statistic	Value	F	Num DF	Den DF	Pr > F
Wilks' Lambda	0.99084	0.037	2	8	0.9639
Pillai's Trace	0.00916	0.037	2	8	0.9639
Hotelling-Lawley Trace	0.009245	0.037	2	8	0.9639
Roy's Greatest Root	0.009245	0.037	2	8	0.9639

General Linear Models Procedure
Repeated Measures Analysis of Variance
Repeated Measures Level Information

Dependent Variable	Y1	Y2	Y3
Level of TEST	1	2	3

TEST.N represents the nth successive difference in TEST

M Matrix Describing Transformed Variables

	Y1	Y2	Y3
TEST.1	1.000000000	-1.000000000	0.000000000
TEST.2	0.000000000	1.000000000	-1.000000000

E = Error SS&CP Matrix

TEST.N represents the nth successive difference in TEST

	TEST.1	TEST.2
TEST.1	12.00000000	-5.00000000
TEST.2	-5.00000000	12.90000000

Test for Sphericity: Mauchly's Criterion = 0.8374058
Chisquare Approximation = 1.4195715 with 2 df
Prob > Chisquare = 0.4917

```
                     Applied to Orthogonal Components:
          Test for Sphericity: Mauchly's Criterion = 0.9833085
              Chisquare Approximation = 0.134659 with 2 df
                        Prob > Chisquare = 0.9349


     Univariate Tests of Hypotheses for Within Subject Effects

  Source: TEST
                                                  Adj  Pr > F
       DF Type III SS Mean Square  F Value  Pr > F  G - G   H - F
        2   0.0666667   0.0333333     0.05  0.9559  0.9540  0.9559

  Source: Error(TEST)

       DF Type III SS Mean Square
       18  13.2666667   0.7370370

                 Greenhouse-Geisser Epsilon = 0.9836
                     Huynh-Feldt Epsilon = 1.2564
```

5.2.2   Testing for Covariance Structures

In the preceding section, we discussed the multivariate approach to comparison of components of the mean vector. An alternative, but not necessarily universally better, approach can be taken by interpreting each subject as a block and the memory test periods as the plots within blocks. This interpretation results in a complete block design structure for the experimental layout in which block (subject) effects are random. Remember, however, that since the plots are memory tests and hence of a temporal nature, they cannot be randomized within blocks. In addition, they may exhibit a certain dependence between the observations within a subject. If there were no such dependence, then assuming no SUBJECT*TEST interaction, a comparison of the memory tests could be made using the usual ANOVA F test. Unfortunately, this ideal situation rarely occurs in practice. There are, however, certain covariance structures modeling the dependence which would still admit the valid F tests for some comparisons. These possible covariance structures should therefore be formally tested for, before assuming them and applying the usual ANOVA F tests. We describe statistical tests for some of the covariance structures here.

A test for sphericity   Let $\mathbf{y}_i$, $i = 1, \ldots, n$ be a random sample of size n from $N_p(\boldsymbol{\mu}, \boldsymbol{\Sigma})$. To test the null hypothesis $H_0 : \boldsymbol{\Sigma} = \sigma^2 \mathbf{I}$, σ^2 unknown, Mauchly (1940) derived the likelihood ratio test statistic $L^{n/2}$, where

$$L = |\mathbf{S}|/(p^{-1}tr\,\mathbf{S})^p, \tag{5.1}$$

and $\mathbf{S}$ is the sample variance-covariance matrix defined by

$$\mathbf{S} = \frac{1}{n-1}\sum_{i=1}^{n}(\mathbf{y}_i - \bar{\mathbf{y}})(\mathbf{y}_i - \bar{\mathbf{y}})', \quad \bar{\mathbf{y}} = \frac{1}{n}\sum_{i=1}^{n}\mathbf{y}_i.$$

For large samples, $-\{(n-1) - (2p^2 + p + 2)/6p\}ln\,L$ has an approximate chi-square distribution with degrees of freedom $\frac{1}{2}p(p+1) - 1$.

Using the REPEATED statement in PROC GLM, we can perform the tests for sphericity on certain sets of contrasts but not on the original data. The use of the REPEATED statement included in Program 5.2 will be illustrated later in Section 5.2.3.

A test for compound symmetry   Given a sample of size n from $N_p(\mu, \Sigma)$, consider the problem of testing the null hypothesis $H_0 : \Sigma = \sigma^2 \mathbf{V}$, where

$$\mathbf{V} = \begin{bmatrix} 1 & \rho & . & . & . & \rho \\ \rho & 1 & . & . & . & \rho \\ . & . & . & . & . & . \\ \rho & \rho & . & . & . & 1 \end{bmatrix}, \quad -(p-1)^{-1} \leq \rho \leq 1$$

and σ^2 unknown. The correlation structure commonly known as compound symmetry or the intraclass correlation structure essentially assumes that the correlations between measurements within a subject do not depend on the ordering of the measurements. The likelihood ratio test statistic for the above null hypothesis is $L^{n/2}$, where

$$L = |\mathbf{S}|/[(s^2)^p (1-r)^{p-1}\{1 + (p-1)r\}],$$

$\mathbf{S} = (s_{ij})$ as defined before, $s^2 = p^{-1} \sum_{i=1}^{p} s_{ii}$, and

$$r = 2[p(p-1)s^2]^{-1} \sum_{i=1}^{p} \sum_{j=i+1}^{p} s_{ij}.$$

For large samples,

$$Q = -[(n-1) - \{p(p+1)^2(2p-3)\}/\{6(p-1)(p^2+p-4)\}]ln\, L \qquad (5.2)$$

follows approximate chi-square distribution with $\frac{1}{2}p(p+1) - 2$ degrees of freedom.

EXAMPLE 1   *Testing Compound Symmetry, Memory Data (continued)*   To test if the variance-covariance matrix of the scores on the three memory tests possesses compound symmetry, we use the likelihood ratio test described in Equation 5.2. The SAS/IML code to perform the calculations appears in Program 5.3. From Output 5.3,

$$s^2 = 8.9963, \; r = 0.9181$$

$$|\mathbf{S}| = 12.5091, \; L = 0.9025, \; \text{and } Q = 0.9873.$$

Corresponding to the observed values 0.9873 of $Q \sim \chi^2_{\frac{1}{2}p(p+1)-2}$ (or χ^2_4) the p value is 0.9117. Therefore we do not reject the null hypothesis of compound symmetry.

```
/* Program  5.3 */

options ls = 64 ps=45 nodate nonumber;
proc iml;
y={
19 18 18,
15 14 14,
13 14 15,
12 11 12,
16 15 15,
17 18 19,
12 11 11,
13 15 12,
19 20 20,
18 18 17};
Title1 'Output 5.3' ;
p=ncol(y);
n=nrow(y);
s=y`*(I(n)-(1/n)*j(n,n))*y;
svar=s/(n-1);
/*
```

```
*Test for sphericity;
const1=-((n-1)-(2*p*p+p+2)/(6*p));
wlam=( det(svar)/((trace(svar)/p)**p) );
llam=wlam**(2/n);
print llam;
test=const1*log(wlam);
print const1 wlam test;
*/
* Test of Compound Symmetry;
detment=det(svar);
square=sum (diag(svar));
sumall=sum(svar);
ssquare=square/p;
correl=(sumall-square)/(p*(p-1)*ssquare);
lrlam=detment /( (ssquare**p)*((1-correl)**(p-1))
*(1+(p-1)*correl) );
correct=(n-1)-( p*(p-1)**2*(2*p-3))/(6*(p-1)*(p*p+p-4));
lrstat=-correct*log(lrlam);
df=p*(p+1)/2-2;
pvalue=probchi(lrstat,df);
pvalue=1-pvalue;
print 'Test of Compound Symmetry for Memory Data' ;
print ssquare correl detment lrlam;
print lrstat pvalue;
run;
```

Output 5.3

```
                         Output 5.3

           Test of Compound Symmetry for Memory Data


             SSQUARE   CORREL   DETMENT    LRLAM
            8.9962963 0.9180733 12.509053 0.9025129


                     LRSTAT    PVALUE
                    0.8846863 0.9267475
```

The usual ANOVA F tests which traditionally assume sphericity for the variance-covariance matrix remain valid under compound symmetry as well. It is so because the compound symmetry of the original variables leads to the sphericity of the orthonormal contrasts on which the F test is based. Thus, an acceptance of the null hypothesis of compound symmetry may be helpful to researchers applying or wishing to apply the analysis of variance techniques in their data analysis.

A test for circular covariance   A useful covariance structure which can naturally occur when repeated measures are taken with spatial instead of time considerations is a circular pattern for the variance-covariance matrix. For example, for $p = 5$ and 6, a circular variance-covariance matrix is

$$
\begin{bmatrix}
\sigma_0 & \sigma_1 & \sigma_2 & \sigma_2 & \sigma_1 \\
\sigma_1 & \sigma_0 & \sigma_1 & \sigma_2 & \sigma_2 \\
\sigma_2 & \sigma_1 & \sigma_0 & \sigma_1 & \sigma_2 \\
\sigma_2 & \sigma_2 & \sigma_1 & \sigma_0 & \sigma_1 \\
\sigma_1 & \sigma_2 & \sigma_2 & \sigma_1 & \sigma_0
\end{bmatrix},
\begin{bmatrix}
\sigma_0 & \sigma_1 & \sigma_2 & \sigma_3 & \sigma_2 & \sigma_1 \\
\sigma_1 & \sigma_0 & \sigma_1 & \sigma_2 & \sigma_3 & \sigma_2 \\
\sigma_2 & \sigma_1 & \sigma_0 & \sigma_1 & \sigma_2 & \sigma_3 \\
\sigma_3 & \sigma_2 & \sigma_1 & \sigma_0 & \sigma_1 & \sigma_2 \\
\sigma_2 & \sigma_3 & \sigma_2 & \sigma_1 & \sigma_0 & \sigma_1 \\
\sigma_1 & \sigma_2 & \sigma_3 & \sigma_2 & \sigma_1 & \sigma_0
\end{bmatrix}
$$

Examples of the situations in which it is natural to assume such a covariance structure are spatial repeated measurements taken on the petals of a flower or on the tentacles of a starfish, or along the perimeter of a circle at equally spaced angles.

Although the assumption of circular covariance does not necessarily simplify the analysis of data, a test for it is important in its own right. A likelihood ratio test statistic for the null hypothesis of circular covariance structure given by Olkin and Press (1969) is

$$L^{2/n} = 2^{2(p-m-1)} |\mathbf{U}| / \prod_{j=1}^{p} v_j, \tag{5.3}$$

where m is such that $p = 2m$ or $p = 2m + 1$, $\mathbf{U} = (u_{ij}) = \mathbf{\Gamma}'(n-1)\mathbf{S}\mathbf{\Gamma}$, $\mathbf{\Gamma} = (\gamma_{ij})$ is a p by p orthogonal (orthonormal) matrix with

$$\gamma_{ij} = p^{-1/2}[\cos\ 2\pi\ p^{-1}\ (i-1)(j-1) + \sin\ 2\pi\ p^{-1}\ (i-1)(j-1)], i, j = 1, \dots, p.$$

Further,

$$v_1 = u_{11},\ \ v_2 = u_{22} + u_{pp},\ \ v_3 = u_{33} + u_{p-1,p-1},$$

$$\dots, v_m = u_{mm} + u_{m+2,m+2},\ \ v_{m+1} = u_{m+1,m+1},$$

for $p = 2m$ and

$$v_1 = u_{11},\ \ v_2 = u_{22} + u_{pp},\ \ v_3 = u_{33} + u_{p-1,p-1},$$

$$\dots, v_{m+1} = u_{m+1,m+1} + u_{m+2,m+2},\ \ \text{for } p = 2m + 1.$$

Also, $v_j = v_{p-j+2}$ for $j = 2, \dots, p$.

Under the null hypothesis, for L defined in Equation 5.3

$$-2\left(1 - \frac{2b}{n}\right)\ln\ L \tag{5.4}$$

follows an approximate chi-square distribution with f degrees of freedom, where for $p = 2m$, $f = (p^2 - 2)/2$ and $b = (2p^3 + 9p^2 - 2p - 18)/[12(p^2 - 2)]$; for $p = 2m + 1$, $f = (p^2 - 1)/2$ and for $b = (2p + 9)/12$. Hence the null hypothesis can be tested using the appropriate χ^2_f cutoff point.

The SAS/IML code to test for circular symmetry is given in Program 5.4 along with an illustrative analysis of cork data presented in Output 5.4.

EXAMPLE 2   ***Testing Circular Covariance, Cork Boring Data***   For the data set of Rao (1948), extensively discussed in Chapters 1, 2, and 3 (cork boring in four directions: North, East, South, and West), we expect that the amount of correlation of a measurement, say taken at the north facing of the tree, with measurements at its immediate neighboring facings, east and west, may be the same while measurement on the opposite direction, south, may be different from these two. This assumption would lead to a circular structure for the 4 by 4 variance-covariance matrix for the cork measurements. Therefore, we want to statistically test the validity of this assumption.

```
/* Program  5.4 */

options ls = 64 ps=45 nodate nonumber;
title1 'Output 5.4';
  /* This program computes the LRT for testing circular
covariance structure vs. general cov. Ref. Olkin & Press,
AMS, 1969,40, 1358-1373.*/
data cork;
infile 'cork.dat';
```

```
                  input y1 y2 y3 y4;
                  run;
                  proc iml;
                  use cork;
                  read all var {y1 y2 y3 y4} into y;
                  p=ncol(y);
                  n=nrow(y);
                  s=y`*(I(n)-(1/n)*j(n,n))*y;
                  svar=s/(n-1);
                  pi=3.1415927;
                  gam=I(p);
                  do k=1 to p;
                   do l=1 to p;
                   gam(|k,l|)=(p**(-0.5))*(cos(2*pi*(k-1)*(l-1)/p)+
                   sin(2*pi*(k-1)*(l-1)/p));
                   end;
                  end;
                  m=floor(p/2);
                  if(m = p/2) then b = (2*p**3+9*p**2-2*p-18)/
                  (12*(p**2-2)) ;
                  else b = (2*p+9)/12 ;
                  if(m = p/2) then f = (p**2-2)/2 ;
                  else f = (p**2-1)/2 ;
                   v=gam*s*gam`;
                  x=j(p);
                  nu=x(|,1|);
                   do k=1 to p;
                   nu(|k|)=v(|k,k|);
                   end;
                   snu=x(|,1|);
                   snu(|1|)=nu(|1|);
                  if (m=p/2) then
                   snu(|m+1|)=nu(|m+1|);
                   else
                   snu(|m+1|)=nu(|m+1|)+nu(|m+2|);
                    do k=2 to m;
                    kp=p+2-k;
                    snu(|k|)=nu(|k|)+nu(|kp|);
                    end;
                    do k=m+2 to p;
                    snu(|k|)=snu(|p-k+2|);
                    end;
                   pdt=1.0;
                    do k=1 to p;
                    pdt=pdt*snu(|k|);
                    end;
                    wlamda=(2**(2*(p-m-1)))*det(v)/pdt;
                    wlamda=wlamda**(n/2);
                    lrstat=-2*(1-2*b/n)*log(wlamda);
                  pvalue=probchi(lrstat,f);
                  pvalue=1-pvalue;
                  print 'Test of Circular Structure for Cork Data' ;
                  print wlamda lrstat pvalue;
                  run;
```

Output 5.4

```
                        Output 5.4

          Test of Circular Structure for Cork Data


              WLAMDA    LRSTAT    PVALUE
            0.0000273  18.820383  0.008769
```

To perform the likelihood ratio test of Olkin and Press, we have, $L = 0.0000273$. Correspondingly, the observed value of the approximate chi-square statistic (Equation 5.4) with $df = 7$ is 18.8204. This leads to a p value of 0.0088 and hence a rejection of the hypothesis of circular covariance structure.

The MIXED procedure described in Chapter 6 uses, among many others, a Toeplitz structure in modeling and data analysis. The circular structure is a special case of the Toeplitz structure. However the MIXED procedure only implements the most general form of the Toeplitz structure. Fortunately the circular covariance structure can still be implemented since it can be expressed as a *linear structure* which can be easily specified in the MIXED procedure. Details will be discussed in Chapter 6.

Covariance structures guaranteeing the sphericity of orthogonal contrasts   Huynh and Feldt (1970) and Rouanet and Lépine (1970) derived a set of necessary and sufficient conditions on the covariance structures under which the usual F tests formed by the ratios of mean squares still follow the exact F distributions. As a result, in the repeated measures context despite the presence of correlation among the repeated measures on the same subject, the usual univariate ANOVA tests can still be used so long as these correlations can be assumed to have a particular structure. Specifically, the vector of all orthogonal contrasts would satisfy the sphericity requirement if for the original variance-covariance matrix $\Sigma = (\sigma_{ij})$, $\sigma_{ii} + \sigma_{jj} - 2\sigma_{ij}$ is a constant for all i, j. As in Huynh and Feldt (1970), we call this structure of a variance-covariance matrix a Type H structure and the condition of having a constant value for $\sigma_{ii} + \sigma_{jj} - 2\sigma_{ij}$ a Type H condition. This condition is automatically satisfied by the covariance matrices with compound symmetry, and hence the class of Type H structure covariance matrices form a slightly more general class. Note, however, that the circular covariance structure and many other important covariance structures including the autoregressive structure *do not* belong to this class.

To test if the variance-covariance matrix can be assumed to have this structure, Huynh and Feldt (1970) suggest applying the sphericity test on the variance-covariance matrix of the set of $(p - 1)$ orthogonal contrasts. If $\mathbf{S}$ is the sample variance-covariance matrix computed from the original data and $\mathbf{C}$ is the matrix defining $(p - 1)$ orthogonal contrasts, then the likelihood ratio test is given by $L^{n/2}$, where

$$L = |\mathbf{CSC}'|/\{(p - 1)^{-1} tr(\mathbf{CSC}')\}^{p-1}. \tag{5.5}$$

The value of L in Equation 5.5 does not depend on the choice of the suborthogonal matrix $\mathbf{C}$, and hence any of the several choices (such as POLYNOMIAL, as specified in the REPEATED statement of the GLM procedure) would serve the purpose. Under the null hypothesis of sphericity of the orthogonal contrasts,

$$-\{(n - 1) - (2(p - 1)^2 + (p - 1) + 2)/(6(p - 1))\}ln\, L$$

approximately follows a chi-square distribution with $p(p - 1)/2 - 1\ df$. When we specify the PRINTE option on the REPEATED statement in PROC GLM, it produces this test and titles it

```
Applied to Orthogonal Components: Test for Sphericity: Mauchly's
Criterion=
```

For the memory data, Mauchly's test on the orthogonal contrasts (see part of Output 5.2) strongly favors the null hypothesis (p value $= 0.9349$).

This suggests, as will be seen later, that the analysis of these data using the univariate ANOVA may be deemed valid. Since certain orthogonal contrasts have meaningful and simple interpretations, accepting the independence of these contrasts and constant variances is very desirable.

5.2.3   Univariate Analysis

Repeated measures data can be analyzed using the univariate techniques applicable for split plot designs (Milliken and Johnson, 1989) under certain assumptions on the covariance structures of within-subject measurements. These requirements on covariance structure are derived by the necessary and sufficient conditions for the usual F test to be valid and hence are rather artificial. It is hard to imagine repeated measures situations where such correlations would naturally occur due to practical considerations. Hence the validity of these assumptions should always be statistically tested before any univariate analysis based on these assumptions is used to draw conclusions about the significance of the effects. If the hypothesis of the validity of these assumptions is rejected, there may still be a way to draw meaningful conclusions from the univariate analysis after making certain adjustments to the degrees of freedom of the F test.

For example, consider the memory data discussed earlier in this chapter. In this experiment, each subject was given three memory tests. If we interpret each subject as a random block (whole plot) containing three treatments (memory tests), then the design resembles a complete block design, except that, within each block, the randomization of the treatments is not possible. In fact, from the very design of this experiment, such a randomization has no meaning. Since the scores on the three memory tests of a given subject are correlated, the design can be treated as a split plot design and can probably be analyzed using the standard univariate analysis of variance techniques for this particular design. However, there is still a subtle difference in that the split plot experiments assume that plots (tests) within a given block (subject) are equicorrelated with each other. This assumption we would seriously doubt for data collected over time, as in the present example. Hence it is necessary to formally test if such an assumption, usually referred to as the *assumption of compound symmetry*, can be made for the data at hand. The approximate chi square test based on the likelihood ratio for compound symmetry has been given in the previous pages, and we can use it for this purpose. If the data pass this test, univariate analysis of these data using the aforementioned techniques may be applicable, for most practical purposes.

Huynh and Feldt (1970) and Rouanet and Lépine (1970) give a weaker requirement for the validity of the ANOVA F test in the split plot design. This requirement, already described in Section 5.2.2, amounts to a condition of sphericity of the variance-covariance matrix of all orthogonal contrasts of repeated measures and, hence, can be tested using Mauchly's sphericity test.

When the sphericity (or compound symmetry) assumption is false, it is still possible to use the ANOVA F test by modifying its degrees of freedom. Box (1954) gave a measure defined as

$$\epsilon = \frac{[tr\,(\mathbf{C'}\mathbf{\Sigma}\mathbf{C})]^2}{(p-1)tr\,(\mathbf{C'}\mathbf{\Sigma}\mathbf{C})^2} = \frac{[\sum_j \theta_j]^2}{(p-1)\sum_j \theta_j^2},$$

where the θ_j, $j = 1, \ldots, p-1$ are the $p-1$ eigenvalues of $\mathbf{C'}\mathbf{\Sigma}\mathbf{C}$. From the above formula, it is evident that $(p-1)^{-1} \leq \epsilon \leq 1$, when p is the number of repeated measures. When the variance-covariance matrix is spherical, all the eigenvalues $\theta_1, \ldots, \theta_{p-1}$ are equal and hence $\epsilon = 1$. The smaller values of ϵ indicate a relatively high degree of departure from sphericity.

Box also suggests that in the case of departure from sphericity, the conventional F test with degrees of freedom $(p-1)$ and $(p-1)(n-1)$ should be replaced by an approximate F test with degrees of freedom $\epsilon(p-1)$ and $\epsilon(p-1)(n-1)$. In practice, since ϵ is unknown, its estimate

$$\hat{\epsilon}_{GG} = \frac{[tr\,(\mathbf{CSC'})]^2}{(p-1)tr\,(\mathbf{C'SC'})^2}$$

can be substituted to obtain the approximate degrees of freedom. This is known as the Greenhouse-Geisser procedure. See Greenhouse and Geisser (1959). As can be easily seen, the estimate $\hat{\epsilon}_{GG}$ of ϵ is obtained by replacing the variances and covariances in the formula of ϵ given above by the corresponding elements of the sample variance-covariance matrix.

Huynh and Feldt (1976) have provided the following estimate of ϵ

$$\hat{\epsilon}_{HF} = \frac{n(p-1)\hat{\epsilon}_{GG} - 2}{(p-1)[n-1-(p-1)\hat{\epsilon}_{GG}]}.$$

Note that the value of this estimate may exceed one. In this case, its value is taken to be one.

EXAMPLE 1   *Testing Type H Structure, Memory Data (continued)*   For the memory data discussed above, $n = 10$ and $p = 3$. Based on the p value in Output 5.3, we have not rejected the null hypothesis of compound symmetry. This implies that the Type H conditions hold, since compound symmetry is a more restrictive condition. Hence the univariate split plot approach is justified. However, for the sake of illustration, let us consider these data again and formally test for the Type H structure. This task can be performed using the REPEATED statement.

Since the group of 10 subjects forms a random sample, SUBJECT is a random effect. The variable TEST, representing the variable with three memory tests as the treatments, is fixed and the interaction SUBJECT*TEST is random. The conventional F test statistic for the null hypothesis of no difference between the memory test is given by

$$F = \frac{MS_{TEST}}{MS_{SUBJECT*TEST}}$$

which under H_0 follows an F distribution with $(p-1) = 2$ and $(n-1)(p-1) = 18$ degrees of freedom. This test is automatically performed when a REPEATED statement is used for the variable TEST. The SAS code for this analysis is given in the latter part of Program 5.2. The resulting output is shown as part of the Output 5.2.

The statement

```
repeated test 3 profile/printe printm;
```

performs a repeated measures analysis, with the variable TEST as the within-subject effect.

Both the univariate as well as multivariate analyses are performed. If desired, the multivariate output can be suppressed by using the NOM option. The matrix **E** is printed if we use the PRINTE option and, if we use PRINTM, the matrix defining the contrasts is printed. The contrasts do not necessarily have to be orthogonal. The default matrix of contrasts is the *contrast* matrix (which corresponds to the option CONTRAST) comparing the last treatment with all the previous ones and is given by

$$\begin{bmatrix} 1 & 0 & -1 \\ 0 & 1 & -1 \end{bmatrix}.$$

However, the contrast matrix specified here corresponds to the comparison of PROFILE and is given by

$$\begin{bmatrix} 1 & -1 & 0 \\ 0 & 1 & -1 \end{bmatrix}$$

which is referred to as the M matrix in Output 5.2. The **E** matrix of the two transformed variables using the above profile matrix is

$$\begin{bmatrix} 12.0 & -5.0 \\ -5.0 & 12.9 \end{bmatrix}.$$

There are many other transformations such as HELMERT or POLYNOMIAL for the **M** matrix. However, in most situations, the main interest may be to compare the responses at various time points. This can be accomplished by choosing the default CONTRAST or by choosing the option PROFILE as done here.

There are two sets of transformed variables on which Mauchly's sphericity test is performed. The first of these corresponds to the sphericity test on the variables obtained by using the transformation matrix **M** specified in the REPEATED statement. As is the case in our example, this may correspond to a nonorthogonal transformation. In the present example, for the profiles Y1-Y2 and Y2-Y3, Mauchly's criterion is 0.8374 with a corresponding observed value of an approximate χ_2^2 as 1.4196. The corresponding p value is 0.4917.

Recall that in general the validity of the F test is subject to the Type H condition which is equivalent to the sphericity of orthogonal contrasts. The above test was conducted on a set of two nonorthogonal contrasts and is not applicable for this purpose.

The test applied to the orthogonal contrasts results in a value of 0.9833 for Mauchly's criterion with an observed approximate $\chi_2^2 = 0.1347$ and a corresponding p value = 0.9349. This appears to support the null hypothesis of Type H structure for the variance-covariance matrix. The choice of which orthogonal matrix to use is immaterial, as the test does not depend on any such choice. Thus, it is not necessary to know the specific choices of orthogonal contrasts, nor are they printed as part of the SAS output.

Note that if a POLYNOMIAL transformation is selected in the repeated statement, for example as in the following

```
repeated test 3 polynomial/printe printm;
```

the two Mauchly's tests that are printed will be identical. This is because the polynomial contrast is an *orthogonal transformation*. Although the rows of HELMERT transformation (as implemented in SAS) are orthogonal to each other, it is not an orthogonal transformation since they have not been scaled to have the same norm.

The univariate F test is performed for the variable TEST. Since the design is balanced, all three types of sums of squares are identical and hence the default choice of Type III sums of squares is used. Under the null hypothesis of no difference between the three memory tests, and with the understanding that the error term is $MS_{SUBJECT*TEST}$,

$$F = \frac{MS_{TEST}}{MS_{SUBJECT*TEST}} = \frac{MS_{TEST}}{MS_{ERROR(TEST)}} = \frac{0.0333}{0.7370} \approx 0.05,$$

is an observed value of $F(2, 18)$. The corresponding p value = 0.9559 is very high.

A few observations and checks need to be made before we decide not to reject H_0 with such a high p value. First of all, the sphericity test resulted in the acceptance of a hypothesized Type H structure. This is confirmed by the values of $\hat{\epsilon}_{GG}$ (= 0.9836) and $\hat{\epsilon}_{HF}$ (= 1.2564 and truncated to 1) both of which are close to 1. Thus, the adjustment in the degrees of freedom is not necessary, and the distribution of the F statistic indicated above can be safely assumed to be $F(2, 18)$. In view of the very high p value, we do not reject the null hypothesis of no treatment effect. However, comparable p values (0.9540 and 0.9559 respectively) would have been obtained had the degrees of freedom been adjusted using $\hat{\epsilon}_{GG}$ and $\hat{\epsilon}_{HF}$, resulting in the same conclusion.

5.2.4   Fitting the Polynomial Curve: Determination of Degree

When the measurements are repeated on the same subject or unit over several time points, we may want to model these responses as a function of time. The model may arise from some theoretical considerations or may be empirically chosen to be simple enough for inferential purposes but, at the same time, to provide sufficient flexibility in fitting the data reasonably well. As an empirical approximation, justified by the Taylor expansion of the function, a model can often be found by fitting polynomials to data.

Suppose n subjects or units receive a treatment and their responses are measured over p time points. The model can be written as

$$\mathbf{y}_j = \boldsymbol{\mu} + \boldsymbol{\epsilon}_j,$$

where $\mathbf{y}_j$ is the p by 1 vector of measurements on the jth subject and $\boldsymbol{\mu}$ is the vector of the true means for these measurements.

To fit an rth order polynomial, we take $\boldsymbol{\mu} = \mathbf{G}\boldsymbol{\beta}$, where $\mathbf{G}$ is the known p by $(r + 1)$ matrix

$$\mathbf{G} = \begin{bmatrix} 1 & t_1 & . & . & . & t_1^r \\ 1 & t_2 & . & . & . & t_2^r \\ . & . & . & . & . & . \\ 1 & t_p & . & . & . & t_p^r \end{bmatrix}$$

and $\boldsymbol{\beta}' = (\beta_0, \beta_1, \ldots, \beta_r)$ is the vector of unknown coefficients in the rth degree polynomial.

To test if the rth order model may indeed be sufficient, we test the hypothesis

$$H_0 : \boldsymbol{\mu} = \mathbf{G}\boldsymbol{\beta} \text{ vs. } H_1 : \boldsymbol{\mu} \neq \mathbf{G}\boldsymbol{\beta},$$

where $\mathbf{G}$ is as defined above. It is known that the rank of $\mathbf{G}$, $Rank(\mathbf{G}) = r + 1$. If $\mathbf{H}'$ is a p by $(p - (r + 1))$ matrix orthogonal to $\mathbf{G}$, with $Rank(\mathbf{H}) = p - (r + 1)$, then since $\mathbf{HG} = \mathbf{0}$, we must have under the null hypothesis $H_0 : \mathbf{H}\boldsymbol{\mu} = \mathbf{HG}\boldsymbol{\beta} = \mathbf{0}$, and hence the null hypothesis can be reduced to a linear hypothesis of the type $H_0 : \mathbf{L}\boldsymbol{\mu} = \mathbf{0}$. The choice of $\mathbf{H}$ is not unique but the resulting Wilks' Λ and other multivariate tests would, however, be invariant of the particular choice of $\mathbf{H}$. The test requires us to reject H_0 if

$$\frac{(n - p + r + 1)n}{(n - 1)(p - r - 1)} \bar{\mathbf{y}}' \mathbf{H}' (\mathbf{HSH}')^{-1} \mathbf{H}\bar{\mathbf{y}} > F_\alpha(p - r - 1, n - p + r + 1),$$

where $\bar{\mathbf{y}}$ is the sample mean vector and $\mathbf{S}$ is the sample variance-covariance matrix.

The matrix $\mathbf{H}$ can conveniently be chosen as a matrix corresponding to the orthogonal polynomials. This is especially helpful since such a matrix can be instantly created by using the function ORPOL in SAS/IML.

EXAMPLE 3   *Polynomial Fitting for Fish Data*   Consider the fish data discussed in Chapter 3. After fish were given various doses of copper (in mg/liter), the study measured the number of fish that were dead after 8, 14, 24, 36, and 48 hours in 25 tanks of 20 trout each. The objective of the study was to model the effects of copper dosage on fish mortality over time. The arcsine transformed data on the number of fish that were dead at various time points were used as the dependent variables. A natural question to ask is, for a fixed level of copper dose, what is the appropriate degree of polynomial in time which can be fitted to describe the death rate?

Suppose we are interested in fitting only the second-degree models. For a fixed dose, the second-degree model for the jth individual then would be

$$y_j(t) = \beta_0 + \beta_1 t + \beta_2 t^2 + \epsilon_j$$

while the biggest model that can be fit would be the fourth degree (since there are five time points), namely,

$$y_j(t) = \beta_0 + \beta_1 t + \beta_2 t^2 + \beta_3 t^3 + \beta_4 t^4 + \epsilon_j.$$

Thus, the null hypothesis $H_0 : \mu = G\beta$ can be expressed as

$$H_0 : \mu = \begin{bmatrix} 1 & 1 & 1 & 1 & 1 \\ 8 & 14 & 24 & 36 & 48 \\ 8^2 & 14^2 & 24^2 & 36^2 & 48^2 \end{bmatrix}' \begin{bmatrix} \beta_0 \\ \beta_1 \\ \beta_2 \end{bmatrix}.$$

Since G is completely specified by the time points (second column of G), we could alternatively choose the 5 by 3 matrix of second-degree orthogonal polynomials. More specifically, if we transform the data on variables $y_1, y_2, y_3, y_4,$ and y_5 to new variables $z_1, z_2, z_3, z_4,$ and z_5 through a 5 by 5 matrix of fourth-degree orthogonal polynomials, then we only need to test that the last two variables z_4 and z_5 which respectively represent the third- and fourth-degree effects have zero means.

In Program 5.5, we used the IML procedure to first obtain the new variables z_i, $i = 1, \ldots, 5$, which are coded as Z1, Z2, Z3, Z4, and Z5. For the vector of time points (8, 14, 24, 36, 48), the matrix of fourth-degree orthogonal polynomials denoted as **OPOLY** is constructed using the IML function ORPOL and is presented below.

$$\mathbf{OPOLY} = \begin{bmatrix} 0.4472136 & -0.5539120 & 0.5034303 & -0.4085320 & 0.2698908 \\ 0.4472136 & -0.3692740 & -0.0575260 & 0.4921961 & -0.6465830 \\ 0.4472136 & -0.0615460 & -0.5507160 & 0.3103943 & 0.6297453 \\ 0.4472136 & 0.3077287 & -0.4136770 & -0.6535700 & -0.3271400 \\ 0.4472136 & 0.6770032 & 0.5184895 & 0.2595117 & 0.0740877 \end{bmatrix}.$$

Subsequently, the data on the transformed variables are obtained by using the matrix transformation

$$(Z1 : Z2 : Z3 : Z4 : Z5)_{n \times 5} = (Y1 : Y2 : Y3 : Y4 : Y5)_{n \times 5} * OPOLY_{5 \times 5}$$

or $\mathbf{Z} = \mathbf{Z}_0 * OPOLY$, where n is the number of data points. The columns of matrix $\mathbf{Z}$ are then assigned the variable names Z1, Z2, Z3, Z4, and Z5 respectively. The variable z_i thus represents the effect resulting from the terms corresponding to the $(i-1)^{th}$ power of time. Thus, if an rth order polynomial is adequate, then $E(z_{r+2}) = \cdots = E(z_p) = 0$. The results appear in Output 5.5. Some output has been suppressed.

```
/* Program 5.5 */

options ls=64 ps=45 nodate nonumber;
data fish;
infile 'fish.dat' firstobs = 1;
input p1 p2 p3 p4 p5 dose wt @@;
y1=arsin(sqrt(p1));
y2=arsin(sqrt(p2));
y3=arsin(sqrt(p3));
y4=arsin(sqrt(p4));
y5=arsin(sqrt(p5));
x1=log(dose);
x2=wt;
Title1 'Output 5.5' ;
title2 'Polynomial Fitting for Fish Data' ;
data growth;
set fish;
keep y1-y5;
proc iml;
```

```
use growth;
read all into z0;
vec={8 14 24 36 48};
opoly=orpol (vec,4);
print 'Orthogonal Polynomial Matrix';
print opoly;
z=z0*opoly;
varnames={z1 z2 z3 z4 z5};
create newdata from z (|colname=varnames|);
append from z;
close newdata;
data newdata;
set newdata fish;
merge newdata fish;
run;
proc sort data=newdata;
by dose ;
run;

proc glm data = newdata;
by dose;
model z4 z5 = /nouni;
manova h=intercept;
title3 'Second Degree Polynomial Fit for Individual Doses';
run;

proc glm data = newdata;
by dose;
model  z5 = /nouni;
manova h=intercept;
title3 'Third Degree Polynomial Fit for Individual Doses';
run;

proc glm data = newdata;
class dose;
model  z5 = dose /nouni;
manova h=intercept;
title3 'Common 3rd degree Polynomial Fit';
run;
```

Output 5.5

```
                         Output 5.5
                 Polynomial Fitting for Fish Data
           Second Degree Polynomial Fit for Individual Doses


------------------------- DOSE=270 -------------------------


            Manova Test Criteria and Exact F Statistics for
              the Hypothesis of no Overall INTERCEPT Effect
    H = Type III SS&CP Matrix for INTERCEPT   E = Error SS&CP Matrix


                       S=1     M=0     N=0.5


Statistic                    Value       F     Num DF  Den DF  Pr > F

Wilks' Lambda               0.678781   0.7098     2       3    0.5592
Pillai's Trace              0.321219   0.7098     2       3    0.5592
Hotelling-Lawley Trace 0.473228   0.7098     2       3    0.5592
Roy's Greatest Root    0.473228   0.7098     2       3    0.5592
```

```
-------------------------- DOSE=410 --------------------------
```

Manova Test Criteria and Exact F Statistics for
the Hypothesis of no Overall INTERCEPT Effect
H = Type III SS&CP Matrix for INTERCEPT   E = Error SS&CP Matrix

S=1   M=0   N=0.5

Statistic	Value	F	Num DF	Den DF	Pr > F
Wilks' Lambda	0.221426	5.2743	2	3	0.1042
Pillai's Trace	0.778574	5.2743	2	3	0.1042
Hotelling-Lawley Trace	3.516189	5.2743	2	3	0.1042
Roy's Greatest Root	3.516189	5.2743	2	3	0.1042

```
-------------------------- DOSE=610 --------------------------
```

Manova Test Criteria and Exact F Statistics for
the Hypothesis of no Overall INTERCEPT Effect
H = Type III SS&CP Matrix for INTERCEPT   E = Error SS&CP Matrix

S=1   M=0   N=0.5

Statistic	Value	F	Num DF	Den DF	Pr > F
Wilks' Lambda	0.078083	17.71	2	3	0.0218
Pillai's Trace	0.921917	17.71	2	3	0.0218
Hotelling-Lawley Trace	11.80692	17.71	2	3	0.0218
Roy's Greatest Root	11.80692	17.71	2	3	0.0218

```
-------------------------- DOSE=940 --------------------------
```

Manova Test Criteria and Exact F Statistics for
the Hypothesis of no Overall INTERCEPT Effect
H = Type III SS&CP Matrix for INTERCEPT   E = Error SS&CP Matrix

S=1   M=0   N=0.5

Statistic	Value	F	Num DF	Den DF	Pr > F
Wilks' Lambda	0.011851	125.07	2	3	0.0013
Pillai's Trace	0.988149	125.07	2	3	0.0013
Hotelling-Lawley Trace	83.37844	125.07	2	3	0.0013
Roy's Greatest Root	83.37844	125.07	2	3	0.0013

```
-------------------------- DOSE=1450 --------------------------
```

Manova Test Criteria and Exact F Statistics for
the Hypothesis of no Overall INTERCEPT Effect
H = Type III SS&CP Matrix for INTERCEPT   E = Error SS&CP Matrix

S=1   M=0   N=0.5

Statistic	Value	F	Num DF	Den DF	Pr > F
Wilks' Lambda	0.010281	144.41	2	3	0.0010
Pillai's Trace	0.989719	144.41	2	3	0.0010

```
Hotelling-Lawley Trace    96.271   144.41      2       3   0.0010
Roy's Greatest Root       96.271   144.41      2       3   0.0010
```

Output 5.5
continued

```
                              Output 5.5
                      Polynomial Fitting for Fish Data
                      Common 3rd degree Polynomial Fit

                      General Linear Models Procedure
                        Class Level Information

              Class    Levels    Values

              DOSE        5       270 410 610 940 1450

              Number of observations in data set = 25


                Manova Test Criteria and Exact F Statistics for
                   the Hypothesis of no Overall INTERCEPT Effect
            H = Type III SS&CP Matrix for INTERCEPT    E = Error SS&CP Matrix

                           S=1    M=-0.5    N=9

            Statistic              Value        F   Num DF  Den DF  Pr > F

            Wilks' Lambda         0.988946   0.2236     1      20   0.6415
            Pillai's Trace        0.011054   0.2236     1      20   0.6415
            Hotelling-Lawley Trace 0.011178  0.2236     1      20   0.6415
            Roy's Greatest Root   0.011178   0.2236     1      20   0.6415
```

To test if the second-degree model is adequate for a given dose level, we have $r = 2$, and hence the transformed variables corresponding to the higher degrees, namely, Z4 and Z5, should have zero means. We thus use multivariate tests to see if the intercepts corresponding to Z4 and Z5 can be assumed to be zero. This is accomplished by first sorting the data by dose levels and then by using the code presented in Program 5.5:

```
proc glm data=newdata;
by dose;
model z4 z5= /nouni;
manova h=intercept;
```

Output 5.5 indicates that this hypothesis can be accepted for lower levels of doses. However, as the doses increase, there is a relatively stronger case against a second-degree model. This is not surprising as, at the higher doses, most of the fish died early in the experiment. A third-degree polynomial, however, seems to fit data for all the five doses. In this case, the corresponding MODEL and MANOVA statements are

```
model z5= /nouni;
manova h=intercept;
```

We have suppressed the output to save space. Having known that a third-degree polynomial may suffice in the case of all five doses, we ask if it is possible to fit a common third-degree polynomial. In other words, if DOSE is taken as a variable at five levels then does the variable Z5, when fitted as a function of DOSE, have a zero intercept? If the entire data set of 25 observations for Z5 is analyzed under the cubic polynomial model with DOSE as a variable, the value of Wilks' Λ for the null hypothesis of no intercept for Z5 is not significant at $\alpha = 0.05$. This suggests that a model of the type

$$y(t) = \beta_0 + \beta_1 t + \beta_2 t^2 + \beta_3 t^3 + \gamma x_1 + \epsilon_t,$$

where x_1 is the logarithm of the dose level can be fitted to this data set. In other words, acceptance of this null hypothesis implies that the cubic polynomial curves for the five dose levels are parallel in that they differ only in the intercepts. The intercepts are determined by the corresponding value of the term $(\beta_0 + \gamma x_1)$. In this case, the number of fish that died over time can be described by cubic polynomials differing only in the intercept terms.

5.2.5 Repeated Measure Designs for Treatment Combinations/Conditions

When various treatment combinations of a factorial experiment with two or more factors are applied to the same group of subjects, then it may not be appropriate to analyze the data from this factorial experiment using the univariate analysis of variance techniques. It is so since the conventional ANOVA assumes the independence of all the measurements (or the Type H covariance structure) as the minimum requirement for the distributional validity of the ANOVA F test. Obviously, if the same unit has been subjected to various treatments one after the other, the corresponding measurements cannot be assumed to be independent and are not guaranteed to have the Type H covariance structure. However, it is possible to formulate the comparison of treatments problem in the Hotelling's T^2 framework.

By appropriately numbering the treatments, say as $1, \ldots, p$, and denoting the measurements on the jth unit as $\mathbf{y}'_j = (y_{1j}, \ldots, y_{pj})$, $j = 1, \ldots, n$, we can write the model as

$$\mathbf{y}_j = \boldsymbol{\mu} + \boldsymbol{\epsilon}_j, \quad j = 1, \ldots, n, \tag{5.6}$$

where $\boldsymbol{\mu}$ is the p by 1 vector of true means of various treatment effects. The effects of individual variables can be expressed as the contrasts of the vector $\boldsymbol{\mu}$, say $\mathbf{c}'_i \boldsymbol{\mu}$, by appropriately choosing the p by 1 vectors $\mathbf{c}_i$. Likewise a simultaneous comparison of several treatments can be accomplished by simultaneously testing for several, say r, linearly independent contrasts $\mathbf{C}\boldsymbol{\mu}$, where $\mathbf{C}$ is an appropriately defined r by p matrix. To test the null hypothesis $H_0 : \mathbf{C}\boldsymbol{\mu} = \mathbf{d}$, where $\mathbf{d}$ is a known vector of order r by 1, Hotelling's T^2 test can be used by defining

$$T^2 = n(\mathbf{C}\bar{\mathbf{y}} - \mathbf{d})'(\mathbf{C}\mathbf{S}\mathbf{C}')^{-1}(\mathbf{C}\bar{\mathbf{y}} - \mathbf{d}). \tag{5.7}$$

The null hypothesis then is rejected if

$$F = \frac{(n-r)}{(n-1)r} T^2 > F_\alpha(r, n-r).$$

The vector $\bar{\mathbf{y}}$ here is the vector of sample means for all treatments and $\mathbf{S}$ is the sample variance-covariance matrix. Chapter 4 showed that this hypothesis can be tested using the M= specification in the MANOVA statement of PROC GLM or PROC ANOVA.

EXAMPLE 4   *A Two-Way Factorial Experiment, Dog Data*   Johnson and Wichern (1998) provide a very fitting example of such a study, where a two-way factorial experiment was conducted as a repeated measures design to explore the possibility of finding improved anesthetics. Two variables, carbon dioxide pressure and the presence or absence of halothane, each at two levels (namely, high, low, and absent, present) respectively were used. The four treatments here referred to as 1, 2, 3, and 4, were administered in the order (high, absent), (low, absent), (high, present) and (low, present) on each of the 19 dogs. The number of milliseconds between the heartbeats was taken as the response variable. These data were collected by Dr. J. Atlee, a physician at Veteran's Hospital, Madison, Wisconsin. If the mean response of the four treatments is represented by a 4 by 1 vector $\boldsymbol{\mu}$, then the effect of halothane is represented by the contrast $(\mu_1 + \mu_2) - (\mu_3 + \mu_4) = \mathbf{c}_1'\boldsymbol{\mu}$ with $\mathbf{c}_1' = (1\ 1\ -1\ -1)$, the effect of carbon dioxide is represented by $(\mu_1 + \mu_3) - (\mu_2 + \mu_4) = \mathbf{c}_2'\boldsymbol{\mu}$ with $\mathbf{c}_2' = (1\ -1\ 1\ -1)$, and the interaction effect by $(\mu_1 + \mu_4) - (\mu_2 + \mu_3) = \mathbf{c}_3'\boldsymbol{\mu}$ with $\mathbf{c}_3' = (1\ -1\ -1\ 1)$. Also to test that there is no difference among any of the four means, that is, $H_0 : \mu_1 = \mu_2 = \mu_3 = \mu_4$, it is possible to simultaneously test that all the three contrasts described above are zero. This is so since the three equations

$$(\mu_1 + \mu_2) - (\mu_3 + \mu_4) = 0$$

$$(\mu_1 + \mu_3) - (\mu_2 + \mu_4) = 0 \tag{5.8}$$

$$(\mu_1 + \mu_4) - (\mu_2 + \mu_3) = 0$$

imply and are implied by the null hypothesis $H_0 : \mu_1 = \mu_2 = \mu_3 = \mu_4$. Thus, the null hypothesis can alternatively be expressed as $H_0 : \mathbf{C}\boldsymbol{\mu} = \mathbf{0}$, where the 3 by 4 matrix $\mathbf{C}$ consists of $\mathbf{c}_1'$, $\mathbf{c}_2'$, and $\mathbf{c}_3'$ as its three rows. It may be noted that there are many other choices of $\mathbf{c}$ to attain this equivalence. However, the value of T^2 and hence of the resulting F statistic remains invariant of the choice of $\mathbf{c}$ matrix.

To carry out these comparisons using SAS, we need to express the problem in the linear model setup. We denote the responses corresponding to four treatments as HIGH_NOH, LOW_NOH, HIGH_H, and LOW_H respectively and collectively represent the data on these responses as a 19 by 4 matrix $\mathbf{Y}$. Thus, we have the linear model

$$\mathbf{Y}_{19 \times 4} = \mathbf{1}_{19 \times 1}\boldsymbol{\mu}_{1 \times 4}' + \mathcal{E}_{19 \times 4}.$$

The linear model above has only the intercept term. The null hypothesis $H_0 : \mathbf{C}\boldsymbol{\mu} = \mathbf{0}$ can be written as $H_0 : \mathbf{BM} = \mathbf{0}$, with $\mathbf{M} = \mathbf{C}'$. This hypothesis can be tested by specifying the MANOVA statement

```
manova h=intercept m=(1  1 -1 -1,
                      1 -1  1 -1,
                      1 -1 -1  1);
```

In Program 5.6, we have used the representation of the $\mathbf{M}$ matrix directly through Equations 5.8 only to illustrate that the two alternatives are equivalent. Also in Program 5.6, the PRINTE and PRINTH options are used to print the error and the hypothesis sums of squares and crossproduct matrices. The results appear in Output 5.6.

```
/* Program 5.6 */

options ls=64 ps=45 nodate nonumber;
data dog ;
input high_noh low_noh high_h low_h ;
y1 = high_noh;
y2 = low_noh ;
y3 = high_h;
y4 = low_h;
z=y1+y2-y3-y4;
lines ;
426 609 556 600
253 236 392 395
359 433 349 357
432 431 522 600
405 426 513 513
324 438 507 539
310 312 410 456
326 326 350 504
```

```
                375 447 547 548
                286 286 403 422
                349 382 473 497
                429 410 488 547
                348 377 447 514
                412 473 472 446
                347 326 455 468
                434 458 637 524
                364 367 432 469
                420 395 508 531
                397 556 645 625
                ;
/* Original Data Source: Dr. J. Atlee, III, M.D.  Reproduced
   with permission from Dr. J. Atlee. */
title1 'Output 5.6 ';
title2 'Two-way Factorial Experiment: Dog Data';
proc glm data = dog ;
model high_noh low_noh high_h low_h = /nouni;
/* Test for Factor halothane;
manova h=intercept m=high_noh + low_noh -high_h -low_h
/printe printh;
*Test for Factor Co2 ;
manova h=intercept m=high_noh - low_noh +high_h -low_h
 /printe printh;

*Test for interaction Co2*halothane;
manova h=intercept m=high_noh - low_noh -high_h +low_h
/printe printh; */

*Testing Both factors and interaction simultaneously:
Comparing all treatments;
manova h=intercept m=high_noh+low_noh -high_h -low_h ,
                   high_noh - low_noh +high_h -low_h ,
                   high_noh - low_noh -high_h +low_h
/printe printh ;
run;
```

Output 5.6

```
                          Output 5.6
                Two-way Factorial Experiment: Dog Data

                  General Linear Models Procedure
                  Multivariate Analysis of Variance

              M Matrix Describing Transformed Variables

                 HIGH_NOH        LOW_NOH        HIGH_H         LOW_H

        MVAR1         1              1            -1            -1
        MVAR2         1             -1             1            -1
        MVAR3         1             -1            -1             1


                      E = Error SS&CP Matrix

                        MVAR1            MVAR2            MVAR3

        MVAR1     169780.10526     -19780.31579     -16696.73684
        MVAR2     -19780.31579     93524.947368     16462.210526
        MVAR3     -16696.73684     16462.210526     136033.15789
```

```
                H = Type III SS&CP Matrix for INTERCEPT


                        MVAR1              MVAR2              MVAR3

        MVAR1      832448.89474       238829.31579       50863.736842
        MVAR2      238829.31579       68520.052632       14592.789474
        MVAR3      50863.736842       14592.789474       3107.8421053



                Manova Test Criteria and Exact F Statistics for
                  the Hypothesis of no Overall INTERCEPT Effect
             on the variables defined by the M Matrix Transformation
        H = Type III SS&CP Matrix for INTERCEPT   E = Error SS&CP Matrix

                      S=1      M=0.5     N=7

        Statistic              Value       F     Num DF  Den DF  Pr > F

        Wilks' Lambda          0.134312   34.375      3      16  0.0001
        Pillai's Trace         0.865688   34.375      3      16  0.0001
        Hotelling-Lawley Trace 6.445351   34.375      3      16  0.0001
        Roy's Greatest Root    6.445351   34.375      3      16  0.0001
```

The output corresponding to this statement is given in the first part of Output 5.6. First the matrix $\mathbf{M} = \mathbf{C}'$ is printed, which is followed by the error SS&CP and the hypothesis SS&CP matrices labeled $\mathbf{E}$ and $\mathbf{H}$ respectively for the three new variables defined as the linear combinations of the four measurements using the $\mathbf{M} = \mathbf{C}'$ matrix. SAS also prints the matrix of partial correlation coefficients between the three linear combinations as well as all the eigenvalues of $\mathbf{E}^{-1}\mathbf{H}$. We have suppressed certain parts of the SAS output in Output 5.6 to save space. Since the design is balanced, all the three types of SS&CP matrices described in Chapter 4 are identical. We have therefore accepted the SAS default option of Type III matrices.

For the null hypothesis described in Equations 5.8, we observe that the value of Wilks' Λ is 0.1343, correspondingly giving the observed value of F(3, 16) as $F = 34.375$. This is highly significant with the corresponding p value $= 0.0001$, indicating that at least one of the equations in Equations 5.8 is possibly not true. Incidentally, the other three multivariate tests are equivalent leading to the identical observed values of (exact) F(3, 16). This is so, since all four test statistics are the functions of

$$T^2 = n(\mathbf{C}\bar{\mathbf{y}} - \mathbf{d})'(\mathbf{C}\mathbf{S}\mathbf{C}')^{-1}(\mathbf{C}\bar{\mathbf{y}} - \mathbf{d})$$
$$= n\,tr\;(\mathbf{C}\mathbf{S}\mathbf{C}')^{-1}(\mathbf{C}\bar{\mathbf{y}} - \mathbf{d})(\mathbf{C}\bar{\mathbf{y}} - \mathbf{d})'$$
$$= tr\;\mathbf{E}^{-1}\mathbf{H}.$$

Since T^2 is a 1 by 1 matrix, the matrix $\mathbf{E}^{-1}\mathbf{H}$ has rank 1 and hence only one nonzero eigenvalue. Consequently, all four multivariate test criteria, which are the functions of the eigenvalues of $\mathbf{E}^{-1}\mathbf{H}$, have a one-to-one correspondences between any pair of tests and hence are all equivalent. As remarked earlier, Hotelling's T^2 statistic and hence also the four multivariate tests listed in the SAS output do not depend on the choice of the $\mathbf{C}$ matrix. Another meaningful choice of the $\mathbf{C}$ matrix is a PROFILE matrix given by

$$\begin{bmatrix} 1 & -1 & 0 & 0 \\ 0 & 1 & -1 & 0 \\ 0 & 0 & 1 & -1 \end{bmatrix}.$$

Although the **E** and **H** matrices are different in this case, the four test statistics are all identical to those obtained in Output 5.6.

Having rejected the null hypothesis given in Equations 5.8, we want to know first if the interaction between carbon dioxide levels and the presence or absence of halothane is nonexistent; second, if halothane's presence has no effect on the response; and third, if the level of carbon dioxide has no effect on the response. The corresponding **M** matrices can respectively be defined by using the third, first, and the second equations in Equations 5.8, as shown in Program 5.6. The corresponding SAS output is not presented here. The hypothesis of no interaction is not rejected with the corresponding p value of 0.5294. The other two null hypotheses are rejected with the respective p values as 0.0001 and 0.0019 concluding that halothane's presence has a strong effect on the response as does the level of carbon dioxide.

We could also analyze these data as a univariate two-way classification, provided that the Type H covariance structure can be assumed for the variance-covariance matrix $D(\mathbf{y}_j) = \mathbf{\Sigma}$. Since the matrix $\mathbf{C}$ used in Hotelling's T^2 statistic in Equation 5.7 is, by construction, orthogonal, we need only to test for the sphericity of $\mathbf{C\Sigma C'}$. Since the **E** matrix printed in Output 5.6 is equal to $[n - (p - 1)]\mathbf{CSC'}$ or $16\mathbf{CSC'}$, the estimated variance-covariance matrix $\mathbf{CSC'}$ for these orthogonal contrasts can be obtained as $\mathbf{E}/[n - (p - 1)] = \mathbf{E}/16$, on which a sphericity test as given in Equation 5.1 can easily be performed. Alternatively, all these computations can be achieved with the use of REPEATED statement as described in the case of memory data. Specifically the REPEATED statement is

```
repeated trtment 4/ printe printm;
```

One can verify that the Type H structure can be assumed for these data as well, thereby validating the ANOVA F tests. Further, the same conclusions about the main effects and their interaction are reached by the univariate analysis. We stress, however, that to test the Type H structure, the transformations for which sphericity would be tested should be orthogonal. That is, the rows of **C** should be orthogonal to each other and should be of the same length. This was the case for the particular **C** we have used in M= option of MANOVA statement but may not be so in several other possible choices such as in the case where **C** is a profile matrix as defined earlier.

Often, the data are collected over time by applying different treatments or are collected under different conditions, and the number of longitudinal observations under various conditions may be different. In such cases, we may want to compare the mean responses under various conditions. Assuming no carryover effects, a simple way to compare various treatments or conditions is to test for the suitable weighted linear combinations of the mean response at different time points. The approach is best illustrated by an example.

EXAMPLE 5   *Comparing Treatments, A Dietary Treatment Study*   A group of 12 patients was subjected to a dietary regime treatment. Two observations before the treatment, three during and two after the conclusion of treatment, all at different time points, were made on a variable representing the level of plasma ascorbic acid (Figure 5.1). The problem is to compare the effectiveness of the treatments by comparing the three sets of responses.

As in the previous example, we have one group of repeated measures with an unequal number of time points for each of the three conditions. If μ_1, $\mu_2, \ldots, \mu_7$ are the mean responses at seven consecutive time points then we may want to test the hypothesis

$$H_0 : \frac{\mu_1 + \mu_2}{2} = \frac{\mu_3 + \mu_4 + \mu_5}{3} = \frac{\mu_6 + \mu_7}{2}$$

or

$$H_0 : 3(\mu_1 + \mu_2) = 2(\mu_3 + \mu_4 + \mu_5) = 3(\mu_6 + \mu_7).$$

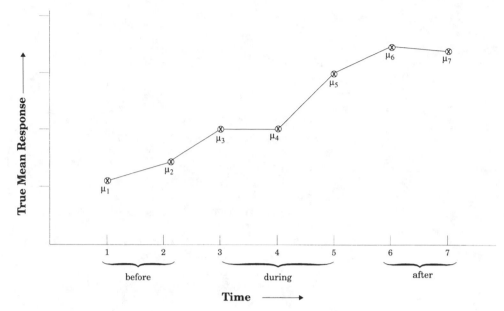

Figure 5.1
Design for Dietary
Treatment Study

In matrix form, it can be written as

$$H_0 : \begin{bmatrix} 3 & 3 & -2 & -2 & -2 & 0 & 0 \\ 0 & 0 & 2 & 2 & 2 & -3 & -3 \end{bmatrix} \begin{bmatrix} \mu_1 \\ \mu_2 \\ \mu_3 \\ \mu_4 \\ \mu_5 \\ \mu_6 \\ \mu_7 \end{bmatrix} = \begin{bmatrix} 0 \\ 0 \end{bmatrix}.$$

With $\mathbf{B} = \boldsymbol{\mu}' = (\mu_1, \ldots, \mu_7)$, , $\mathbf{X} = \mathbf{1}_{12}$, and

$$\mathbf{M}' = \begin{bmatrix} 3 & 3 & -2 & -2 & -2 & 0 & 0 \\ 0 & 0 & 2 & 2 & 2 & -3 & -3 \end{bmatrix}$$

and $\mathbf{Y}$ a 12 by 7 matrix of observed responses on twelve patients at seven time points, the above hypothesis testing problem can be formulated as

$$H_0 : \mathbf{BM} = \mathbf{0}$$

under the linear model setup

$$\mathbf{Y} = \mathbf{XB} + \mathcal{E} = \mathbf{1}\boldsymbol{\mu}' + \mathcal{E}.$$

Taking normality as the underlying assumption, we can test the null hypothesis as in the previous example by appropriately defining the $\mathbf{M}$ matrix in the MANOVA statement in PROC GLM.

The SAS code and the selected parts of the output are respectively presented in Program 5.7 and Output 5.7. The evidence is strongly against the null hypothesis (p value $= 0.0001$) for all multivariate tests which are equivalent in this case and we consequently reject H_0. We thereby conclude that the dietary regime treatment is indeed effective. However, we may still want to know if the effect of the treatment lasts for the latter periods. This hypothesis can be tested by defining the new $\mathbf{M}$ matrix consisting only of the second column of the $\mathbf{M}$ matrix indicated above, that is, $(0\ 0\ 2\ 2\ 2\ -3\ -3)'$. We can verify that this hypothesis is also rejected (p value $= 0.0002$), leading us not to reject the well-known fact that the effect of dietary treatment does not continue after the diet ends.

```
/* Program 5.7 */

options ls=64 ps=45 nodate nonumber;
title1 'Output 5.7';
data react;
input patient y1-y7;
lines;
1   0.22 0.00 1.03 0.67 0.75 0.65 0.59
2   0.18 0.00 0.96 0.96 0.98 1.03 0.70
3   0.73 0.37 1.18 0.76 1.07 0.80 1.10
4   0.30 0.25 0.74 1.10 1.48 0.39 0.36
5   0.54 0.42 1.33 1.32 1.30 0.74 0.56
6   0.16 0.30 1.27 1.06 1.39 0.63 0.40
7   0.30 1.09 1.17 0.90 1.17 0.75 0.88
8   0.70 1.30 1.80 1.80 1.60 1.23 0.41
9   0.31 0.54 1.24 0.56 0.77 0.28 0.40
10 1.40 1.40 1.64 1.28 1.12 0.66 0.77
11 0.60 0.80 1.02 1.28 1.16 1.01 0.67
12 0.73 0.50 1.08 1.26 1.17 0.91 0.87
;
/* Source: Crowder and Hand (1990, p. 32). */
proc glm data = react;
model y1-y7= /nouni;
manova h=intercept
m=3*y1+3*y2-2*y3-2*y4-2*y5, 2*y3+2*y4+2*y5-3*y6-3*y7;
title2 'Comparison of Dietary Regime Treatments';
run;
```

Output 5.7

```
                        Output 5.7
              Comparison of Dietary Regime Treatments


              Manova Test Criteria and Exact F Statistics for
                 the Hypothesis of no Overall INTERCEPT Effect
             on the variables defined by the M Matrix Transformation
        H = Type III SS&CP Matrix for INTERCEPT    E = Error SS&CP Matrix


                         S=1      M=0      N=4

        Statistic              Value       F    Num DF  Den DF  Pr > F

        Wilks' Lambda          0.136226  31.704    2       10   0.0001
        Pillai's Trace         0.863774  31.704    2       10   0.0001
        Hotelling-Lawley Trace 6.340768  31.704    2       10   0.0001
        Roy's Greatest Root    6.340768  31.704    2       10   0.0001
```

5.3 *k* Populations

Suppose there are k different treatments and the ith treatment, say A_i, is applied to a group of n_i subjects which are observed over time for p time points. In such a situation, the p dimensional observations on these groups can be thought of as the respective random samples from the k populations corresponding to various treatments. As a first step, we can look at the problem of comparing various treatments as a problem of multivariate one-way classification. Clearly, the p dependent variables, namely, the observations taken over time, are correlated and may have different means for different time points as well as for the different treatments.

For the jth dependent variable (that is, response at j^{th} time point) $j = 1, \ldots, p$ the (univariate) one-way classification model is

$$y_{ijl} = \nu_j + \tau_{ij} + \epsilon_{ijl}, \ \ i = 1, \ldots, k; l = 1, \ldots, n_i,$$

which, by stacking $y_{1j1}, y_{1j2}, \ldots, y_{kjn_k}$ one below the other, can be written in matrix form as

$$\mathbf{y}_j = \mathbf{X}\boldsymbol{\beta}_j + \boldsymbol{\epsilon}_j, \tag{5.9}$$

with $\mathbf{X}$ and $\boldsymbol{\beta}_j$ respectively as

$$\mathbf{X} = \begin{bmatrix} 1 & 1 & 0 & . & . & . & 0 \\ 1 & 1 & 0 & . & . & . & 0 \\ . & . & . & . & . & . & . \\ 1 & 1 & 0 & . & . & . & 0 \\ 1 & 0 & 1 & . & . & . & 0 \\ . & . & . & . & . & . & . \\ 1 & 0 & 1 & . & . & . & 0 \\ . & . & . & . & . & . & . \\ . & . & . & . & . & . & . \\ 1 & 0 & 0 & . & . & . & 1 \\ . & . & . & . & . & . & . \\ 1 & 0 & 0 & . & . & . & 1 \end{bmatrix}_{n \times (k+1)} \quad \boldsymbol{\beta}_j = \begin{bmatrix} \nu_j \\ \tau_{1j} \\ \tau_{2j} \\ . \\ . \\ . \\ \tau_{kj} \end{bmatrix}_{(k+1) \times 1},$$

where $n = \sum_{i=1}^{k} n_i$. The vector $\boldsymbol{\epsilon}_j$ has been arranged in essentially the same way as the vector $\mathbf{y}_j$, and $\boldsymbol{\epsilon}_j$ has zero mean and the variance-covariance matrix $\sigma_{jj}\mathbf{I}_n$.

The multivariate model is obtained by arranging the univariate linear models given in Equation 5.9 side by side as columns. That is,

$$(\mathbf{y}_1 : \mathbf{y}_2 : \cdots : \mathbf{y}_p) = (\mathbf{X}\boldsymbol{\beta}_1 : \mathbf{X}\boldsymbol{\beta}_2 : \cdots : \mathbf{X}\boldsymbol{\beta}_p) + (\boldsymbol{\epsilon}_1 : \boldsymbol{\epsilon}_2 : \cdots : \boldsymbol{\epsilon}_p)$$

or

$$\mathbf{Y} = \mathbf{X}\mathbf{B} + \mathcal{E},$$

where the jth columns of $\mathbf{Y}$, $\mathcal{E}$, and $\mathbf{B}$ respectively are $\mathbf{y}_j$, $\boldsymbol{\epsilon}_j$, and $\boldsymbol{\beta}_j$. Since the observations taken at different time points on the same subject may be correlated, we assume that each row of $\mathcal{E}$ has zero mean and a variance-covariance matrix $\boldsymbol{\Sigma}$. On the other hand since the measurements on different subjects are uncorrelated, we assume, that for $j, j' = 1, \ldots, p,$

$$Cov\,(\mathbf{y}_j, \mathbf{y}_{j'}) = \sigma_{jj'}\mathbf{I}_n,$$

so that a nonzero covariance between two responses is assumed only when they are observed on the same subject at possibly two different time points j and j'. In other words, the variance-covariance matrix of the elements of $\mathbf{Y}$ stacked as a column vector by taking row after row is an np by np matrix

$$\begin{bmatrix} \boldsymbol{\Sigma} & & & \\ & \boldsymbol{\Sigma} & & \\ & & \ddots & \\ & & & \boldsymbol{\Sigma} \end{bmatrix}$$

or $diag\,(\boldsymbol{\Sigma}, \boldsymbol{\Sigma}, \ldots, \boldsymbol{\Sigma})$.

The statistical problems of interest can be broadly classified into two classes: first, the comparison of various treatment groups and second, the comparison within the repeated measures. The hypothesis concerning the latter aspects are often referred to as *within-subject hypotheses*, while the former are termed *between-subject hypotheses*.

5.3.1   Comparison of Treatments

The multivariate approach to the analysis and the hypothesis testing for between-subject analysis has already been described in Chapter 4. Specifically, the hypothesis of interest is the comparison of the treatment mean vectors, namely, μ_1, $\mu_2, \ldots, \mu_k$, where $\mu_i = \nu + \tau_i$, $i = 1, \ldots, k$ and $\nu = (\nu_1, \ldots, \nu_p)'$. This hypothesis is written as $H_0 : \mu_1 = \mu_2 = \cdots = \mu_k$ or $\tau_1 = \tau_2 = \cdots = \tau_k$, where $\tau_i' = (\tau_{i1}, \tau_{i2}, \ldots, \tau_{ip})$, $i = 1, \ldots, k$. As discussed in Chapter 4, the testing of H_0 can be accomplished by MANOVA partitioning of the corrected total sums of squares and a crossproducts (SS&CP) matrix into the model SS&CP and error SS&CP matrices and then by using the appropriate Wilks' Λ or any other multivariate test statistic derived from these quantities. We will illustrate this using an example of drug comparison.

EXAMPLE 6   *A Three-Population Study, Heart Rate Data*   Spector (1987) presented a heart rate study which was carried out to compare the effects of two drugs on human heart rate. The twenty-four subjects were randomly assigned to one of the three groups (eight to each group): two receiving the experimental drugs and one receiving the control. The heart rate measurements were observed at four different time points five minutes apart after administering the drug. Consequently, for this data set $k = 3$, $n_1 = n_2 = n_3 = 8$, and $p = 4$. We test the null hypothesis of no differences between the three drugs, that is, $\tau_1 = \tau_2 = \tau_3$, by using the Wilks' Λ defined as

$$\Lambda = \frac{|\mathbf{E}|}{|\mathbf{E} + \mathbf{H}|}$$

or any of the other three multivariate tests defined in Chapters 3 and 4. Denoting the measurements at four time points as the variables y_1, y_2, y_3, and y_4 respectively, and the levels of variable DRUG as the three drug names AX23, BWW9, and CONTROL, the corresponding SAS statements are

```
proc glm;
class drug;
model y1 y2 y3 y4=drug/nouni;
manova h=drug;
```

These statements are specified in the beginning of Program 5.8. Correspondingly, Output 5.8 shows that there are significant differences between the three drugs, including placebo ($\Lambda = 0.0790$ leading to F = 11.51 at (8, 36) degrees of freedom with a corresponding p value = 0.0001). All three types of analyses, namely, Type I, II, and III, are identical in this case. Also, all multivariate test criteria lead to the same conclusions.

5.3.2   Profile Analysis

Profile analysis is a collection of statistical hypothesis testing procedures used to explore any possible similarities between the treatment effects. We have briefly touched upon the profile analysis of the population mean in Section 5.2. Profile analysis is especially relevant for the longitudinal data on a given response variable or in the situations where responses on several dependent variables are measured on the same experimental unit. A population profile is a plot of the components of the population mean vector versus the order in which these means are arranged. In such cases, the order usually represents the time, especially in cases such as longitudinal studies or clinical trials. Such an order may also occur when a battery of treatments is sequentially applied on a group of subjects. Since the population means are usually unknown, the sample profile plots in which the population means are replaced by the sample means are in some sense the graphical estimates of the population profile plots. A typical sample profile plot has already been shown in Output 5.1.

The hypothesis of equality of means $H_0 : \boldsymbol{\mu}_1 = \boldsymbol{\mu}_2 = \cdots = \boldsymbol{\mu}_k$ implies that the treatments have the same average effects. Though useful, mere acceptance or rejection of such a hypothesis does not provide adequate insight into the type of similarities and dissimilarities that may exist among the treatments. In order to gain more understanding, we can formulate the above hypothesis as three hypotheses to be tested sequentially and subjected to the acceptance of the hypothesis at the previous stage. Specifically, we can ask: Are the profiles parallel? If so, are they coincidental? and finally, If so, are they all horizontal?

Are the k population profiles parallel?   The question here is, for the k populations, are the profile curves all identical except for the constant shifts in the levels? At a given time point j, the mean responses for i^{th} and i'^{th} treatments are μ_{ij} and $\mu_{i'j}$ and their difference is $\mu_{ij} - \mu_{i'j}$. Thus, the profiles for the ith and i'^{th} treatments are parallel if this difference remains constant for all time points $j = 1, \ldots, p$, that is, if $\mu_{i1} - \mu_{i'1} = \mu_{i2} - \mu_{i'2} = \cdots = \mu_{ip} - \mu_{i'p}$. Thus, the null hypothesis of all treatment profiles being parallel is equivalent to the null hypothesis (see Figure 5.2)

$$H_0^{(a)} : \mu_{11} - \mu_{21} = \mu_{12} - \mu_{22} = \cdots = \mu_{1p} - \mu_{2p}$$

$$\mu_{11} - \mu_{31} = \mu_{12} - \mu_{32} = \cdots = \mu_{1p} - \mu_{3p}$$

$$\vdots$$

$$\mu_{11} - \mu_{k1} = \mu_{12} - \mu_{k2} = \cdots = \mu_{1p} - \mu_{kp}.$$

The first equation in this set can be written as

$$\begin{bmatrix} 1 & -1 & 0 & \ldots & 0 \end{bmatrix} \begin{bmatrix} \boldsymbol{\mu}_1' \\ \boldsymbol{\mu}_2' \\ \vdots \\ \boldsymbol{\mu}_k' \end{bmatrix}_{k \times p} \begin{bmatrix} 1 & 1 & \ldots & 1 \\ -1 & 0 & \ldots & 0 \\ 0 & -1 & \ldots & 0 \\ \vdots & & & \\ 0 & 0 & \ldots & -1 \end{bmatrix}_{p \times (p-1)} = \mathbf{0}.$$

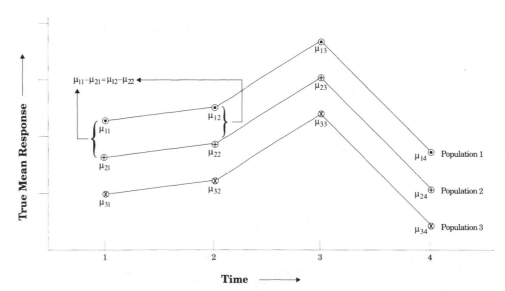

Figure 5.2
Profile Analysis,
Parallel Profiles

Similarly for the jth equation, $j = 1, \ldots, k-1$, the left-most row vector would have 1 at the first place and -1 at the jth place. As a result, the system given in $H_0^{(a)}$ can be conveniently expressed as

$$
\begin{bmatrix} 1 & -1 & 0 & \ldots & 0 \\ 1 & 0 & -1 & \ldots & 0 \\ \vdots & & & & \\ 1 & 0 & 0 & \ldots & -1 \end{bmatrix}_{(k-1) \times k}
\begin{bmatrix} \boldsymbol{\mu}_1' \\ \boldsymbol{\mu}_2' \\ \vdots \\ \boldsymbol{\mu}_k' \end{bmatrix}_{k \times p}
\begin{bmatrix} 1 & 1 & \ldots & 1 \\ -1 & 0 & \ldots & 0 \\ 0 & -1 & \ldots & 0 \\ \vdots & & & \\ 0 & 0 & \ldots & -1 \end{bmatrix}_{p \times p-1}
= \begin{bmatrix} \mathbf{0} \\ \mathbf{0} \\ \vdots \\ \mathbf{0} \end{bmatrix}.
$$

Since $\boldsymbol{\mu}_i = \boldsymbol{\nu} + \boldsymbol{\tau}_i, i = 1, \ldots, k$, the above set of equations has an alternative representation as

$$
\begin{bmatrix} 0 & 1 & -1 & 0 & \ldots & 0 \\ 0 & 1 & 0 & -1 & \ldots & 0 \\ \vdots & & & & & \\ 0 & 1 & 0 & 0 & \ldots & -1 \end{bmatrix}_{(k-1) \times (k+1)}
\begin{bmatrix} \boldsymbol{\nu}' \\ \boldsymbol{\tau}_1' \\ \boldsymbol{\tau}_2' \\ \vdots \\ \boldsymbol{\tau}_k' \end{bmatrix}_{(k+1) \times p}
\begin{bmatrix} 1 & 1 & \ldots & 1 \\ -1 & 0 & \ldots & 0 \\ 0 & -1 & \ldots & 0 \\ \vdots & & & \\ 0 & 0 & \ldots & -1 \end{bmatrix}
= \begin{bmatrix} \mathbf{0} \\ \mathbf{0} \\ \vdots \\ \mathbf{0} \end{bmatrix}.
$$

or

$$
H_0^{(a)} : \mathbf{LBM} = \mathbf{0}, \tag{5.10}
$$

where the definitions of $\mathbf{L}$, $\mathbf{B}$, and $\mathbf{M}$ are obvious. Thus, the hypothesis of parallel profiles can be formulated as a general linear hypothesis. As indicated in Section 3.4, the matrices $\mathbf{L}$ and $\mathbf{M}$ respectively have some special interpretations. The ith row of matrix $\mathbf{L}$ forms the appropriate linear functions of the regression coefficients (in matrix $\mathbf{B}$) within the ith model, $i = 1, \ldots, p$, whereas the jth column of the post-multiplied matrix $\mathbf{M}$ creates the desired linear combination of the coefficients from different models but corresponding to the same independent variable. In other words, in the present context, the specific comparisons between the population means are indicated by using the $\mathbf{L}$ matrix. Further, the longitudinal comparisons (or the comparisons involving the parameters of the models for various dependent variables) for a fixed population are specified through the $\mathbf{M}$ matrix. Note, however, that although there are several equivalent choices of $\mathbf{L}$ and $\mathbf{M}$, the final test statistics and the conclusions do not depend on these choices.

The general linear hypothesis given in Equation 5.10 is tested using the appropriate multivariate test. All the tests described in Section 3.4 apply, with some specific modifications. This is so since the general linear model

$$
\mathbf{Y} = \mathbf{XB} + \mathcal{E},
$$

when post-multiplied by a p by s matrix $\mathbf{M}$, reduces to the model $\mathbf{YM} = \mathbf{XBM} + \mathcal{E}\mathbf{M}$ or $\mathbf{Y}^* = \mathbf{XB}^* + \mathcal{E}^*$ which is essentially the same model with p replaced by s. In the present context of the parallel profile hypothesis, the matrix $\mathbf{M}$ is of order p by $(p-1)$ and hence s is equal to $(p-1)$.

EXAMPLE 6   ***Heart Rate Data (continued)***   The sample profile plots for three drugs are given in part of Output 5.8. The SAS code used to obtain this plot is given in Program 5.8. Chapter 2 also discussed similar SAS code. Friendly (1991) provides an explanation as well. The sample profile plots reveal the differences between the drugs in several respects, and we consider some of them here.

```
/* Program 5.8 */

options ls=64 ps=45 nodate nonumber;
data heart;
infile 'heart.dat';
input drug $ y1 y2 y3 y4;
title1 ' Output 5.8';
title2 'Comparison of Drugs: Profile Analysis';
proc glm data = heart;
class drug;
model y1 y2 y3 y4 = drug/nouni;
manova h = drug/printe printh ;
run;

filename gsasfile "prog58.graph";
goptions gaccess=gsasfile dev=pslmono;
goptions horigin=1in vorigin=2in;
goptions hsize=5in vsize=7in;
title1 h=1.5 'Comparison of Drugs: Profiles of the Means';
title2 j=l ' Output 5.8';
proc summary nway data=heart;
class drug;
var y1 y2 y3 y4;
output out=new mean=my1-my4;
data plot;
set new;
array my{4} my1-my4;
do test =1 to 4;
Response=my(test);
output;
end;
drop my1-my4;
  proc gplot data = plot;
plot response*test=drug /vaxis=axis1 haxis=axis2
vminor=3 legend=legend1 ;
axis1 label =(a=90 h=1.2 'Response');
axis2 offset=(2) label=(h=1.2 'Test');
symbol1 v=+ i = join;
symbol2 v=x i=join;
symbol3 v=* i=join;
legend1 across = 3;
run;

title1 ' Output 5.8';
title2 'Comparison of Drugs: Profile Analysis';
proc glm data = heart ;
class drug;
model y1 y2 y3 y4 =drug/nouni;
contrast '"bww9 vs. control"' drug 0 1  -1;
contrast '"ax23 vs. the rest"'  drug 2 -1 -1;
contrast '"parallel?"' drug 1 0 -1, drug  0 1 -1;
contrast '"horizontal?"' intercept 1;
manova h=drug
        m= (1 -1 0 0,   1 0 -1 0,  1 0 0 -1)/printe printh;
contrast '"coincidental?"' drug  1 0 -1, drug  0 1 -1;
manova h=drug m=(1 1 1 1) /printe printh;
run;
```

Output 5.8

Comparison of Drugs: Profiles of the Means

Output 5.8

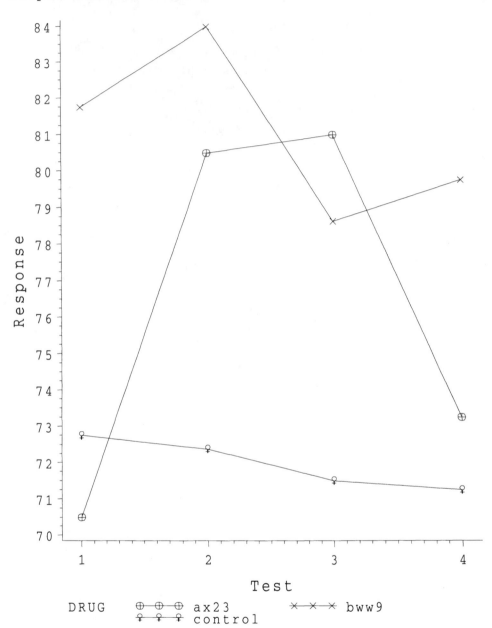

Output 5.8
continued

Output 5.8
Comparison of Drugs: Profile Analysis

General Linear Models Procedure
Multivariate Analysis of Variance

Manova Test Criteria and F Approximations for
the Hypothesis of no Overall DRUG Effect
H = Type III SS&CP Matrix for DRUG   E = Error SS&CP Matrix

S=2   M=0.5   N=8

Statistic	Value	F	Num DF	Den DF	Pr > F
Wilks' Lambda	0.079007	11.51	8	36	0.0001
Pillai's Trace	1.283456	8.5081	8	38	0.0001
Hotelling-Lawley Trace	7.069384	15.022	8	34	0.0001
Roy's Greatest Root	6.346509	30.146	4	19	0.0001

NOTE: F Statistic for Roy's Greatest Root is an upper bound.
NOTE: F Statistic for Wilks' Lambda is exact.

Comparison of Drugs: Profile Analysis
M Matrix Describing Transformed Variables

	Y1	Y2	Y3	Y4
MVAR1	1	-1	0	0
MVAR2	1	0	-1	0
MVAR3	1	0	0	-1

E = Error SS&CP Matrix

	MVAR1	MVAR2	MVAR3
MVAR1	261.375	119.5	147.5
MVAR2	119.5	226.375	235
MVAR3	147.5	235	463.5

H = Contrast SS&CP Matrix for "parallel?"

	MVAR1	MVAR2	MVAR3
MVAR1	465.58333333	593.54166667	212.25
MVAR2	593.54166667	872.58333333	308.25
MVAR3	212.25	308.25	109

```
              Manova Test Criteria and F Approximations for
                the Hypothesis of no Overall "parallel?" Effect
            on the variables defined by the M Matrix Transformation
                 H = Contrast SS&CP Matrix for "parallel?"
                         E = Error SS&CP Matrix


                      S=2      M=0     N=8.5

       Statistic                 Value       F     Num DF  Den DF  Pr > F

       Wilks' Lambda            0.110286   12.738      6      38   0.0001
       Pillai's Trace           1.089171    7.972      6      40   0.0001
       Hotelling-Lawley Trace   6.258785   18.776      6      36   0.0001
       Roy's Greatest Root      5.955089   39.701      3      20   0.0001

          NOTE: F Statistic for Roy's Greatest Root is an upper bound.
              NOTE: F Statistic for Wilks' Lambda is exact.




              H = Contrast SS&CP Matrix for "bww9 vs. control"


                        MVAR1              MVAR2              MVAR3

       MVAR1          27.5625           -19.6875             -5.25
       MVAR2         -19.6875            14.0625              3.75
       MVAR3          -5.25              3.75                1




              Manova Test Criteria and Exact F Statistics for
              the Hypothesis of no Overall "bww9 vs. control" Effect
            on the variables defined by the M Matrix Transformation
              H = Contrast SS&CP Matrix for "bww9 vs. control"
                         E = Error SS&CP Matrix


                      S=1      M=0.5    N=8.5

       Statistic                 Value       F     Num DF  Den DF  Pr > F

       Wilks' Lambda            0.743569   2.1841      3      19   0.1233
       Pillai's Trace           0.256431   2.1841      3      19   0.1233
       Hotelling-Lawley Trace   0.344864   2.1841      3      19   0.1233
       Roy's Greatest Root      0.344864   2.1841      3      19   0.1233




              H = Contrast SS&CP Matrix for "ax23 vs. the rest"


                        MVAR1              MVAR2              MVAR3

       MVAR1       438.02083333       613.22916667          217.5
       MVAR2       613.22916667       858.52083333          304.5
       MVAR3          217.5              304.5              108
```

```
          Manova Test Criteria and Exact F Statistics for
      the Hypothesis of no Overall "ax23 vs. the rest" Effect
     on the variables defined by the M Matrix Transformation
        H = Contrast SS&CP Matrix for "ax23 vs. the rest"
                    E = Error SS&CP Matrix


              S=1      M=0.5     N=8.5

Statistic                  Value        F    Num DF  Den DF  Pr > F

Wilks' Lambda           0.144636   37.455        3      19  0.0001
Pillai's Trace          0.855364   37.455        3      19  0.0001
Hotelling-Lawley Trace  5.913921   37.455        3      19  0.0001
Roy's Greatest Root     5.913921   37.455        3      19  0.0001
```

First of all, we want to see if the three drugs have parallel profiles. Accepting such a hypothesis would mean that for the three drugs the changes in the heart rate measurements are in the same direction and have similar patterns. We call this the hypothesis of no interaction between the drug and time, if time itself is taken as a factor. In notation, we want to test

$$H_0^{(a)} : \mu_{11} - \mu_{21} = \mu_{12} - \mu_{22} = \mu_{13} - \mu_{23} = \mu_{14} - \mu_{24},$$

$$\mu_{11} - \mu_{31} = \mu_{12} - \mu_{32} = \mu_{13} - \mu_{33} = \mu_{14} - \mu_{34}$$

or equivalently,

$$\begin{bmatrix} 1 & -1 & 0 \\ 1 & 0 & -1 \end{bmatrix} \begin{bmatrix} \mu_{11} & \mu_{12} & \mu_{13} & \mu_{14} \\ \mu_{21} & \mu_{22} & \mu_{23} & \mu_{24} \\ \mu_{31} & \mu_{32} & \mu_{33} & \mu_{34} \end{bmatrix} \begin{bmatrix} 1 & 1 & 1 \\ -1 & 0 & 0 \\ 0 & -1 & 0 \\ 0 & 0 & -1 \end{bmatrix} = \begin{bmatrix} 0 \\ 0 \end{bmatrix}$$

or

$$\begin{bmatrix} 0 & 1 & -1 & 0 \\ 0 & 1 & 0 & -1 \end{bmatrix} \begin{bmatrix} \boldsymbol{\nu}' \\ \boldsymbol{\tau}'_1 \\ \boldsymbol{\tau}'_2 \\ \boldsymbol{\tau}'_3 \end{bmatrix} \begin{bmatrix} 1 & 1 & 1 \\ -1 & 0 & 0 \\ 0 & -1 & 0 \\ 0 & 0 & -1 \end{bmatrix} = \begin{bmatrix} 0 \\ 0 \end{bmatrix}.$$

Thus, with

$$\mathbf{L} = \begin{bmatrix} 0 & 1 & -1 & 0 \\ 0 & 1 & 0 & -1 \end{bmatrix} \text{ and } \mathbf{M} = \begin{bmatrix} 1 & 1 & 1 \\ -1 & 0 & 0 \\ 0 & -1 & 0 \\ 0 & 0 & -1 \end{bmatrix},$$

we have $H_0^{(a)}$: $\mathbf{LBM} = \mathbf{0}$. Note that another alternative choice (among many others) for the matrices $\mathbf{L}$ and $\mathbf{M}$ may be

$$\mathbf{L} = \begin{bmatrix} 0 & 1 & 0 & -1 \\ 0 & 0 & 1 & -1 \end{bmatrix} \text{ and } \mathbf{M} = \begin{bmatrix} 1 & 0 & 0 \\ -1 & 1 & 0 \\ 0 & -1 & 1 \\ 0 & 0 & -1 \end{bmatrix}.$$

In PROC GLM, the matrix $\mathbf{M}$ is specified by the M= specification on the MANOVA statement. The matrix $\mathbf{L}$, however, is specified in the CONTRAST statement which should appear before the MANOVA statement. The following statements (see Program 5.8) are used to test the hypothesis of parallel profiles.

```
proc glm;
class drug;
model y1 y2 y3 y4=drug;
contrast '"parallel?"' drug 1 0 -1,
                       drug 0 1 -1;
manova h=drug m=(1 -1 0 0,
                 1 0 -1 0,
                 1 0 0 -1)/printe printh;
```

A few comments are in order. As mentioned earlier, the matrix **L** is specified in the CONTRAST statement. Since the comparison involves μ_1, μ_2, and μ_3 or equivalently τ_1, τ_2, and τ_3 the coefficients for the intercept are zero and hence need not be specified. As a result, the matrix **L** is shortened in the SAS code by deleting the first column. Further, we have used the second of the two equivalent choices for **L** stated earlier. The **M** matrix is specified in the MANOVA statement, where the corresponding hypothesis is on the equality of the treatment effects for the variable DRUG. The specification of **M** is column by column, and hence it would resemble **M**′ rather than **M**. In the actual program, we have also used the NOUNI option to suppress the univariate analysis.

The selected pieces of the output resulting from these statements are given as part of Output 5.8. First the matrix **M** is printed. The transformed (by post-multiplication of **M**) variables are referred as MVAR1 MVAR2 MVAR3 (the output can be cosmetically improved by using the PREFIX option to assign the appropriate names to the transformed variables). The error SS&CP matrix **E** as well as the corresponding hypothesis SS&CP matrix **H** are listed next. Based on these two matrices, the four multivariate test statistics are calculated for $H_0 : \mathbf{L}^*\mathbf{BM} = \mathbf{0}$ with **M** indicated above and $\mathbf{L}^* = (\mathbf{0} : \mathbf{I})$ with **0** as a vector of zeros. However, the null hypothesis $H_0 : \mathbf{L}^*\mathbf{BM} = \mathbf{0}$ is not the null hypothesis of just the parallel profiles. What this particular null hypothesis states is that for all three treatments, profiles are horizontally aligned. This, of course would imply the parallelism of the profiles but as a hypothesis it is considerably more restrictive and demanding than the requirement of just the parallel profiles. Not being relevant in the present context, this part of the output is not included here.

The null hypothesis of parallel profiles gives the corresponding hypothesis SS&CP matrix **H** as

$$\mathbf{H} = \begin{bmatrix} 465.5833 & 593.5417 & 212.2500 \\ 593.5417 & 872.5833 & 308.2500 \\ 212.2500 & 308.2500 & 109.0000 \end{bmatrix},$$

and the error SS&CP matrix as

$$\mathbf{E} = \begin{bmatrix} 261.3750 & 119.5000 & 147.5000 \\ 119.5000 & 226.3750 & 235.0000 \\ 147.5000 & 235.0000 & 463.5000 \end{bmatrix}.$$

The value of Wilks' Λ is 0.1103 leading to the observed value of the (exact) F(6, 38) random variable as 12.738. This is highly significant with a very low p value of 0.0001. As a result, we reject the null hypothesis of parallel profiles. The other three multivariate tests are also in agreement with this conclusion.

Are the profiles coincidental, given that profiles are parallel?   If the hypothesis of parallel profiles is not rejected, the next thing to ask may be if they are all identical. Since parallelism of profiles guarantees that profiles do not intersect each other and are consistently one below the other, the profiles will all be identical only if sums of components in each profile vector μ_i are all equal. The corresponding null hypothesis is

$$H_0^{(b)} : \mathbf{1}_p'\mu_1 = \mathbf{1}_p'\mu_2 = \cdots = \mathbf{1}_p'\mu_k$$

$$\text{or} \quad \mathbf{LBM} = \mathbf{0}$$

with the choice of **L** and **M** respectively as

$$
\mathbf{L} = \begin{bmatrix} 0 & 1 & 0 & \ldots & 0 & -1 \\ 0 & 0 & 1 & \ldots & 0 & -1 \\ \vdots & & & & & \\ 0 & 0 & 0 & \ldots & 1 & -1 \end{bmatrix}_{(k-1)\times(k+1)} \quad \text{and } \mathbf{M} = \mathbf{1}_p.
$$

Two sample profiles suggesting the possibility of coincidental population profiles are shown in Figure 5.3. Note that even when the hypothesis of possible parallel population profiles has been accepted, the sample profiles may still intersect with each other due to the sampling variability in computing the sample means.

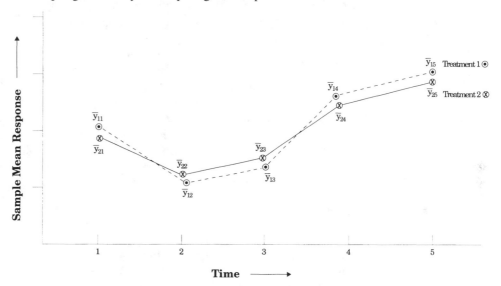

Figure 5.3
Two Sample Profiles
Suggesting the Possible
Coincidentalness

The hypothesis of parallel profile was rejected for the heart data in Example 6. In view of this rejection, the null hypothesis of coincidental profiles has no meaning there. However, if such a hypothesis were to be tested for this data set, the following **L** and **M** matrices would need to be used.

$$
\mathbf{L} = \begin{bmatrix} 0 & 1 & 0 & -1 \\ 0 & 0 & 1 & -1 \end{bmatrix} \quad \text{and } \mathbf{M}' = \begin{bmatrix} 1 & 1 & 1 & 1 \end{bmatrix}.
$$

As earlier, omitting the first column of **L**, which corresponds to the coefficients of the intercepts, the corresponding SAS statements would be

```
contrast '"coincidental?"' drug 1 0 -1,
                          drug 0 1 -1;
manova h=drug m=(1 1 1 1)/printe printh;
```

Are the profiles horizontal?   If we did not reject the null hypothesis of coincidental profiles, then the k populations supposedly have a common mean vector. It is natural to ask if the components in this common mean vector are also all equal. This means that profiles can be represented by a common horizontal line. Statistically speaking, this amounts to testing the hypothesis $H_0^{(c)} : \mathbf{LBM} = \mathbf{0}$ with

$$
\mathbf{L} = (1, 0, \ldots, 0) \quad \text{and } \mathbf{M} = \begin{bmatrix} 1 & 1 & \ldots & 1 \\ -1 & 0 & \ldots & 0 \\ 0 & -1 & \ldots & 0 \\ \vdots & \vdots & \vdots & \vdots \\ 0 & 0 & \ldots & -1 \end{bmatrix}.
$$

In this case, the coefficient of intercept vector in the **L** matrix is 1, while all other coefficients are zero. The **M** matrix is the same as that used to test the hypothesis of parallel profiles. If this hypothesis were to be tested for the heart data in Example 6, the corresponding part of the PROC GLM statements would be

```
contrast '"horizontal?"' intercept 1;
manova h=drug m=(1 -1 0 0,
                 1 0 -1 0,
                 1 0 0 -1)/printe printh;
```

EXAMPLE 6   *Heart Rate Data (continued)*   We recall that in Example 6, the hypothesis of parallel profiles was rejected. To gain further insight and identify the possible causes for nonparallelism, we may want to test for the significance of individual orthogonal contrasts. What we need to do first is to identify an appropriate **L** matrix in which rows are orthogonal to each other and then perform statistical tests for the significance of these individual contrasts. Although technically not necessary, one should attempt to use an **L** matrix so that its rows correspond to some intuitive and meaningful subhypotheses.

The sample profile plots of the three drugs are given in the beginning of Output 5.8. It appears that the profile for the drug AX23 is very different from the other two and falls between the profiles for BWW9 and CONTROL. These observations seem to suggest that such comparisons should also be part of the analysis. Specifically, we should first compare the profiles of the other two drugs, namely, BWW9 and CONTROL, with each other and then their average profile with the profile of drug AX23. If the 2 by 4 matrix **L** is chosen as

$$\mathbf{L} = \begin{bmatrix} 0 & 0 & 1 & -1 \\ 0 & 2 & -1 & -1 \end{bmatrix},$$

the two rows of **L** are mutually orthogonal and the corresponding functions **LB** are estimable. This is an alternative choice of **L** in addition to the two nonorthogonal choices previously indicated (see $H_0^{(a)}$), namely,

$$\begin{bmatrix} 0 & 1 & -1 & 0 \\ 0 & 1 & 0 & -1 \end{bmatrix} \text{ and } \begin{bmatrix} 0 & 1 & 0 & -1 \\ 0 & 0 & 1 & -1 \end{bmatrix}.$$

With matrix **M** the same as earlier, the rows of this new matrix **L**, namely,

$$\ell_1' = (0\ 0\ 1\ -1) \quad \text{and} \quad \ell_2' = (0\ 2\ -1\ -1)$$

respectively, can be used to test the hypotheses $H_0^{(a1)} : \ell_1'\mathbf{BM} = 0$, which tests for the parallelism of the profiles of drug BWW9 and CONTROL, and $H_0^{(a2)} : \ell_2'\mathbf{BM} = 0$ tests for the parallelism of the profiles of drug AX23 to the average profile of drugs BWW9 and CONTROL. These were the two hypotheses we have found to be of interest by looking at the profile plots given in Output 5.8. However, we hasten to add that the practice of developing hypotheses after looking at the data may be of questionable value at least from the classical statistics point of view.

The Wilks' Λ and other tests, all of which are exact and equivalent, support the hypothesis $H_0^{(a1)}$. The observed value of F(3, 19) corresponding to all of these tests is 2.1841 with a p value of 0.1233. Thus, we conclude that, although it may be different and superior, drug BWW9 behaves similarly to CONTROL over time.

The null hypothesis given in $H_0^{(a2)}$ is rejected and in view of the very small p value (= 0.0001), there is a strong argument for doing so. It indicates that as a function of time, the drug AX23 has a behavior very different from the other two. A look at the profile plots is visually convincing. Clearly, the major reason for the nonparallelism of profiles and hence the major component in the rejection of $H_0^{(a)}$ is the fact that drug AX23 acts very differently from the average of other two over time. CONTROL and drug BWW9 appear to be similar in their behavior.

5.3.3   A Univariate Approach

As in the one population case, a univariate approach to the analysis of repeated measure data for k populations can be devised by treating the experiments as a split plot design in which the units in the whole plots (whole plots are subjects in the present context) are correlated. By first testing for the Type H covariance structure and then accordingly taking the appropriate action (to adjust or not adjust the degrees of freedom for the test statistics), we can make inferences using the univariate approach as well. Of course each of the two approaches has its own shortcomings, and we should not think of these approaches as interchangeable. The split plot experiment leads to the linear model

$$y_{iju} = \mu + \alpha_i + \beta_j + (\alpha\beta)_{ij} + \delta_{iu} + \epsilon_{iju}, \tag{5.11}$$

$u = 1, \ldots, n_i,\ i = 1, \ldots, k,\ j = 1, \ldots, p$, in which μ represents the intercept or the general mean, α_i is the effect of the ith treatment, β_j is the effect of the jth time point, $(\alpha\beta)_{ij}$ is the interaction effect between the ith treatment and j^{th} time point, δ_{iu} represents the random error for the uth subject in the ith treatment group, and ϵ_{iju} is the random error corresponding to the uth subject in the ith treatment group at the jth time point.

We assume that δ_{iu} and ϵ_{iju} are both independently and normally distributed with zero means and respective variances σ_δ^2 and σ^2. As a consequence of this assumption and the assumed split plot model, the repeated measurements on the subjects have the variance-covariance matrix possessing the property of compound symmetry. This is a mixed effect model in which the appropriate F tests for treatment effect, time effect, and their interaction can be constructed using ANOVA partitioning of the corrected total sum of squares. However, this model is seldom realistic. The assumption of compound symmetry of the variance-covariance matrix is artificial and should not be accepted on face value. If the appropriate statistical tests on the data suggest this assumption to be acceptable, then under this assumption the usual split plot ANOVA F ratio tests for all the variables and the interaction are valid, exact F tests.

However, when the data do not conform to compound symmetry, only the test for the whole plot or between-subject treatment variable (DRUG) is a valid exact F test. As far as the other two tests, that is, for the time variable and the interaction between the drug and time, the F tests will be valid provided the Type H conditions, which are less stringent than compound symmetry assumption on the covariance structure, are satisfied. These are discussed in Section 5.2.2 and also in Section 5.2.3. If Type H conditions do not hold, then, the distribution of usual F ratios for TIME and DRUG*TIME interaction can be approximated by an F distribution for which the degrees of freedom are appropriately adjusted. The two adjustments suggested by Greenhouse and Geisser (1959) and Huynh and Feldt (1976) have been discussed in Section 5.2.3. Huynh and Feldt's (1976) adjustment has been shown to maintain the desired level of significance to a higher degree than Greenhouse and Geisser's.

EXAMPLE 6   ***Heart Rate Data (continued)***   Let us analyze the heart rate data of Spector (1987) using the univariate approach. To fit the model in Equation 5.11, for the dependent variable heart rate (Y) on variables DRUG, TIME, and their interaction with SUBJECT as the whole plot using PROC GLM, we use the SAS statements

```
proc glm;
class subject drug time;
model y= drug subject (drug) time drug*time;
random subject (drug)/test;
```

In the MODEL statement we have specified the model given in Equation 5.11. The RANDOM statement indicates that subjects are a random sample, and hence SUBJECT is a random effect *nested* within DRUG. It is necessary to account for such facts in constructing the appropriate F tests. The RANDOM statement prints the table of expected values of various mean squares. The TEST option computes the values of the appropriate test statistics. Of course, since it is a univariate analysis, all values of Y1, Y2, Y3, and Y4 are to be stacked one below the other as Y with corresponding levels of the variable TIME appropriately identified. This is done in Program 5.9. The ANOVA part of the resulting output is presented in Output 5.9. The results are identical to certain parts of Output 5.10 shown later.

```
/* Program 5.9 */

title1 'Output 5.9';
title2 'Univariate Analysis of Heart Rate Data';
options ls=64 ps=45 nodate nonumber;
data heart;
infile 'heart.dat';
input drug $ y1 y2 y3 y4;
data split;
set heart ;
array t{4} y1-y4;
subject+1;
do time=1 to 4;
y=t{time};
output;
end;
drop y1-y4;
proc glm data = split;
class subject drug time;
model y = drug subject(drug) time time*drug;
random subject(drug)/test;
run;
```

Output 5.9

```
                        Output 5.9
            Univariate Analysis of Heart Rate Data


              General Linear Models Procedure
      Tests of Hypotheses for Mixed Model Analysis of Variance


Dependent Variable: Y


Source: DRUG *
Error: MS(SUBJECT(DRUG))
                    Denominator   Denominator
   DF    Type III MS       DF          MS      F Value  Pr > F
    2      657.40625        21    111.08630952   5.9180  0.0092
* - This test assumes one or more other fixed effects are zero.



Source: SUBJECT(DRUG)
Error: MS(Error)
                    Denominator   Denominator
   DF    Type III MS       DF          MS      F Value  Pr > F
   21   111.08630952       63    7.3402777778  15.1338  0.0001

Source: TIME *
Error: MS(Error)
                    Denominator   Denominator
   DF    Type III MS       DF          MS      F Value  Pr > F
    3   93.069444444       63    7.3402777778  12.6793  0.0001
* - This test assumes one or more other fixed effects are zero.
```

```
Source: DRUG*TIME
Error: MS(Error)
                       Denominator  Denominator
     DF     Type III MS      DF            MS   F Value  Pr > F
      6    88.059027778      63   7.3402777778  11.9967  0.0001
```

The analysis performed in Program 5.9 can be more efficiently achieved by using the REPEATED statement. It eliminates the need for arranging the data on Y1, Y2, Y3, and Y4 in a vector **Y**. The other added advantage is the availability of the test for sphericity for the prescribed contrasts and the Type H structure as well as the computation of the estimates of ϵ, to adjust the degrees of freedom, in case these conditions are not met. To indicate that the repeated measures are taken across TIME, which in this case has four levels, the SAS statement is

```
repeated time 4;
```

If we want, we can specify the levels of TIME (5, 10, 15, and 20 in our example) within parentheses following 4, the number of levels. Further, any other relevant variable name to represent the repeated measures is also applicable. For example, in spatial data the measurements may be named TRANSECT.

Tests on various contrasts of $E(Y1), \ldots, E(Y4)$ can also be performed. While the default choice is the difference between the means of the responses at various time points and the last time point, we can choose any other time point instead of the last for these contrasts. For example, to obtain the contrasts $E(Y1) - E(Y2)$, $E(Y1) - E(Y3)$ and $E(Y1) - E(Y4)$ we specify the CONTRAST(1) transformation after TIME 4 in the REPEATED statement given above. The number 1 within parentheses ( ) indicates that time point 1 is taken as the base or reference point for CONTRAST comparisons. Many other choices for contrasts may be used. For example, PROFILE can be used if we want the contrasts to consist of successive differences such as $E(Y1) - E(Y2)$, $E(Y2) - E(Y3)$, $E(Y3) - E(Y4)$. The contrasts using the orthogonal polynomials (of degree $p - 1$ if there are p time points) can be obtained by specifying the POLYNOMIAL transformation in the REPEATED statement. This choice is especially useful if we want to study time trends. In fact, we will utilize this option later on in this example.

Returning to the univariate split plot analysis of heart rate data, suppose we want to fit the model in Equation 5.11 and examine if this split plot model is appropriate for the analysis. The statement

```
repeated time 4 profile/summary printm printe;
```

given in Program 5.10 first fits the split plot model given in Equation 5.11 and examines its appropriateness. If $\boldsymbol{\Sigma}$ is the p by p variance-covariance matrix of responses collected over time, then, program tests for the sphericity of $\mathbf{C\Sigma C'}$, where $\mathbf{C}$ is the profile matrix given by

$$\mathbf{C} = \begin{bmatrix} 1 & -1 & 0 & \ldots & 0 & 0 \\ 0 & 1 & -1 & \ldots & 0 & 0 \\ \vdots & & & & & \\ 0 & 0 & 0 & \ldots & 1 & -1 \end{bmatrix}_{(p-1) \times p}.$$

In the case of heart rate data it simply is

$$\begin{bmatrix} 1 & -1 & 0 & 0 \\ 0 & 1 & -1 & 0 \\ 0 & 0 & 1 & -1 \end{bmatrix}.$$

In addition, the sphericity test is also applied on the matrix $\mathbf{D\Sigma D}'$, where $\mathbf{D}$ is a $(p-1)$ by p matrix of orthogonal contrasts. This test is invariant of the choice of suborthogonal matrix $\mathbf{D}$ and is used to test for the Type H condition. If the matrix $\mathbf{C}$ (specified in the REPEATED statement) itself is the matrix of orthogonal contrasts, then the two tests result in identical output. Observe that PROFILE or CONTRAST transformations are not the matrices of orthogonal contrasts. Therefore, we use Mauchly's test described in Section 5.2.2 to test the sphericity of $\mathbf{C\Sigma C}'$ and $\mathbf{D\Sigma D}'$.

Program 5.10 results in Output 5.10. With the choice of contrast matrix $\mathbf{C}$ as PROFILE matrix, the null hypothesis of the sphericity of $\mathbf{C\Sigma C}'$ is rejected. The observed value of approximate χ_5^2 is 7.7411 with a large p value of 0.1711. This part of the output has been suppressed. To test if the Type H structure can be assumed, we apply Mauchly's test to $\mathbf{D\Sigma D}'$ (see Section 5.2.2). In view of the observed value of approximate $\chi_5^2 = 8.0703$ with a p value of 0.1524, we do not reject this null hypothesis. This suggests that we can probably analyze these data using the split plot model, and we may not need to adjust the degrees of freedom for certain resulting F ratios, using $\hat{\epsilon}_{GG}$ or $\hat{\epsilon}_{HF}$.

```
/* Program 5.10 */

options ls=64 ps=45 nodate nonumber;
title1 'Output 5.10';
title2 'Repeated Measures Analysis of Heart Rate Data';
data heart;
infile 'heart.dat';
input drug $ y1 y2 y3 y4;
proc glm data = heart;
class drug ;
model y1 y2 y3 y4 = drug/ nouni ;
repeated time 4 profile/ printe;
run;
proc glm data = heart;
class drug ;
model y1 y2 y3 y4 = drug/ nouni ;
repeated time 4 polynomial/summary printm printe;
run;
```

Output 5.10

```
                            Output 5.10
              Repeated Measures Analysis of Heart Rate Data

                    General Linear Models Procedure
                  Repeated Measures Analysis of Variance

        Test for Sphericity: Mauchly's Criterion = 0.6641817
           Chisquare Approximation = 8.0703245 with 5 df
                    Prob > Chisquare = 0.1524


                    Applied to Orthogonal Components:
        Test for Sphericity: Mauchly's Criterion = 0.6641817
           Chisquare Approximation = 8.0703245 with 5 df
                    Prob > Chisquare = 0.1524



          Tests of Hypotheses for Between Subjects Effects

     Source              DF        Type III SS   F Value    Pr > F

     DRUG                 2       1314.8125000      5.92    0.0092

     Error               21       2332.8125000
```

```
     Univariate Tests of Hypotheses for Within Subject Effects

Source: TIME
                                                    Adj  Pr > F
     DF  Type III SS  Mean Square  F Value  Pr > F   G - G    H - F
      3  279.208333    93.069444    12.68   0.0001   0.0001   0.0001

Source: TIME*DRUG
                                                    Adj  Pr > F
     DF  Type III SS  Mean Square  F Value  Pr > F   G - G    H - F
      6  528.354167    88.059028    12.00   0.0001   0.0001   0.0001

Source: Error(TIME)

     DF  Type III SS  Mean Square
     63  462.437500    7.340278


          Greenhouse-Geisser Epsilon = 0.7947
              Huynh-Feldt Epsilon = 0.9887

         Analysis of Variance of Contrast Variables

 TIME.N represents the nth degree polynomial contrast for TIME

Contrast Variable: TIME.1

Source               DF        Type III SS  F Value   Pr > F

MEAN                  1          8.53333333    0.72   0.4042
DRUG                  2         85.40416667    3.63   0.0444

Error                21        247.26250000

Contrast Variable: TIME.2

Source               DF        Type III SS  F Value   Pr > F

MEAN                  1        234.37500000   46.30   0.0001
DRUG                  2        398.31250000   39.34   0.0001

Error                21        106.31250000

Contrast Variable: TIME.3

Source               DF        Type III SS  F Value   Pr > F

MEAN                  1         36.30000000    7.00   0.0151
DRUG                  2         44.63750000    4.31   0.0271

Error                21        108.86250000
```

Next, note that the remaining part of Output 5.10 essentially includes all the information provided by Output 5.9. The usual ANOVA F test for the interaction DRUG*TIME is significant (p value = 0.0001). No adjustments are needed since the null hypothesis of Type H structure has been accepted, and hence we assume this structure. If we also want to conduct individual testing for certain specific contrasts, we can do it using the CONTRAST statement as shown in Program 5.8.

5.3.4   A Study of Time Trends

If an interaction between the treatment and time exists, we are interested in studying the time trends for individual treatments and in identifying the differences in such trends if any. The sample profile plots provide some visual insight about such trends and the differences between them, but we need a formal study. This can be done by partitioning the sum of squares corresponding to the variable TIME as well as the interaction TREAT-MENT*TIME into the independent sums of squares corresponding to various degrees of orthogonal polynomials up to $(p-1)^{th}$ degree. When orthogonal polynomial contrasts are used, the sums of squares corresponding to successive contrasts provide us with necessary information. In SAS, we can accomplish this by using the POLYNOMIAL transformation in the REPEATED statement as shown in Program 5.10.

EXAMPLE 6   *Heart Rate Data (continued)*   Since $p = 4$, we can fit up to a third-degree polynomial and we can accordingly partition the sum of squares corresponding to the variable TIME as well as TIME*DRUG into three orthogonal components, namely, linear, quadratic, and cubic. For the heart rate data in Output 5.10, these components are respectively labeled TIME.1, TIME.2, and TIME.3. The three sums of squares corresponding to the MEAN transformation provide the orthogonal partitioning of the sum of squares for the variable TIME. The sums of squares corresponding to DRUG give the orthogonal decomposition of the sum of squares corresponding to the interaction TIME*DRUG (say, INT). We perform similar partitioning for the error sum of squares. The results are given in Table 5.1. The values are taken from Output 5.10.

The analysis of variance is carried out separately for each of the three contrasts. If $z^{(h)}$ represents the hth contrast, then the linear model for the hth contrast $h = 1, 2, 3$ is a one-way classification model in the drug effect $\alpha_i^{(h)}$, that is,

$$z_{ij}^{(h)} = \mu^{(h)} + \alpha_i^{(h)} + \epsilon_{ij}^{(h)}, \ i = 1, 2, 3, \ j = 1, \dots, 8.$$

All the usual assumptions for this model are made. There are two questions that could be asked:

Are the contrast means zero?   If all contrast means are zero, then it implies that there is no treatment effect.

Is there an interaction between the drug and trend?   If there is no interaction between drug and trend, the situation corresponds to the case when all $\alpha_i^{(h)}$ are zero.

Thus, we may individually test

$$H_0^{(a)} : \mu^{(h)} = 0$$

$$\text{and } H_0^{(b)} : \alpha_1^{(h)} = \alpha_2^{(h)} = \alpha_3^{(h)} = 0$$

for $h = 1, 2, 3$. We recall that in the split plot experiment, we rejected the null hypothesis of no DRUG effect and of no TIME*DRUG interaction. The tests performed here

TABLE 5.1 Partitioning of SS into Othogonal Polynomial Contrasts

Contrast	SS(TIME)	df	SS(INT)	df	SS(ERROR)	df
Linear	8.5333	1	85.4042	2	247.2625	21
Quadratic	234.3750	1	398.3125	2	106.3125	21
Cubic	36.3000	1	44.6375	2	108.8625	21
Total	279.2083	3	528.3542	6	462.4375	63

would provide the cause of rejection in each of the two cases. Output 5.10 shows that the DRUG for TIME.1 (linear contrast) is marginally significant (p value $= 0.0444$). It is strongly significant (p value $= 0.0001$) for TIME.2 (quadratic contrast) and is reasonable (p value $= 0.0271$) in the case of TIME.3 (the cubic contrast). This indicates that there is sufficient evidence to assume that the response curves are quite different for the three drugs. Since the response curves are not the same for the three drugs, it is currently of no interest to test if the contrast means are all zero, because it is equivalent to testing that there is no treatment effect. These hypotheses would have been of interest if $H_0^{(b)}$ had been accepted for all the three contrasts.

A few comments are in order with respect to the analysis of contrasts outlined here. First of all, since two or more hypotheses are being tested to draw the simultaneous inference, we can use a Bonferroni approach to choose the appropriate levels of significance for each individual hypothesis (we have used this approach in the context of confidence intervals in Chapter 3). Secondly, the analysis of contrasts may be subject to serious error if the error variances for these contrasts are not nearly equal. Thus, we should be especially careful in the interpretation of these results if the Type H conditions are not met. Finally, the split plot analysis may be more useful than the present analysis of orthogonal contrasts. The reason for this is that when the Type H conditions are not met the degrees of freedom of the F tests in the former analysis can be adjusted to get approximate tests. Finally a multivariate approach to trend analysis can be taken using the growth curve models.

Given the two different approaches to the repeated measure data, namely, multivariate and univariate, a natural question to ask is, which of the two approaches is superior? This question cannot be answered in a clear-cut way. Due to the data collection scheme and design, the problem is multivariate in nature and hence, theoretically, the multivariate approach is the "correct" approach for the problem. However, the multivariate tests usually have lower power than the corresponding univariate tests. This is more so when the sphericity assumptions are satisfied. Thus, the power of the tests can possibly be increased by following the univariate approach when the assumptions on the covariance structure (such as compound symmetry or Type H conditions) can be made.

Thus, although the multivariate approach is always a legitimate approach, we may be able to obtain more conclusive results and possibly gain more insight by using the univariate split plot models. Crowder and Hand (1990), however, state that in the absence of sphericity of orthogonal contrasts, there is little to choose between the multivariate tests and the univariate tests where the degrees of freedom of univariate F tests have been modified using either $\hat{\epsilon}_{GG}$ or $\hat{\epsilon}_{HF}$. Of course, in certain cases circumstances such as small sample size permit only univariate analysis of data.

There are certain precautions that we must take in analyzing repeated measures data especially when using the univariate approach. The failure of the user or the software to appropriately identify the fixed and random effects in the model and to form appropriate F ratios for the tests may result in incorrect analyses and hence in invalid conclusions. Schaefer (1994) provides an excellent account, with special reference to SAS, of possible pitfalls in using the default packaged analysis: in PROC GLM, all the effects are by default considered fixed unless indicated otherwise by using a RANDOM statement and then using the TEST option to perform the appropriate test. Using simulation studies, he found that Type I and Type II errors in the testing of hypotheses can both be greatly affected if inappropriate tests are performed.

5.4 Factorial Designs

When the treatment combinations are made up of various levels of several factors, then both the multivariate and univariate approaches are generalized in a straightforward way. It is so because all these situations can still be dealt within the general framework of multivariate linear model. Also, the assumption of compound symmetry of covariance structure needed in the univariate analysis is on the error. Therefore, what design has been used for data collection does not affect the univariate approach. For the multivariate approach, the MANOVA partitioning of the total sums of squares and crossproduct matrix has been seen to be the straightforward generalization of the univariate ANOVA and therefore the multivariate approach essentially parallels that for the univariate ANOVA. This similarity has already been discussed in Chapters 3 and 4. We illustrate the approach for the factorial designs using a three-factor experiment originally discussed by Box (1950).

EXAMPLE 7   *A Two-Way Factorial Experiment, Abrasion Data*   Box (1950) presented data on fabric weight loss due to abrasion. Some of the fabrics were given a surface treatment (SURFTRT) and some were not, leading to two levels, YES and NO, for SURFTRT. Two fillers (FILL) A and B were used at three proportions (PROP), namely, 25%, 50%, and 75%, and the weight losses were recorded at three successive periods, each after 1000 revolutions of the machine which tested the abrasion resistance. Two replicates for each treatment combination in this $2 \times 2 \times 3$ factorial experiment were obtained. Program 5.11 performs the multivariate as well as the univariate analysis of the data.

The repeated measures are taken with respect to the increasing number of revolutions (REVOLUTN). The corresponding values of the dependent variable weight loss are denoted by Y1, Y2, and Y3. We first fit the multivariate model with all main variables and two- and three-factor interactions. The standard MANOVA (not shown) indicates that SURFTRT, FILL, SURFTRT*FILL, PROP, and FILL*PROP are all highly significant. For illustration we examine only the profiles corresponding to variable PROP. Thus, all the comparisons are about the average weight losses due to various proportions of the fillers. The three sequential hypotheses to be tested are, are the PROP profiles parallel? given that they are parallel, are the PROP profiles coincidental? given that they are coincidental, are the PROP profiles horizontal?

Each of these three hypotheses is specified as **LBM = 0** in the linear model setup with the specific choices of matrices of **L** and **M** in the three hypotheses. Since the three hypotheses are about the variable PROP only, the **L** matrix is specified by using the CONTRAST statement with PROP as the variable of interest. The **M** matrices are appropriately defined with the M= specification on the MANOVA statement. Specifically, the SAS statements specifying **L** and **M** for each hypothesis are listed below.

Are the PROP profiles parallel?

```
contrast '"prop-parallel?"' prop 1 0 -1,
                            prop 0 1 -1;
manova h=prop m=(1 -1 0,
                 1 0 -1);
```

Given that they are parallel, are the PROP profiles coincidental?

```
contrast '"prop-coincidental?"' prop 1 0 -1
                                prop 0 1 -1;
manova h=prop m=(1 1 1);
```

Given that they are coincidental, are the PROP profiles horizontal?

```
contrast '"prop-horizontal?"' intercept 1;
manova h=prop m=(1 -1 0,
                 1 0 -1);
```

```
/* Program 5.11 */

options ls=64 ps=45 nodate nonumber;
title1 ' Output 5.11';
data box;
infile 'box.dat';
input surftrt $ fill $ prop y1 y2 y3 ;
title2 'Repeated Measures in Factorials: Tire Wear Data';
proc glm data = box;
class surftrt fill prop;
model y1 y2 y3 =surftrt|fill|prop/nouni;
contrast '"prop-parallel?"' prop 1 0 -1,
                            prop  0 1 -1;
contrast '"prop-horizontal?"' intercept 1;
manova h=prop m= (1 -1 0, 1 0 -1)/printe printh;
contrast '"prop-concidental?"'  prop  1 0 -1,
                                prop 0 1 -1 ;
manova h = prop m=(1 1 1) /printe printh;
run;
proc glm data = box;
class surftrt fill prop;
model y1 y2 y3  = surftrt|fill|prop/ nouni ;
repeated revolutn 3 polynomial/summary printm printe ;
title2 'Univariate Split Plot Analysis of Tire Wear Data';
run;
data boxsplit;
set box;
array yy{3} y1-y3;
subject+1;
do time=1 to 3;
y=yy(time);
output;
end;
run;
proc mixed data = boxsplit  method = reml ;
class surftrt fill prop subject;
model y =  surftrt fill prop surftrt*fill surftrt*prop
fill*prop  surftrt*time  fill*time  prop*time
surftrt*fill*time surftrt*prop*time fill*prop*time/chisq;
repeated /type = ar(1) subject = subject r ;
title2 'Analysis of Tire Wear Data Using PROC MIXED';
run;
```

Output 5.11

```
                      Output 5.11
       Repeated Measures in Factorials: Tire Wear Data

               General Linear Models Procedure
               Multivariate Analysis of Variance

           Manova Test Criteria and F Approximations for
          the Hypothesis of no Overall "prop-parallel?" Effect
         on the variables defined by the M Matrix Transformation
             H = Contrast SS&CP Matrix for "prop-parallel?"
                     E = Error SS&CP Matrix

                 S=2    M=-0.5    N=4.5
```

Statistic	Value	F	Num DF	Den DF	Pr > F
Wilks' Lambda	0.469985	2.5227	4	22	0.0701
Pillai's Trace	0.550475	2.2786	4	24	0.0904
Hotelling-Lawley Trace	1.084195	2.7105	4	20	0.0594
Roy's Greatest Root	1.042434	6.2546	2	12	0.0138

NOTE: F Statistic for Roy's Greatest Root is an upper bound.
NOTE: F Statistic for Wilks' Lambda is exact.

Manova Test Criteria and Exact F Statistics for
the Hypothesis of no Overall "prop-concidental?" Effect
on the variables defined by the M Matrix Transformation
H = Contrast SS&CP Matrix for "prop-concidental?"
E = Error SS&CP Matrix

S=1   M=0   N=5

Statistic	Value	F	Num DF	Den DF	Pr > F
Wilks' Lambda	0.216071	21.769	2	12	0.0001
Pillai's Trace	0.783929	21.769	2	12	0.0001
Hotelling-Lawley Trace	3.628115	21.769	2	12	0.0001
Roy's Greatest Root	3.628115	21.769	2	12	0.0001

Univariate Split Plot Analysis of Tire Wear Data

General Linear Models Procedure
Repeated Measures Analysis of Variance
Univariate Tests of Hypotheses for Within Subject Effects

Source: REVOLUTN

| | | | | | Adj Pr > F | |
DF	Type III SS	Mean Square	F Value	Pr > F	G - G	H - F
2	60958.5278	30479.2639	160.68	0.0001	0.0001	0.0001

Source: REVOLUTN*SURFTRT

| | | | | | Adj Pr > F | |
DF	Type III SS	Mean Square	F Value	Pr > F	G - G	H - F
2	8248.0278	4124.0139	21.74	0.0001	0.0001	0.0001

Source: REVOLUTN*FILL

| | | | | | Adj Pr > F | |
DF	Type III SS	Mean Square	F Value	Pr > F	G - G	H - F
2	18287.6944	9143.8472	48.20	0.0001	0.0001	0.0001

Source: REVOLUTN*SURFTRT*FILL

| | | | | | Adj Pr > F | |
DF	Type III SS	Mean Square	F Value	Pr > F	G - G	H - F
2	2328.0833	1164.0417	6.14	0.0070	0.0111	0.0070

Source: REVOLUTN*PROP

| | | | | | Adj Pr > F | |
DF	Type III SS	Mean Square	F Value	Pr > F	G - G	H - F
4	1762.8056	440.7014	2.32	0.0857	0.1002	0.0857

```
Source: REVOLUTN*SURFTRT*PROP
```

					Adj Pr > F	
DF	Type III SS	Mean Square	F Value	Pr > F	G - G	H - F
4	685.9722	171.4931	0.90	0.4772	0.4658	0.4772

```
Source: REVOLUTN*FILL*PROP
```

					Adj Pr > F	
DF	Type III SS	Mean Square	F Value	Pr > F	G - G	H - F
4	1415.6389	353.9097	1.87	0.1493	0.1633	0.1493

```
Source: REVOLUTN*SURFTRT*FILL*PROP
```

					Adj Pr > F	
DF	Type III SS	Mean Square	F Value	Pr > F	G - G	H - F
4	465.9167	116.4792	0.61	0.6566	0.6308	0.6566

```
Source: Error(REVOLUTN)
```

DF	Type III SS	Mean Square
24	4552.6667	189.6944

$$\text{Greenhouse-Geisser Epsilon} = 0.8384$$
$$\text{Huynh-Feldt Epsilon} = 1.8522$$

Analysis of Variance of Contrast Variables

REVOLU.N represents the nth degree
polynomial contrast for REVOLUTN

Contrast Variable: REVOLU.1

Source	DF	Type III SS	F Value	Pr > F
MEAN	1	60705.18750000	565.91	0.0001
SURFTRT	1	7676.02083333	71.56	0.0001
FILL	1	9436.02083333	87.96	0.0001
SURFTRT*FILL	1	1938.02083333	18.07	0.0011
PROP	2	1035.12500000	4.82	0.0290
SURFTRT*PROP	2	191.54166667	0.89	0.4350
FILL*PROP	2	260.54166667	1.21	0.3309
SURFTRT*FILL*PROP	2	255.79166667	1.19	0.3371
Error	12	1287.25000000		

Contrast Variable: REVOLU.2

Source	DF	Type III SS	F Value	Pr > F
MEAN	1	253.34027778	0.93	0.3536
SURFTRT	1	572.00694444	2.10	0.1727
FILL	1	8851.67361111	32.53	0.0001
SURFTRT*FILL	1	390.06250000	1.43	0.2543
PROP	2	727.68055556	1.34	0.2991
SURFTRT*PROP	2	494.43055556	0.91	0.4292
FILL*PROP	2	1155.09722222	2.12	0.1625
SURFTRT*FILL*PROP	2	210.12500000	0.39	0.6879
Error	12	3265.41666667		

```
              Analysis of Tire Wear Data Using PROC MIXED
                         The MIXED Procedure

                        R Matrix for SUBJECT 1

       Row          COL1           COL2           COL3

         1    368.56581667   -166.1922475    74.93875414
         2   -166.1922475    368.56581667   -166.1922475
         3     74.93875414   -166.1922475    368.56581667
```

Output 5.11 shows that while the null hypothesis of parallel profiles for PROP is not rejected at 5% level of significance (with a p value for Wilks' Λ equal to 0.0701), the hypothesis of coincidental profiles is rejected with $\Lambda = 0.2161$ (leading to an observed value of F (2,12) as 21.769 which corresponds to a very small p value of 0.0001).

To perform the univariate split plot analysis, we use the REPEATED statement. The corresponding time variable is defined here as REVOLUTN as the repeated measures are taken for the three increasing numbers of revolutions, namely, 1,000, 2,000, and 3,000. If we are interested in studying the effects of various orthogonal polynomial contrasts, we must choose the POLYNOMIAL transformation for the type of orthogonal contrasts. The corresponding SAS statement is

```
repeated revolutn 3 polynomial;
```

In Program 5.11 we have made certain other choices such as printing the corresponding **E** and **M** matrices and only summary output. The **M** matrix here is a 2 by 3 matrix of first- and second-degree orthogonal polynomial coefficients and the 2 by 2 matrix **E** is the matrix of the corresponding error SS&CP.

Output 5.11 shows that the interaction REVOLUTN*SURFTRT*FILL is highly significant (with a p value $= 0.0070$) and respective interactions of REVOLUTN with SURFTRT and FILL are very highly significant (both p values $= 0.0001$). The interaction of REVOLUTN with PROP is only marginally significant (p value $= 0.0857$). This indicates that the abrasion curves for various levels of the variables SURFTRT, FILL, and PROP are not parallel. The univariate F tests adjusted by using $\hat{\epsilon}_{GG}(= 0.8384)$ or $\hat{\epsilon}_{HF}(= 1.0$, since it cannot exceed 1) for various interactions with REVOLUTN lead to the same conclusions. These adjustments, however, may not be necessary as Mauchly's test for the sphericity of orthogonal contrasts provides sufficient evidence (p value $= 0.3079$) to assume sphericity. This part has not been included in Output 5.11.

The tests on the linear and quadratic contrasts respectively denoted by REVOLUTN.1 and REVOLUTN.2 reveal that the quadratic contrast is not significant for any effect or interaction except FILL (p value $= 0.0001$). However, the linear contrast is significant for all main variables and the SURFTRT*FILL interaction. The data can also be analyzed under various covariance structures, other than compound symmetry using the MIXED procedure which will be discussed in detail in Chapter 6. While most of the output has been suppressed, the appropriate statements under AR(1) covariance structure and REML estimation procedure are included in Program 5.11. One especially interesting observation is that some of the off-diagonal elements of the estimates of the **R** matrix (referred to as $\Sigma_{subject}$ in the previous example) are negative. Normally we would not expect the negative correlations between the repeated measures of these weight losses. Lindsey (1993, p. 83) interprets such an occurrence as evidence of a situation where there is a greater variability within the experiment or subject than among the experiments or subjects.

EXAMPLE 8   *Two-Factor Experiment with Both Repeated Measures Factors*   In a factorial design, if there are repeated measures on more than two variables then the analysis of the previous section can be applied in a straightforward manner. For example, consider this example:

three participants in an experiment were given a large amount of a sleep-inducing drug on the day before the experiment. The next day, they were given placebos. The participants were tested in the morning (AM) and afternoon (PM) of the two different days. Each participant was given a stimulus, and his or her reaction (the response variable) was timed. See Cody and Smith (1991, p. 182). The problem considered here is to determine whether the drug had any effect on the reaction time and the effects are the same for AM and PM.

Since each subject is measured under two levels of TIME (AM, PM) and for two levels of DRUG (DRUG1 (Placebo), DRUG2), it is a 2 by 2 factorial experiment with both TIME and DRUG as *repeated measures variables* or *repeated measures factors*. The data are given below and the SAS code for analyzing these data is given in Program 5.12.

	DRUG1	DRUG1	DRUG2	DRUG2
SUBJECT	AM	PM	AM	PM
1	77	67	82	72
2	84	76	90	80
3	102	92	109	97

Both the univariate and multivariate analysis are performed using the SAS statement

```
repeated drug 2, time 2;
```

One important thing to remember is that the order in which the variables are written in the REPEATED statement and the order in which data are presented in the INPUT statement must correspond. For every level of variable DRUG there are two levels of the variable TIME and the REPEATED statement given above reads exactly that way. That is, level 1 of variable DRUG and levels 1 and 2 of variable TIME are selected first, level 2 of variable DRUG and levels 1 and 2 of variable TIME are selected next, and so on. The logic behind writing the variables in the REPEATED statement is the same as the logic behind nested DO loops.

```
/*  Program 5.12 */

option ls=64 ps=45 nodate nonumber;
title1 'Output 5.12';
title2 'Analysis with Two Repeated Factors';
data react;
input y1 y2 y3 y4;
lines;
77 67 82 72
84 76 90 80
102 92 109 97
;
proc glm;
model y1-y4= /nouni;
repeated drug 2, time 2;
run;
```

Output 5.12

```
                         Output 5.12
                Analysis with Two Repeated Factors

                  General Linear Models Procedure
                Repeated Measures Analysis of Variance

             Manova Test Criteria and Exact F Statistics for
                  the Hypothesis of no DRUG Effect
       H = Type III SS&CP Matrix for DRUG    E = Error SS&CP Matrix
```

S=1   M=-0.5   N=0

Statistic	Value	F	Num DF	Den DF	Pr > F
Wilks' Lambda	0.007752	256	1	2	0.0039
Pillai's Trace	0.992248	256	1	2	0.0039
Hotelling-Lawley Trace	128	256	1	2	0.0039
Roy's Greatest Root	128	256	1	2	0.0039

Manova Test Criteria and Exact F Statistics for
the Hypothesis of no TIME Effect
H = Type III SS&CP Matrix for TIME   E = Error SS&CP Matrix

S=1   M=-0.5   N=0

Statistic	Value	F	Num DF	Den DF	Pr > F
Wilks' Lambda	0.006623	300	1	2	0.0033
Pillai's Trace	0.993377	300	1	2	0.0033
Hotelling-Lawley Trace	150	300	1	2	0.0033
Roy's Greatest Root	150	300	1	2	0.0033

Manova Test Criteria and Exact F Statistics for
the Hypothesis of no DRUG*TIME Effect
H = Type III SS&CP Matrix for DRUG*TIME   E = Error SS&CP Matrix

S=1   M=-0.5   N=0

Statistic	Value	F	Num DF	Den DF	Pr > F
Wilks' Lambda	0.333333	4	1	2	0.1835
Pillai's Trace	0.666667	4	1	2	0.1835
Hotelling-Lawley Trace	2	4	1	2	0.1835
Roy's Greatest Root	2	4	1	2	0.1835

An examination of Output 5.12 reveals that there is no interaction between the variables TIME and DRUG (p value $= 0.1835$). However, the variable DRUG with a p value of 0.0039 and the variable TIME with a p value of 0.0033 are both highly significant.

EXAMPLE 9   *Three-Factor Experiment with Two Repeated Measures Factors*   In an educational testing program, students from two groups, namely, those with relatively higher socioeconomic status (GP1) and with lower socioeconomic status (GP2), were tested and their scores in the tests recorded. See Cody and Smith (1991, pp. 194–195). The study was conducted for three years (denoted by 1, 2, and 3) during each of the two seasons, FALL and SPRING, for a group of 10 students. Each group consisted of five subjects. In this setup there are three variables, namely, GROUP, YEAR, and SEASON. Repeated measures on each of the ten subjects are available for the variables YEAR and SEASON. In that sense, YEAR and SEASON are two repeated measures variables.

The three problems of interest are to test

- whether students do better on a reading comprehension test in the Spring than in the Fall,
- whether the differences in the mean scores become negligible as students get older,
- whether these differences in the mean scores are more prominent for one socio-economic group than the other.

In order to find answers to these problems, we first determine whether there are significant main effects and interactions. Once that is determined, if needed, we can perform an analysis of the means, applying multiple comparison techniques, to conduct pairwise differences between the variables. However, we restrict ourselves to the determination of the significance of the main effects and their interactions. We assume that the underlying variances of responses for both seasons and for all three years are the same. The data and SAS code are given in Program 5.13 and the corresponding output in Output 5.13. The REPEATED statement is used to perform both the multivariate and univariate analyses.

```
/* Program 5.13 */

options ls=64 ps=45 nodate nonumber;
title1 'Output 5.13';
title2 'Three Factors Case with Two Repeated Factors';
data read;
input group y1-y6;
lines;
1 61 50 60 55 59 62
1 64 55 62 57 63 63
1 59 49 58 52 60 58
1 63 59 65 64 67 70
1 62 51 61 56 60 63
2 57 42 56 46 54 50
2 61 47 58 48 59 55
2 55 40 55 46 57 52
2 59 44 61 50 63 60
2 58 44 56 49 55 49
;
/* Source: Cody, R. P./Smith, J. K. APPLIED STATISTICS AND SAS
PROGRAMMING LANGUAGE, 3/E, 1991, p. 194.  Reprinted by permission of
Prentice-Hall, Inc. Englewood Cliffs, N.J. */
proc glm data=read;
class group;
model y1-y6=group/nouni;
repeated year 3, season 2;
run;
proc glm data=read;
class group;
model y1-y6=group/nouni;
repeated year 3(1 2 3) polynomial,
season 2/summary nom nou;
run;
```

Output 5.13

```
                         Output 5.13
            Three Factors Case with Two Repeated Factors

                 General Linear Models Procedure
                 Repeated Measures Analysis of Variance

             Manova Test Criteria and Exact F Statistics for
                    the Hypothesis of no YEAR Effect
        H = Type III SS&CP Matrix for YEAR   E = Error SS&CP Matrix

                   S=1     M=0     N=2.5

   Statistic             Value       F    Num DF  Den DF  Pr > F

   Wilks' Lambda       0.195582   14.395     2       7    0.0033
   Pillai's Trace      0.804418   14.395     2       7    0.0033
```

```
Hotelling-Lawley Trace 4.112941    14.395        2        7   0.0033
Roy's Greatest Root    4.112941    14.395        2        7   0.0033
```

Manova Test Criteria and Exact F Statistics for
the Hypothesis of no YEAR*GROUP Effect
H = Type III SS&CP Matrix for YEAR*GROUP
E = Error SS&CP Matrix

S=1   M=0   N=2.5

Statistic	Value	F	Num DF	Den DF	Pr > F
Wilks' Lambda	0.96176	0.1392	2	7	0.8724
Pillai's Trace	0.03824	0.1392	2	7	0.8724
Hotelling-Lawley Trace	0.03976	0.1392	2	7	0.8724
Roy's Greatest Root	0.03976	0.1392	2	7	0.8724

Manova Test Criteria and Exact F Statistics for
the Hypothesis of no SEASON Effect
H = Type III SS&CP Matrix for SEASON   E = Error SS&CP Matrix

S=1   M=-0.5   N=3

Statistic	Value	F	Num DF	Den DF	Pr > F
Wilks' Lambda	0.034362	224.82	1	8	0.0001
Pillai's Trace	0.965638	224.82	1	8	0.0001
Hotelling-Lawley Trace	28.10193	224.82	1	8	0.0001
Roy's Greatest Root	28.10193	224.82	1	8	0.0001

Manova Test Criteria and Exact F Statistics for
the Hypothesis of no SEASON*GROUP Effect
H = Type III SS&CP Matrix for SEASON*GROUP
E = Error SS&CP Matrix

S=1   M=-0.5   N=3

Statistic	Value	F	Num DF	Den DF	Pr > F
Wilks' Lambda	0.177593	37.047	1	8	0.0003
Pillai's Trace	0.822407	37.047	1	8	0.0003
Hotelling-Lawley Trace	4.630854	37.047	1	8	0.0003
Roy's Greatest Root	4.630854	37.047	1	8	0.0003

Manova Test Criteria and Exact F Statistics for
the Hypothesis of no YEAR*SEASON Effect
H = Type III SS&CP Matrix for YEAR*SEASON
E = Error SS&CP Matrix

S=1   M=0   N=2.5

Statistic	Value	F	Num DF	Den DF	Pr > F
Wilks' Lambda	0.037144	90.727	2	7	0.0001
Pillai's Trace	0.962856	90.727	2	7	0.0001
Hotelling-Lawley Trace	25.92199	90.727	2	7	0.0001
Roy's Greatest Root	25.92199	90.727	2	7	0.0001

```
            Manova Test Criteria and Exact F Statistics for
             the Hypothesis of no YEAR*SEASON*GROUP Effect
            H = Type III SS&CP Matrix for YEAR*SEASON*GROUP
                      E = Error SS&CP Matrix

                    S=1      M=0      N=2.5

Statistic                    Value       F     Num DF  Den DF  Pr > F

Wilks' Lambda              0.916038   0.3208      2       7    0.7357
Pillai's Trace            0.083962   0.3208      2       7    0.7357
Hotelling-Lawley Trace 0.091658   0.3208      2       7    0.7357
Roy's Greatest Root      0.091658   0.3208      2       7    0.7357
```

```
              Repeated Measures Analysis of Variance
            Tests of Hypotheses for Between Subjects Effects

Source                 DF        Type III SS  F Value   Pr > F

GROUP                   1       680.0666667    13.54    0.0062

Error                   8       401.6666667
```

```
    Univariate Tests of Hypotheses for Within Subject Effects

Source: YEAR
                                                   Adj  Pr > F
    DF Type III SS Mean Square  F Value  Pr > F   G - G    H - F
     2  252.033333  126.016667    26.91  0.0001  0.0002   0.0001

Source: YEAR*GROUP
                                                   Adj  Pr > F
    DF Type III SS Mean Square  F Value  Pr > F   G - G    H - F
     2    1.033333    0.516667     0.11  0.8962  0.8186   0.8700

Source: Error(YEAR)

    DF Type III SS Mean Square
    16   74.933333    4.683333

              Greenhouse-Geisser Epsilon = 0.6757
                 Huynh-Feldt Epsilon = 0.8658

Source: SEASON
                                                   Adj  Pr > F
    DF Type III SS Mean Square  F Value  Pr > F   G - G    H - F
     1  680.066667  680.066667   224.82  0.0001    .        .

Source: SEASON*GROUP
                                                   Adj  Pr > F
    DF Type III SS Mean Square  F Value  Pr > F   G - G    H - F
     1  112.066667  112.066667    37.05  0.0003    .        .

Source: Error(SEASON)

    DF Type III SS Mean Square
     8   24.200000    3.025000
```

Source: YEAR*SEASON

DF	Type III SS	Mean Square	F Value	Pr > F	Adj Pr > F G - G	Adj Pr > F H - F
2	265.433333	132.716667	112.95	0.0001	0.0001	0.0001

Source: YEAR*SEASON*GROUP

DF	Type III SS	Mean Square	F Value	Pr > F	Adj Pr > F G - G	Adj Pr > F H - F
2	0.433333	0.216667	0.18	0.8333	0.7592	0.8168

Source: Error(YEAR*SEASON)

DF	Type III SS	Mean Square
16	18.800000	1.175000

Greenhouse-Geisser Epsilon = 0.7073
Huynh-Feldt Epsilon = 0.9221

Analysis of Variance of Contrast Variables

YEAR.N represents the nth degree polynomial contrast for YEAR

Contrast Variable: YEAR.1

Source	DF	Type III SS	F Value	Pr > F
MEAN	1	490.05000000	31.06	0.0005
GROUP	1	1.25000000	0.08	0.7855
Error	8	126.20000000		

Contrast Variable: YEAR.2

Source	DF	Type III SS	F Value	Pr > F
MEAN	1	14.01666667	4.74	0.0612
GROUP	1	0.81666667	0.28	0.6135
Error	8	23.66666667		

SEASON.N represents the contrast between the
nth level of SEASON and the last

Contrast Variable: SEASON.1

Source	DF	Type III SS	F Value	Pr > F
MEAN	1	4080.40000000	224.82	0.0001
GROUP	1	672.40000000	37.05	0.0003
Error	8	145.20000000		

YEAR.N represents the nth degree polynomial contrast for YEAR
SEASON.N represents the contrast between the
nth level of SEASON and the last

Contrast Variable: YEAR.1*SEASON.1

Source	DF	Type III SS	F Value	Pr > F
MEAN	1	530.45000000	157.17	0.0001
GROUP	1	0.05000000	0.01	0.9061
Error	8	27.00000000		

Contrast Variable: YEAR.2*SEASON.1

Source	DF	Type III SS	F Value	Pr > F
MEAN	1	0.41666667	0.31	0.5903
GROUP	1	0.81666667	0.62	0.4550
Error	8	10.60000000		

The multivariate tests indicate that the YEAR*SEASON*GROUP and YEAR*GROUP interactions are not significant (p values are 0.7357 and 0.8724 respectively) whereas, YEAR and SEASON have significant effects with respective p values 0.0033 and 0.0001. Also significant are the interactions YEAR*SEASON (p value = 0.0001) and SEASON* GROUP (p value = 0.0003). The univariate tests also support these findings. Note from the output that there is a significant difference between the two socio-economic groups (p value = 0.0062). Note also that the values of $\hat{\epsilon}_{GG}$ and $\hat{\epsilon}_{HF}$ are quite large, indicating that Type H structure for the covariance may be satisfied.

In order to understand the nature of the significant effect of the repeated measures variable YEAR, since the levels of it are quantitative (1, 2, 3), it may be useful to analyze the variables obtained by using the POLYNOMIAL transformation. The statement

```
repeated year 3 polynomial, season/summary nom nou;
```

can be used for this purpose. Both NOM and NOU options suppress redundant output. If the time points were not equidistant then the transformation "POLYNOMIAL(t_1, t_2, t_3)" can be used in the REPEATED statement to indicate the time points. The PRINTM option is used to print the contrast transformation. The output for this part of the analysis is shown in Output 5.13. The variable YEAR.N represents the nth degree orthogonal polynomial contrast for the variable YEAR as indicated in the output. Thus, in the output, YEAR.1 represents the first degree (linear) polynomial contrast and YEAR.2 represents the quadratic contrast. The line MEAN, listed under column SOURCE in the output, tests the hypothesis that the linear component of the variable YEAR is zero. Since the mean effect for YEAR.1 is significant (p value = 0.0005), the linear component of YEAR is significantly different from zero. The variable GROUP listed under SOURCE is used to test the hypothesis that the first-order polynomial for the variable YEAR is the same for different levels of the variable GROUP. This hypothesis is not rejected (p value = 0.7855). Similar interpretations are applicable for the contrast YEAR.2. Since the p value for the MEAN effect in this case is 0.0612 there is slight evidence that the quadratic component is different from zero. See Section 5.3.4 for a more detailed description and interpretation of the analysis of the orthogonal polynomial contrasts.

5.5   Analysis in the Presence of Covariates

When data contain covariates that might affect repeated measures tests, it is important to design the analysis to account for the covariate effects. Consider these examples. First, suppose the initial measurements on the subjects *before* the treatments are applied provide the information relevant to the subsequent measurements. Second, suppose various blood pressure drugs are administered to groups of subjects, and blood pressure measurements are taken weekly. Also measured is the amount of sodium consumed by each individual the previous day. See Milliken (1989). Since the amount of sodium in the diet is known to affect blood pressure measurements, a better analysis may be possible by incorporating the sodium content in the previous day's diet as a covariate.

However, the two examples described above are different. In the first case, the value of the covariate is the same for all time points and is specific only to the particular subject; in the second case, the values of covariates may be different for various time points for every subject. In other words, if the subjects are visualized as the whole plots with subplots represented by different time points, then the two situations differ in whether the value of the covariate is the same for all subplots within a whole plot or not. We will contrast these two scenerios by respectively naming them as *subject specific* and *time specific* covariates. Unfortunately, only the first case (subject specific) can be expressed in terms of a standard multivariate linear model. The second case, although quite common, requires that the matrix of regression coefficients corresponding to the covariates be block diagonal, thereby imposing restrictions on the matrix of unknown parameters.

5.5.1   A Multivariate Analysis of Covariance: Subject Specific Covariates

If the values of covariates are the same for all time points for a given subject then the multivariate linear model can be written as

$$\mathbf{Y} = \mathbf{XB} + \mathbf{Z\Gamma} + \mathcal{E},$$

where the matrix $\mathbf{Z}$ represents the matrix of the values of covariates and possibly interaction between the treatments and covariates. The matrix $\mathbf{\Gamma}$ stands for the corresponding unknown coefficients. This model has been described in Chapter 4 along with an appropriate approach to the analysis.

The effect of covariates may be different in various treatment groups. Also, they may influence the measurements taken at different time points very differently, and these two possibilities should be carefully investigated. The first of these possibilities can be examined by testing for the statistical significance of the interactions of the treatment with the covariates. The second possibility can be tested by examining the significance of the interaction between the covariates and time and also possibly that between treatment, covariates, and time. If the covariates have different effects on responses at various time points then the comparison of treatment profiles adjusted for the covariates does not have much meaning. If, however, the covariate*time and the treatment*covariate*time interactions can be assumed to be zero, we can perform the profile analysis with only a slight modification. We will illustrate this in the following example.

EXAMPLE 10   *Subject-Specific Covariates, Diabetic Patients Study Data*   Three groups of diabetic patients, without complications (DINOCOM), with hypertension (DIHYPER), and with postural hypotension (DIHYPOT) respectively and a control (CONTROL) of healthy subjects were asked to perform a small physical task at time zero. A particular response was observed at times $-30, -1, 1, 2, 3, 4, 5, 6, 8, 10, 12$, and 15 minutes. The corresponding variables are denoted by X1, X2, and Y1 through Y10 respectively and the pre-performance responses X1 and X2 are used as the covariates for the repeated measures Y1 through Y10. The data set has certain missing values and in such cases the entire row of the data set on the corresponding subjects is discarded in the SAS analysis. As a result, it is not possible to perform all the desirable analyses on this data set. Hence we illustrate the analysis only for variables Y1, Y2, Y3, and Y4.

The first step in the analysis is to examine if the effects of the two covariates are the same in all treatment groups. To do so, we need to include X1, X2 and their interactions with GROUP, namely, GROUP*X1, and GROUP*X2, in the multivariate linear model. Thus, the corresponding MODEL statement is

```
model y1-y4 =x1 x2 group group*x1 group*x2;
```

and the corresponding MANOVA statement is

```
manova h=group x1 x2 group*x1 group*x2;
```

The NOUNI option in Program 5.14 suppresses the univariate output and the SS3 option produces the Type III SS&CP matrices. The PRINTE and PRINTH options in the MANOVA statement print the **E** and corresponding **H** matrices. The multivariate tests show that neither of the two interactions, GROUP*X1 and GROUP*X2, is significant (respective p values for Wilks' Λ are 0.9420 and 0.7343). Also, there is no overall GROUP effect. The corresponding Wilks' Λ test statistic has a value of 0.6131 with a p value of 0.9248. Further, X1 is not significant (p value $= 0.4753$) but X2 appears to have a very significant effect ($\Lambda = 0.0609$ with a p value of 0.0001). This indicates that X2 should be included in the model for any treatment comparison. Also, since GROUP*X2 was found not to be significant, we can assume that the effect of covariate X2 is same in the four treatment groups.

```
/* Program 5.14 */

options ls=64 ps=45 nodate nonumber;
title1 'Output 5.14';
data task;
infile 'task.dat';
input group$ x1 x2 y1-y10;
title2 'Multivariate Analysis of Covariance (MANCOVA)';
proc glm data=task;
class group;
model y1-y4=x1 x2 group group*x1 group*x2/nouni ss3;
manova  h=group x1 x2 group*x1 group*x2/printe printh;
run;
title2 'MANCOVA: Profile Analysis';
proc glm data = task ;
class group;
model y1-y4 = x1 x2 group/nouni;
contrast '"parallel?"' group 1 -1 0 0, group 1 0 -1 0,
group 1 0 0 -1;
*contrast '"horizontal?"' intercept 1;
manova h=group m=y1-y2,y1-y3,y1-y4/printe printh;
run;
proc glm data = task;
class group;
model y1-y4 = group/nouni;
contrast '"coincidental?"' group 1 -1 0 0,
group 1 0 -1 0, group 1 0 0 -1;
manova h = group m=(1 1 1 1)/printe printh;
run;
title2 'Analysis as in Srivastava and Carter (1983)';
data task;
set task;
z1=y2-y1;
z2=y3-y2;
z3=y4-y3;
```

```
ybar = (y1+y2+y3+y4)/4;
proc glm data = task;
class group;
model ybar = z1 z2 z3 x1 x2 group;
run;
title2 'MANCOVA with Repeated Measures: Using repeated statement';
proc glm data=task;
class group;
model y1-y4=x1 x2 group x1*group x2*group/nouni ss3;
repeated time 4 (1 2 3 4) polynomial/printe;
run;
```

Output 5.14

Output 5.14
Multivariate Analysis of Covariance (MANCOVA)

General Linear Models Procedure
Multivariate Analysis of Variance

Manova Test Criteria and F Approximations for
the Hypothesis of no Overall GROUP Effect
H = Type III SS&CP Matrix for GROUP   E = Error SS&CP Matrix

S=3   M=0   N=4

Statistic	Value	F	Num DF	Den DF	Pr > F
Wilks' Lambda	0.613076	0.4528	12	26.749	0.9248
Pillai's Trace	0.440286	0.516	12	36	0.8903
Hotelling-Lawley Trace	0.546654	0.3948	12	26	0.9529
Roy's Greatest Root	0.307716	0.9231	4	12	0.4822

NOTE: F Statistic for Roy's Greatest Root is an upper bound.

Manova Test Criteria and F Approximations for
the Hypothesis of no Overall X1*GROUP Effect
H = Type III SS&CP Matrix for X1*GROUP   E = Error SS&CP Matrix

S=3   M=0   N=4

Statistic	Value	F	Num DF	Den DF	Pr > F
Wilks' Lambda	0.633684	0.4195	12	26.749	0.9420
Pillai's Trace	0.409494	0.4742	12	36	0.9167
Hotelling-Lawley Trace	0.511456	0.3694	12	26	0.9631
Roy's Greatest Root	0.326124	0.9784	4	12	0.4552

NOTE: F Statistic for Roy's Greatest Root is an upper bound.

Manova Test Criteria and F Approximations for
the Hypothesis of no Overall X2*GROUP Effect
H = Type III SS&CP Matrix for X2*GROUP   E = Error SS&CP Matrix

S=3   M=0   N=4

Statistic	Value	F	Num DF	Den DF	Pr > F
Wilks' Lambda	0.483896	0.7037	12	26.749	0.7343
Pillai's Trace	0.58756	0.7307	12	36	0.7126
Hotelling-Lawley Trace	0.920692	0.6649	12	26	0.7680
Roy's Greatest Root	0.725555	2.1767	4	12	0.1335

NOTE: F Statistic for Roy's Greatest Root is an upper bound.

Manova Test Criteria and Exact F Statistics for
the Hypothesis of no Overall X1 Effect
H = Type III SS&CP Matrix for X1   E = Error SS&CP Matrix

S=1   M=1   N=4

Statistic	Value	F	Num DF	Den DF	Pr > F
Wilks' Lambda	0.724797	0.9492	4	10	0.4753
Pillai's Trace	0.275203	0.9492	4	10	0.4753
Hotelling-Lawley Trace	0.379697	0.9492	4	10	0.4753
Roy's Greatest Root	0.379697	0.9492	4	10	0.4753

Manova Test Criteria and Exact F Statistics for
the Hypothesis of no Overall X2 Effect
H = Type III SS&CP Matrix for X2   E = Error SS&CP Matrix

S=1   M=1   N=4

Statistic	Value	F	Num DF	Den DF	Pr > F
Wilks' Lambda	0.060864	38.575	4	10	0.0001
Pillai's Trace	0.939136	38.575	4	10	0.0001
Hotelling-Lawley Trace	15.43015	38.575	4	10	0.0001
Roy's Greatest Root	15.43015	38.575	4	10	0.0001

Output 5.14
continued

Output 5.14
MANCOVA: Profile Analysis

Manova Test Criteria and F Approximations for
the Hypothesis of no Overall "parallel?" Effect
on the variables defined by the M Matrix Transformation
H = Contrast SS&CP Matrix for "parallel?"
E = Error SS&CP Matrix

S=3   M=-0.5   N=7.5

Statistic	Value	F	Num DF	Den DF	Pr > F
Wilks' Lambda	0.719521	0.6682	9	41.524	0.7324
Pillai's Trace	0.301037	0.7064	9	57	0.7005
Hotelling-Lawley Trace	0.361292	0.6289	9	47	0.7665
Roy's Greatest Root	0.246755	1.5628	3	19	0.2312

NOTE: F Statistic for Roy's Greatest Root is an upper bound.

```
          Manova Test Criteria and Exact F Statistics for
         the Hypothesis of no Overall "coincidental?" Effect
      on the variables defined by the M Matrix Transformation
          H = Contrast SS&CP Matrix for "coincidental?"
                    E = Error SS&CP Matrix


                 S=1      M=0.5      N=10

Statistic                 Value        F    Num DF  Den DF  Pr > F

Wilks' Lambda            0.871497   1.0813      3      22   0.3776
Pillai's Trace           0.128503   1.0813      3      22   0.3776
Hotelling-Lawley Trace   0.147451   1.0813      3      22   0.3776
Roy's Greatest Root      0.147451   1.0813      3      22   0.3776
```

Output 5.14
continued

```
                         Output 5.14
             Analysis as in Srivastava and Carter (1983)

                    General Linear Models Procedure

Dependent Variable: YBAR

Source                   DF   Sum of Squares  F Value   Pr > F

Model                     8     254.29864123    37.00   0.0001

Error                    16      13.74575877

Corrected Total          24     268.04440000

                R-Square           C.V.        YBAR Mean

                0.948718        14.46671        6.40700000


Source                   DF      Type I SS   F Value   Pr > F

Z1                        1      0.41503092      0.48   0.4970
Z2                        1      4.03021154      4.69   0.0458
Z3                        1      0.66998345      0.78   0.3903
X1                        1     63.73020368     74.18   0.0001
X2                        1    177.09538387    206.14   0.0001
GROUP                     3      8.35782776      3.24   0.0498

Source                   DF    Type III SS   F Value   Pr > F

Z1                        1      3.06301372      3.57   0.0773
Z2                        1      0.29590402      0.34   0.5655
Z3                        1      1.13016529      1.32   0.2683
X1                        1      0.38955026      0.45   0.5103
X2                        1    177.85893877    207.03   0.0001
GROUP                     3      8.35782776      3.24   0.0498
```

Output 5.14
continued

General Linear Models Procedure
Repeated Measures Analysis of Variance

Manova Test Criteria and Exact F Statistics for
the Hypothesis of no TIME*X1 Effect
H = Type III SS&CP Matrix for TIME*X1   E = Error SS&CP Matrix

S=1   M=0.5   N=4.5

Statistic	Value	F	Num DF	Den DF	Pr > F
Wilks' Lambda	0.928425	0.2827	3	11	0.8369
Pillai's Trace	0.071575	0.2827	3	11	0.8369
Hotelling-Lawley Trace	0.077092	0.2827	3	11	0.8369
Roy's Greatest Root	0.077092	0.2827	3	11	0.8369

Manova Test Criteria and Exact F Statistics for
the Hypothesis of no TIME*X2 Effect
H = Type III SS&CP Matrix for TIME*X2   E = Error SS&CP Matrix

S=1   M=0.5   N=4.5

Statistic	Value	F	Num DF	Den DF	Pr > F
Wilks' Lambda	0.9661	0.1287	3	11	0.9411
Pillai's Trace	0.0339	0.1287	3	11	0.9411
Hotelling-Lawley Trace	0.035089	0.1287	3	11	0.9411
Roy's Greatest Root	0.035089	0.1287	3	11	0.9411

Manova Test Criteria and F Approximations for
the Hypothesis of no TIME*X1*GROUP Effect
H = Type III SS&CP Matrix for TIME*X1*GROUP
E = Error SS&CP Matrix

S=3   M=-0.5   N=4.5

Statistic	Value	F	Num DF	Den DF	Pr > F
Wilks' Lambda	0.8163	0.2602	9	26.922	0.9802
Pillai's Trace	0.191514	0.2955	9	39	0.9718
Hotelling-Lawley Trace	0.215503	0.2315	9	29	0.9869
Roy's Greatest Root	0.156091	0.6764	3	13	0.5818

NOTE: F Statistic for Roy's Greatest Root is an upper bound.

Manova Test Criteria and F Approximations for
the Hypothesis of no TIME*X2*GROUP Effect
H = Type III SS&CP Matrix for TIME*X2*GROUP
E = Error SS&CP Matrix

S=3   M=-0.5   N=4.5

Statistic	Value	F	Num DF	Den DF	Pr > F
Wilks' Lambda	0.640487	0.6009	9	26.922	0.7849
Pillai's Trace	0.37655	0.622	9	39	0.7710
Hotelling-Lawley Trace	0.534956	0.5746	9	29	0.8066
Roy's Greatest Root	0.481813	2.0879	3	13	0.1513

NOTE: F Statistic for Roy's Greatest Root is an upper bound.

Suppose we can assume that the covariates X1 and X2 do not influence differently the measurements taken at different time points (this assumption will be tested later). This assumption means that the covariates X1 and X2 affect the average of Y1, Y2, Y3, and Y4 for a given subject rather than the temporal changes in response. In that case, we want to perform the profile analysis for this data set. Under the assumption stated above, the hypotheses of parallel profiles and of coincidental profiles, given parallelism, will be unchanged. The SAS code for these hypotheses is given as part of Program 5.14. The two hypotheses are not rejected in view of their large p values. These p values for Wilks' Λ are 0.7324 and 0.3776 respectively.

Given that the profiles are coincidental, the other hypothesis of interest is that all the profiles are horizontal. To test this hypothesis, we made a transformation by defining the variables YBAR = (Y1 + Y2 + Y3 + Y4)/4, Z1 = Y2 − Y1, Z2 = Y3 − Y2, and Z3 = Y4 − Y3. The variable YBAR measures the average response for the subject and Z1, Z2, Z3 measure the changes in response from one time point to the next. For parallel profiles, the expected values of Z1, Z2, and Z3 are the same for all four groups. The hypothesis of horizontal profiles could be tested using the method described in the previous section. However, YBAR is correlated with Z1, Z2, and Z3 and hence these can be taken as the additional covariates in the model. In view of this, the hypothesis of horizontal profiles can be tested by testing the hypothesis of no GROUP effect in the univariate analysis of covariance model for YBAR given by the following statement

```
model ybar =z1 z2 z3 x1 x2 group;
```

Output 5.14 indicates that the GROUP effect is marginally significant (p value = 0.0498). Care must be taken to ensure that the sum of squares corresponding to GROUP have been corrected for all the covariates Z1, Z2, Z3, X1, and X2. Thus, we can either list GROUP last in the MODEL statement, if the sequential sums of squares (Type I) are used or alternatively specify the Type III sums of squares (SS3) as an option in the MODEL statement.

One assumption that was made in the profile analysis using YBAR presented above is that the covariates do not differently influence the measurements taken at different times. This hypothesis can be examined by testing for the significance of the covariate*time and treatment*covariate*time interaction in the corresponding univariate analysis. This can be easily achieved by using the REPEATED statement

```
repeated time 4 (1 2 3 4) polynomial;
```

where the POLYNOMIAL option has often been chosen for further analyses (such as, to examine the polynomial trend etc.) in repeated measures data. The analysis presented in this section does not, however, depend on any such choice. The above REPEATED statement is included in Program 5.14 and the corresponding output appears in Output 5.14. Only the relevant parts of the output are presented. None of the multivariate tests for TIME*X1, TIME*X2, TIME*X1*GROUP, and TIME*X2*GROUP are significant. The respective p values for Wilks' Λ are 0.8369, 0.9411, 0.9802, and 0.7849. All other multivariate test statistics also confirm this conclusion. Thus, the hypothesis that the covariates X1 and X2 do not differently affect the measurements taken at different time points cannot be rejected.

5.5.2   A Univariate Approach

The alternative univariate approach of a split plot design with covariates can also be adopted. Specifically the observed value on the uth subject under the ith treatment at the jth time point y_{iju} can be modeled as (assuming only one covariate)

$$y_{iju} = \mu + \alpha_i + \beta_j + (\alpha\beta)_{ij} + \delta_{iu} + \beta_{ij}x_{iu} + \epsilon_{iju}, \tag{5.12}$$

where all the symbols except β_{ij} and x_{iu} have been defined in Equation 5.11. The covariate x_{iu} represents the value for the uth subject under the ith treatment and β_{ij} are the slope parameters for the jth time period and the ith treatment. The model in Equation 5.12 is a time specific covariate model. The generalization of the above model for more than one covariate is straightforward.

If for a given subject x_{iu} are the same for all time points j, $j = 1, \ldots, p$, a *whole plot model* can be obtained by averaging the above model over suffix j. See Milliken and Johnson(1989). It leads to

$$\bar{y}_{i \cdot u} = \mu + \alpha_i + \bar{\beta}. + (\alpha\beta)_i. + \delta_{iu} + \bar{\beta}_i.x_{iu} + \epsilon_{i \cdot u}$$

or

$$\bar{y}_{i \cdot u} = \bar{\mu}_i. + \bar{\beta}_i.x_{iu} + \epsilon^*_{i \cdot u} \tag{5.13}$$

which is a one-way classification model with a covariate. We prefer to call it a *subject specific model* rather than a *whole plot model* to agree with our present context of repeated measures. Note that unlike the standard analysis of covariance model, "errors" $\epsilon^*_{i \cdot u}$ are not independently distributed but have a compound symmetric structure for the variance-covariance matrix. However, the analysis of variance tests are still valid under compound symmetry. The two hypotheses of interest are

$$H_0^{(1)} : \bar{\beta}_1. = \cdots = \bar{\beta}_k. = 0$$

and

$$H_0^{(2)} : \bar{\beta}_1. = \cdots = \bar{\beta}_k.$$

The null hypothesis $H_0^{(1)}$ tests if the average slopes are all zero, whereas, $H_0^{(2)}$ tests if the regression lines for various treatment groups are parallel. If $H_0^{(2)}$ is true then the common slope can be estimated.

Both of these hypotheses can be tested using the subject model in Equation 5.13 and its appropriate reduction under the corresponding null hypothesis. Specifically, if $H_0^{(1)}$ is true, then Equation 5.13 reduces to a one-way classification model. Although errors are not independent in view of their compound symmetric covariance structure, the usual partial F test can still be applied. The corresponding F statistic

$$F = \frac{(SSE_{Reduced} - SSE_{Full})/k}{SSE_{Full}/f} \tag{5.14}$$

follows an F distribution with (k, f) df under the null hypothesis $H_0^{(1)}$. The quantity SSE_{Full} corresponds to the error sum of squares of the subject model given in Equation 5.13. Its degrees of freedom, for convenience, are denoted by f. We may add that the value of f is in part also determined by the number of missing values in the multivariate data. $SS_{Reduced}$ is the error sum of squares for the reduced model; that is, the model given in

Equation 5.13 without $\bar{\beta}_i.x_{iu}$ terms. To test $H_0^{(2)}$, we observe that under $H_0^{(2)}$, Equation 5.13 becomes

$$\bar{y}_{i\cdot u} = \bar{\mu}_{i\cdot} + \bar{\beta}x_{iu} + \epsilon^*_{i\cdot u}. \qquad (5.15)$$

A partial F test for $H_0^{(2)}$ can also be obtained using SSE_{Full} and $SSE_{Reduced}$ which now correspond to models in Equations 5.13 and 5.15 respectively. Since the null hypothesis $H_0^{(2)}$ specifies only $(k-1)$ linear restrictions, the divisor of the $(SSE_{Reduced} - SSE_{Full})$ is $(k-1)$ and not k. When there are two or more covariates, the hypothesis $H_0^{(1)}$ and $H_0^{(2)}$ can be generalized in a straightforward way. Also the corresponding F tests can be appropriately modified. All this can be best illustrated through the continuation of Example 10. Also see Example 13 in Section 5.6.3 for an alternative analysis of testing the homogeneity of regression coefficients and intercepts.

EXAMPLE 10   ***Example 10: Diabetic Patients Study Data (continued)***   For the fitness data of Crowder and Hand, we first want to test if the covariates X1 and X2 have any effect on the average responses. Since $k = 4$ and there are two covariates, the quantity k in Equation 5.14 is replaced by $2k = 8$ here. The SSE_{Full} and $SSE_{Reduced}$ are the two sums of squares which can be respectively obtained by fitting the appropriate models using two MODEL statements

```
model y=group z11 z12 z13 z14 z21 z22 z23 z24;
```

and

```
model y=group;
```

While the second MODEL statement is straightforward, the first one needs further explanation, especially since Z11, Z12, etc., have not been introduced yet. Since x_{1iu} and x_{2iu} values are available only for the subjects in the ith group (CONTROL) and not others, and since the slopes for each group are different, we need to have (two covariates times four groups =) eight slope parameters in the model and correspondingly eight independent regression variables (apart from the treatment effect). The new regression variables are defined as z_{li^*}, $l = 1, 2$, $i^* = 1, \ldots, 4$ where the value of z_{li^*} for u^{th} subject in the ith group, say z_{li^*u}, $l = 1, 2$, $i^* = 1, \ldots, 4$, is

$$z_{li^*u} = \begin{cases} x_{liu} & \text{if } i = i^* \\ 0 & \text{if } i \neq i^* \end{cases}$$

The hypotheses $H_0^{(1)}$ of no covariate effect is equivalent to testing that all z_{li}, $l = 1, 2$, $i = 1, 2, 3, 4$ are unimportant. Hence to assess their contribution, we use the corresponding hypothesis sum of squares which is computed as the difference between the SSE from two models indicated above in the MODEL statement.

The SAS program to calculate z_{li} and then to fit two models is given as the first part of Program 5.15. The output presented as Output 5.15 gives $SSE_{Reduced} = 243.3190$ and $SSE_{Full} = 11.9889$ and the value of f as the degree of freedom of SSE_{Full}, which is 14 for this data set. Note even though the covariates do not appear in the corresponding model, one observation with a missing X1 value was discarded for this calculation, to have the sum of squares comparable in the two models. Also $k = 8$, and hence the observed value of the F statistic with $(8, 14)$ df is

$$F = \frac{(243.3190 - 11.9889)/8}{11.9889/14} = 33.7669$$

which is highly significant with a p value (calculated from an independent computation using the SAS function PROBF, not shown) almost zero. As a result, we reject the null hypothesis $H_0^{(1)}$ and conclude that the average response is indeed affected by the covariates.

```
/* Program 5.15 */

options ls=64 ps=45 nodate nonumber;
title1 'Output 5.15';
title2 'Analysis of Diabetic Patients Study';
data task;
infile 'task.dat';
input group$ x1 x2 y1-y10;
data task;
set task;
ybar = mean(y1,y2,y3,y4);
if group = 'control' then z11=x1;
if group = 'control' then z12 =0;
if group = 'control' then z13 =0;
if group = 'control' then z14 =0;
if group = 'control' then z21=x2;
if group = 'control' then z22 =0;
if group = 'control' then z23 =0;
if group = 'control' then z24 =0;
if group = 'dinocom' then z11 =0;
if group = 'dinocom' then z12=x1;
if group = 'dinocom' then z13 =0;
if group = 'dinocom' then z14 =0;
if group = 'dinocom' then z21 =0;
if group = 'dinocom' then z22=x2;
if group = 'dinocom' then z23 =0;
if group = 'dinocom' then z24 =0;
if group = 'dihypot' then  z11=0;
if group = 'dihypot' then z12 =0;
if group = 'dihypot' then z13=x1;
if group = 'dihypot' then z14 =0;
if group = 'dihypot' then z21 =0;
if group = 'dihypot' then z22=0 ;
if group = 'dihypot' then z23=x2;
if group = 'dihypot' then z24 =0;
if group = 'dihyper' then  z11=0;
if group = 'dihyper' then  z12=0;
if group = 'dihyper' then  z13=0;
if group = 'dihyper' then z14=x1;
if group = 'dihyper' then  z21=0;
if group = 'dihyper' then  z22=0;
if group = 'dihyper' then  z23=0;
if group = 'dihyper' then z24=x2;
data nomiss;
set task ;
if x1 > 0;
title3 'Subject Model';
proc glm data =nomiss;
classes group ;
model ybar = group z11 z12 z13 z14 z21 z22 z23 z24;
run;
proc glm data =nomiss;
classes group ;
model ybar = group;
title3 'Subject Model: No Covariates';
```

```
run;
proc glm data =nomiss;
classes group ;
model ybar = group x1 x2/solution;
title3 'Subject Model: With Covariates';
run;
```

Output 5.15

```
                          Output 5.15
                Analysis of Diabetic Patients Study
                           Subject Model


                 General Linear Models Procedure

Dependent Variable: YBAR

Source              DF    Sum of Squares   F Value    Pr > F

Model               11      256.43272193    27.22     0.0001

Error               14       11.98888330

Corrected Total     25      268.42160524




                Subject Model: No Covariates


Dependent Variable: YBAR

Source              DF    Sum of Squares   F Value    Pr > F

Model                3       25.10256299     0.76     0.5304

Error               22      243.31904225

Corrected Total     25      268.42160524




                Subject Model: With Covariates

Dependent Variable: YBAR

Source              DF    Sum of Squares   F Value    Pr > F

Model                5      246.79399930    45.64     0.0001

Error               20       21.62760594

Corrected Total     25      268.42160524


                                T for H0:   Pr > |T|   Std Error of
Parameter         Estimate    Parameter=0                Estimate

X1             -0.032229071       -0.30      0.7690      0.10825051
X2              1.017385670       12.61      0.0001      0.08066019
```

Next, to test if the slopes in different groups for the covariate X1 are all equal and those for X2 are all equal, the MODEL statement for the reduced model would be

```
model y=group x1 x2;
```

The corresponding SSE would be used as $SSE_{Reduced}$ while SSE_{Full} is the same as earlier. In Output 5.15, we have eliminated the corresponding calculations to save the space. We may verify that the hypothesis $H_0^{(2)}$ is not rejected (p value $= 0.1157$). Thus, a common slope model can be used for all four groups. The estimated value of the two common slopes corresponding to X1 and X2 are -0.0322 and 1.0174 respectively.

Two other hypotheses of interest from the original time specific model in Equation 5.12 are

$$H_0^{(3)} : \beta_{11} = \cdots = \beta_{1p} = \beta_{21} = \cdots = \beta_{2p} = \cdots = \beta_{k1} = \cdots = \beta_{kp} = 0,$$

which states that the covariate has no effect on the response variable and

$$H_0^{(4)} : \beta_{11} = \cdots = \beta_{1p},$$
$$\beta_{21} = \cdots = \beta_{2p},$$
$$\vdots = \cdots = \vdots$$
$$\beta_{k1} = \cdots = \beta_{kp},$$

which states that for any given treatment group, the effect of the covariate on the response variable is unaffected by the particular time point, and is the same for all time points. The hypothesis $H_0^{(4)}$ can be tested using the model in Equation 5.12 but $H_0^{(3)}$ cannot be tested using the original model due to certain confounding difficulties. However, the acceptance of orthogonal hypotheses $H_0^{(1)}$ and $H_0^{(4)}$ implies the acceptance of $H_0^{(3)}$; rejection of either $H_0^{(1)}$ or $H_0^{(4)}$ implies the rejection of $H_0^{(3)}$. Hence, $H_0^{(3)}$ can be tested by testing both $H_0^{(1)}$ and $H_0^{(4)}$. Milliken and Johnson (1989) provide an extensive discussion of the methodology involved and the related intricacies.

When the covariates are available for each subject and at every time point, the appropriate analysis using the univariate split plot design is given by Milliken (1990), who also warns that the use of the REPEATED statement does not provide the correct sums of squares for within-subject effects. See Milliken (1990) for more information. The analysis of covariance will again be considered in Chapter 6 under more complex covariance structures.

5.6   The Growth Curve Models

Suppose a growth process is observed on a set of experimental subjects over a period of time. We want to build an appropriate growth function for this process. We may also want to compare the growth functions of groups of several sets of individuals or subjects. A generalization of MANOVA models can be used to fit certain polynomial growth curves to the growth process and hence, using the standard theory of multivariate analysis, we can compare the growths of several groups. We consider this model next. For comprehensive treatment of the growth curve models, see Kshirsagar and Smith (1995).

5.6.1   Polynomial Growth

In Chapters 3 and 4 the linear model

$$\mathbf{Y} = \mathbf{XB} + \mathcal{E} \tag{5.16}$$

was considered and applications of this model to analyze various types of data were given. A slightly more general model than Equation 5.16 developed by Potthoff and Roy (1964) follows. It is needed to analyze certain types of measurements on growth curves. In the model

$$\mathbf{Y}_{0_{n \times p}} = \mathbf{X}_{n \times k} \mathbf{B}_{k \times q} \mathbf{A}_{q \times p} + \mathcal{E}_{n \times p},$$

where $\mathbf{X}$ and $\mathbf{A}$ are known full rank (as was also observed in the previous chapters, this assumption is really not needed) matrices of ranks $k(< n)$ and $q(< p)$ respectively. The matrix $\mathbf{B}$ is the matrix of unknown parameters. As in the previous discussion, p represents the number of time points at which the measurements are taken on each of the n experimental units. The degree of the polynomial curve that is being fit for the p measurements over time is $q - 1$. Accordingly $\mathbf{A}$ is a q by p matrix of the coefficients of the polynomials of various degrees up to $q - 1$ (or orthogonal polynomials, if the measurements are taken at equidistant time points) and $\mathbf{X}$ is the design matrix representing k different groups. In general, the matrices $\mathbf{A}$ and $\mathbf{X}$ can be any matrices of known quantities. Assume that the rows of error matrix $\mathcal{E}$ are independently distributed as $N_p(\mathbf{0}, \boldsymbol{\Sigma})$, where $\boldsymbol{\Sigma}$ is a p by p positive definite matrix.

Consider the general linear hypothesis

$$H_0 : \mathbf{LBM} = \mathbf{0},$$

where $\mathbf{L}_{r \times k}$ and $\mathbf{M}_{q \times s}$ are full rank matrices with ranks r and s respectively. There are several approaches for testing H_0. However, we will adopt what is termed the Rao-Khatri approach (Seber, 1984, p. 480), since it reduces the present problem to a testing problem under the usual analysis of covariance model. Hence the standard SAS procedures, like PROC GLM, can be applied to test H_0 given above.

5.6.2   Rao-Khatri Reduction

Let $\mathbf{C}_1$ of order p by q and $\mathbf{C}_2$ of order p by $p - q$ be any two matrices such that *Rank* $(\mathbf{C}_1) = q$, *Rank* $(\mathbf{C}_2) = p - q$, $\mathbf{A}\mathbf{C}_1 = \mathbf{I}_q$, and $\mathbf{A}\mathbf{C}_2 = \mathbf{0}_{p-q}$ and that $\mathbf{C} = (\mathbf{C}_1 : \mathbf{C}_2)$ is a nonsingular matrix. For example, $\mathbf{C}_1$ and $\mathbf{C}_2$ can be taken as $\mathbf{A}'(\mathbf{A}\mathbf{A}')^{-1}$ and any $p - q$ linearly independent columns of $\mathbf{I} - \mathbf{A}'(\mathbf{A}\mathbf{A}')^{-1}\mathbf{A}$ respectively. An easy way of choosing $\mathbf{C}_1$ and $\mathbf{C}_2$ when $\mathbf{A}$ is the matrix of orthogonal polynomials is to take $\mathbf{C}_1 = \mathbf{A}'$ and $\mathbf{C}_2$ such that $\mathbf{C} = (\mathbf{C}_1 : \mathbf{C}_2)$ is an orthogonal matrix. Specifically, choose $\mathbf{C}_1$ to be the matrix of normalized orthogonal polynomials of degree 0 to $q - 1$ and $\mathbf{C}_2$ to be the similar matrix for degrees q to $p - 1$. Once $\mathbf{C}_1$ and $\mathbf{C}_2$ are selected, define $\mathbf{Y}_1 = \mathbf{Y}_0\mathbf{C}_1$ and $\mathbf{Y}_2 = \mathbf{Y}_0\mathbf{C}_2$ as the transformed data matrices of order n by q and n by $p - q$ respectively. This leads to $\mathbf{Y} = (\mathbf{Y}_1 : \mathbf{Y}_2) = \mathbf{Y}_0(\mathbf{C}_1 : \mathbf{C}_2) = \mathbf{Y}_0\mathbf{C}$. Also, $E(\mathbf{Y}_1) = \mathbf{XBAC}_1 = \mathbf{XB}$ and $E(\mathbf{Y}_2) = \mathbf{XBAC}_2 = \mathbf{0}$. Since $\mathbf{Y}_1$ and $\mathbf{Y}_2$ are correlated, even though $E(\mathbf{Y}_2)$ does not involve $\mathbf{B}$, the set of variables $\mathbf{Y}_2$ can be taken as covariates for estimating or testing the hypotheses about $\mathbf{B}$. That is, we consider the conditional model

$$E(\mathbf{Y}_1|\mathbf{Y}_2) = \mathbf{XB} + \mathbf{Y}_2\boldsymbol{\Gamma}, \tag{5.17}$$

where $(p - q)$ by q matrix $\boldsymbol{\Gamma}$ is the parameter matrix representing the effects of various covariates on the conditional mean of $\mathbf{Y}_1$ given $\mathbf{Y}_2$. The rows of $\mathbf{Y}_1$ are independent and are conditionally distributed as multivariate normal with the means as the rows of the right-hand side matrix of Equation 5.17 and a common variance-covariance matrix. Hence the method used in Section 4.7 can be used here. By writing Equation 5.17 as

$$\mathbf{Y}_1 = (\mathbf{X}\ \mathbf{Y}_2) \begin{bmatrix} \mathbf{B} \\ \boldsymbol{\Gamma} \end{bmatrix} + \mathcal{E},$$

it can be shown that the maximum likelihood estimate of **B** is given by

$$\hat{\mathbf{B}} = (\mathbf{X}'\mathbf{X})^{-1}\mathbf{X}'\mathbf{Y}_0\mathbf{S}^{-1}\mathbf{A}'(\mathbf{A}\mathbf{S}^{-1}\mathbf{A}')^{-1}$$

where

$$\mathbf{S} = \mathbf{Y}_0'[\mathbf{I} - \mathbf{X}(\mathbf{X}'\mathbf{X})^{-1}\mathbf{X}']\mathbf{Y}_0/(n - p).$$

EXAMPLE 11   *Modeling Cubic Growth, Dog Response Time Data*   These data were first analyzed by Grizzle and Allen (1969) using the Rao-Khatri method. Also see Seber (1984, p. 487) for the analysis of these data. The data set includes observations on four groups of dogs showing the response of each dog at times 1, 3, 5, 7, 9, 11, and 13 minutes after a coronary occlusion. The four groups, the first being the control group, have respective sample sizes 9, 10, 8, and 9. The three experimental groups respectively include dogs with extrinsic cardiac denervation three weeks prior to coronary occlusion, with extrinsic cardiac denervation immediately prior to coronary occlusion, and with bilateral thoracic sympathectomy and stellectomy three weeks prior to coronary occlusion. A profile plot of the sample mean vectors of the four group is presented as a part of Output 5.16. The SAS code for plotting this profile plot is included in Program 5.16. The null hypothesis is that the third-degree polynomial is adequate.

```
/* Program 5.16 */

/* This is a growth curve anlysis program where
dog data is used (Grizzle and Allen, 1969). */
options ls=64 ps=45 nodate nonumber;
title1 'Output 5.16';
title2 'Growth Curves Analysis of Dog Data';
data dog;
infile "dog.dat";
input d1 d2 d3 d4 d5 d6 d7;
dog+1;
if (dog<10 and dog>0) then group='control';
if (dog<19 and dog>9) then group='treat1';
if (dog<28 and dog>18) then group='treat2';
if (dog<37 and dog>27) then group='treat3';
output;
drop dog;
run;

filename gsasfile "prog516.graph";
goptions gaccess=gsasfile dev=pslmono;
goptions horigin=1in vorigin=2in;
goptions hsize=5in vsize=7in;
title1 h=1.5 'Modeling Growth: Profiles of the Means';
title2 j=l 'Output 5.16';
proc summary nway data=dog;
class group;
var d1 d2 d3 d4 d5 d6 d7;
output out=new mean=md1-md7;
data plot;
set new;
array md{7} md1-md7;
do i =1 to 7;
meanresp=md(i);
time=i*2-1;
output;
end;
```

```
                keep group time meanresp;
                run;
                proc gplot data = plot;
                plot meanresp*time=group/ vaxis=axis1 haxis=axis2;
                axis1 label =(a=90 h=1.2 'Dog Response Time');
                axis2 offset=(2) label=(h=1.2 'Time of Test');
                symbol1 v=+ i = join;
                symbol2 v=x i=join;
                symbol3 v=* i=join;
                symbol4 v=- i=join;
                run;

                proc iml;
                use dog;
                read all into y0;
                /*Generating the Orthogonal Polynomial of degree p-1=6*/
                vec1={1 3 5 7 9 11 13};
                c=orpol(vec1,6);
                y=y0*c;
                /* Converting Y matrix to a data set Trandata*/
                varnames={y1 y2 y3 y4 y5 y6 y7};
                create trandata from y (|colname=varnames|);
                append from y;
                close trandata;
                /* Creating the independent variable named group*/
                data trandata;
                set trandata;
                dog+1;
                if (dog<10 and dog>0) then group='control';
                if (dog<19 and dog>9) then group='treat1';
                if (dog<28 and dog>18) then group='treat2';
                if (dog<37 and dog>27) then group='treat3';
                output;
                run;
                /* Testing the adequacy of a 3rd degree polynomial */
                proc glm data=trandata;
                model y5 y6 y7= /nouni;
                manova h=intercept;
                run;
                /* Fitting a 3rd degree polynomial using
                Rao-Khatri method*/
                /* Contrast statement defines a 3 by 4 L matrix.*/
                proc glm data=trandata;
                classes group;
                model y1 y2 y3 y4=group y5 y6 y7/nouni;
                contrast 'growth curves' group 1 -1 0 0,
                                         group 1 0 -1 0,
                                         group 1 0 0 -1/E;
                manova h=group;
                run;
                *To obtain the estimates use one of the following
                two programs;
                proc glm data=trandata;
                classes group;
                model y1 y2 y3 y4=group y5 y6 y7;
                estimate 'es1' intercept 1 group 1 0 0 0 ;
                estimate 'es2' intercept 1 group 0 1 0 0 ;
                estimate 'es3' intercept 1 group 0 0 1 0 ;
                estimate 'es4' intercept 1 group 0 0 0 1 ;
                run;
```

```
proc glm data=trandata;
classes group;
model y1 y2 y3 y4=group y5 y6 y7/noint;
estimate 'es1' group 1 0 0 0 ;
estimate 'es2' group 0 1 0 0 ;
estimate 'es3' group 0 0 1 0 ;
estimate 'es4' group 0 0 0 1 ;
run;
```

Output 5.16

Output 5.16   Modeling Growth: Profiles of the Means

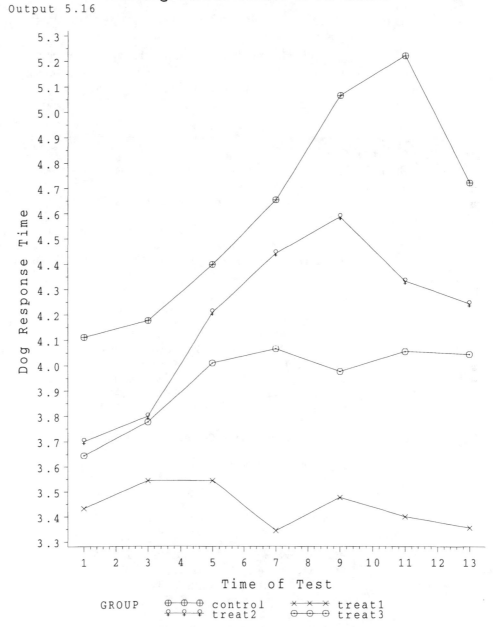

The profile plots of the sample mean vectors for each group seem to suggest a third-degree polynomial growth curve. Alternatively, a third-degree orthogonal polynomial may instead be used. The model then is

$$\mathbf{Y}_{0_{36 \times 7}} = \mathbf{X}_{36 \times 4} \mathbf{B}_{4 \times 4} \mathbf{A}_{4 \times 7} + \mathcal{E}_{36 \times 7},$$

where

$$\mathbf{X} = \begin{bmatrix} \mathbf{1}_9 & \mathbf{0} & \mathbf{0} & \mathbf{0} \\ \mathbf{0} & \mathbf{1}_{10} & \mathbf{0} & \mathbf{0} \\ \mathbf{0} & \mathbf{0} & \mathbf{1}_8 & \mathbf{0} \\ \mathbf{0} & \mathbf{0} & \mathbf{0} & \mathbf{1}_9 \end{bmatrix}$$

and $\mathbf{A}$ is the matrix with the first four columns of a 7 by 7 orthogonal polynomial matrix. Thus $\mathbf{A}$ is known. In SAS the matrix of orthogonal coefficients, $\mathbf{C}$, is generated using the ORPOL function of PROC IML. Then

$$\mathbf{Y} = \mathbf{Y}_0\mathbf{C} = (\mathbf{Y}1 : \mathbf{Y}2 : \mathbf{Y}3 : \mathbf{Y}4 : \mathbf{Y}5 : \mathbf{Y}6 : \mathbf{Y}7).$$

A formal test for the adequacy of the third-degree polynomial is performed by testing whether $E(Y5) = E(Y6) = E(Y7) = 0$. The SAS code is given in Program 5.16, and the output is presented in Output 5.16. The p value using Wilks' Λ is 0.4280. Hence we do not reject the null hypothesis that the third-degree polynomial is adequate.

Output 5.16
continued

```
                          Output 5.16
                 Growth Curves Analysis of Dog Data

                   General Linear Models Procedure

              Manova Test Criteria and Exact F Statistics for
                the Hypothesis of no Overall INTERCEPT Effect
        H = Type III SS&CP Matrix for INTERCEPT    E = Error SS&CP Matrix

                      S=1     M=0.5     N=15.5

   Statistic                  Value       F     Num DF  Den DF  Pr > F

   Wilks' Lambda            0.920548    0.9494      3      33    0.4280
   Pillai's Trace           0.079452    0.9494      3      33    0.4280
   Hotelling-Lawley Trace   0.08631     0.9494      3      33    0.4280
   Roy's Greatest Root      0.08631     0.9494      3      33    0.4280


              Manova Test Criteria and F Approximations for
                the Hypothesis of no Overall GROUP Effect
          H = Type III SS&CP Matrix for GROUP    E = Error SS&CP Matrix

                      S=3     M=0     N=12

   Statistic                  Value       F     Num DF  Den DF  Pr > F

   Wilks' Lambda            0.403257    2.3576     12    69.081  0.0132
   Pillai's Trace           0.675205    2.0331     12      84    0.0309
   Hotelling-Lawley Trace   1.29045     2.6526     12      74    0.0052
   Roy's Greatest Root      1.131816    7.9227      4      28    0.0002

   NOTE: F Statistic for Roy's Greatest Root is an upper bound.
```

The model for further analysis of the data can therefore be taken as the one in which $Y5$, $Y6$, and $Y7$ are covariates for the dependent variables $Y1$, $Y2$, $Y3$, $Y4$, and the basic independent variable is GROUP. For this, the null hypothesis of no difference between the groups can be written as $H_0 : \mathbf{LBM} = \mathbf{0}$ with $\mathbf{M} = \mathbf{I}_4$ and

$$\mathbf{L} = \begin{bmatrix} 0 & 1 & -1 & 0 & 0 \\ 0 & 0 & 1 & -1 & 0 \\ 0 & 0 & 0 & 1 & -1 \end{bmatrix}.$$

Wilks' Λ for this hypothesis is 0.4033 leading to an approximate F = 2.3576 on 12 and 69 degrees of freedom. This yields a p value of 0.0132 thereby leading us to reject the null hypothesis that all the groups have the same third-degree polynomial curves. All the other three tests also lead to the same conclusion. Knowing that the groups may have different growth curves, we may want to estimate the matrix $\mathbf{B}$ consisting of parameters of these curves. To estimate $\mathbf{B}$ we use the ESTIMATE statement of PROC GLM. The columns of $\hat{\mathbf{B}}$, the estimate of $\mathbf{B}$, are obtained as the estimates of the regression coefficients from the individual outputs from the corresponding univariate analyses. The SAS code for finding the estimates is given at the end of Program 5.16 and the corresponding output is suppressed to save space. Another way to get these estimates is by using the REG procedure. See Example 12.

The model in Equation 5.17 is applicable in many other designed experimental situations. For example, suppose data measured at p time points on n experimental units comes from a randomized block design. Then, by choosing the $\mathbf{X}$ matrix appropriately and including all the parameters of this model in $\mathbf{B}$ we can express the model as Equation 5.17.

It is not necessarily true that the inclusion of covariates would improve the efficiency of the estimates. In some instances, including only a few covariates from $\mathbf{Y}_2$ (that is, only a few columns of $\mathbf{Y}_2$) may improve the efficiency of the estimates more than using the entire $\mathbf{Y}_2$. We illustrate a way of selecting a set of covariates from $\mathbf{Y}_2$ to improve the inference, using the mice data of Izenman and Williams (1989). Also see Williams and Izenman (1981).

EXAMPLE 12   *Choosing Covariates, Mice Data*   The data, given in Program 5.17, are a part of those given in Izenman and Williams (1989). Complete details of the data collection scheme are given in Izenman (1987). The observations represent the weights from birth until weaning of 14 male mice measured at 2, 5, 8, 11, 14, 17, and 20 days after birth. A graphical representation of the data along with the profile of the means is presented as Output 5.17. Other output is suppressed. The SAS code to plot this graph is provided in Program 5.17.

```
/* Program 5.17 */

options ls=64 ps=45 nodate nonumber;
data mice;
input d1 d2 d3 d4 d5 d6 d7;
lines;
0.190 0.388 0.621 0.823 1.078 1.132 1.191
0.218 0.393 0.568 0.729 0.839 0.852 1.004
0.141 0.260 0.472 0.662 0.760 0.885 0.878
0.211 0.394 0.549 0.700 0.783 0.870 0.925
0.209 0.419 0.645 0.850 1.001 1.026 1.069
0.193 0.362 0.520 0.530 0.641 0.640 0.751
0.201 0.361 0.502 0.530 0.657 0.762 0.888
0.202 0.370 0.498 0.650 0.795 0.858 0.910
0.190 0.350 0.510 0.666 0.819 0.879 0.929
0.219 0.399 0.578 0.699 0.709 0.822 0.953
0.225 0.400 0.545 0.690 0.796 0.825 0.836
0.224 0.381 0.577 0.756 0.869 0.929 0.999
0.187 0.329 0.441 0.525 0.589 0.621 0.796
0.278 0.471 0.606 0.770 0.888 1.001 1.105
;
```

```
/* Source: Izenman and Williams (1989).  Reproduced by
permission of the International Biometric Society. */

filename gsasfile "prog517.graph";
goptions gaccess=gsasfile dev=pslmono;
goptions horigin=1in vorigin=2in;
goptions hsize=5in vsize=7in;
title1 h=1.5 'Mice Data with Profile of the Mean';
title2 j=l 'Output 5.17';
data plot1;
set mice;
array d{7} d1-d7;
do i=1 to 7;
weight=d(i);
day=i*3-1;
output;
end;
keep day weight;
run;
proc summary nway data=mice;
var d1 d2 d3 d4 d5 d6 d7;
output out=new mean=md1-md7;
data plot2;
set new;
array md{7} md1-md7;
do i =1 to 7;
mean_wt=md(i);
output;
end;
keep mean_wt;
run;
data plot;
merge plot1 plot2;
run;
proc gplot data = plot;
plot (weight mean_wt)*day/overlay vaxis=axis1 haxis=axis2;
axis1 label =(a=90 h=1.2 'Weights of Mice');
axis2 offset=(2) label=(h=1.2 'Days after Birth');
symbol1 v=square i = none;
symbol2 v=dot i=join;
run;

proc iml;
use mice;
read all into y0;
print y0;
/* Generating the Ortho. Poly. of degree p-1=6*/
vec1={2 5 8 11 14 17 20};
c=orpol(vec1,6);
y=y0*c;
/* Convert Y matrix to a data set Trandata*/
varnames={y1 y2 y3 y4 y5 y6 y7};
create trandata from y (|colname=varnames|);
append from y;
close trandata;
proc glm data=trandata;
model y1 y2 y3=y4 y5 y6 y7/nouni;
manova h=intercept/printe;
proc reg data=trandata;
```

```
model y1 y2 y3= ;
model y1 y2 y3=y4;
model y1 y2 y3=y4 y5;
model y1 y2 y3=y4 y5 y6;
model y1 y2 y3=y4 y5 y6 y7;
title1 'Output 5.17';
title2 'Growth Curve Analysis: Mice Data';
run;
```

Output 5.17

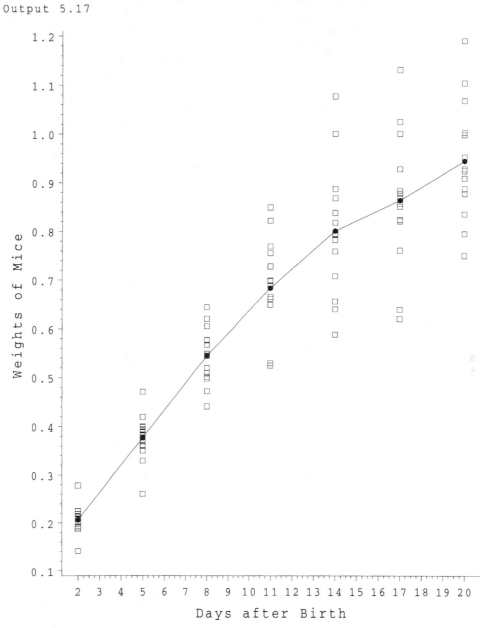

Output 5.17   Mice Data with Profile of the Mean

We analyze this data set to illustrate how a set of covariates may increase the efficiency of the inference in growth curve modeling. Suppose that a second-degree polynomial is the correct model for these data. Rao (1987) has utilized a number of methods of selecting variables to arrive at this model.

A second degree polynomial model for these data can be written in matrix form as

$$\mathbf{Y_0}_{14\times7} = \mathbf{1}_{14\times1}\mathbf{B}_{1\times3}\mathbf{A}_{3\times7} + \mathcal{E}_{14\times7}.$$

As in the previous example, by transforming $\mathbf{Y_0}$ using an orthogonal polynomial matrix we obtain $\mathbf{Y} = \mathbf{Y_0}\mathbf{C}$. Let $\mathbf{Y} = (\mathbf{Y}1 : \dots : \mathbf{Y}7)$. The adequacy of the second-degree polynomial model is established by testing the null hypothesis $H_0 : E(Y4) = \dots = E(Y7) = 0$. The variables $Y4, \dots, Y7$ can thus be used as the covariates for the dependent variables $Y1$, $Y2$, and $Y3$. The problem is to estimate $\mathbf{B} = (\beta_0, \beta_1, \beta_2)$, elements of which represent the expected values of $Y1$, $Y2$, and $Y3$ respectively. In Table 5.2 we present the standard errors (SE) of the least squares estimates of $\hat{\beta}_0$, $\hat{\beta}_1$, and $\hat{\beta}_2$ for the five models; namely, with no covariates, only $Y4$ as a covariate, $Y4$ *and* $Y5$ as covariates, $\dots$, and $Y4, \dots, Y7$ are all covariates. The standard errors presented in the table are taken from the outputs of PROC REG for each of the five models. For example, the values corresponding to the covariate $Y4$ in the table below respectively are the standard errors corresponding to intercepts for dependent variables Y1, Y2, and Y3 from the output of PROC REG when the code

```
proc reg;
model y1 y2 y3 = y4;
```

is used.

TABLE 5.2 Standard Errors for the Estimates

Covariates	$SE(\hat{\beta}_0)$	$SE(\hat{\beta}_1)$	$SE(\hat{\beta}_2)$
None	.0573	.0332	.0138
$Y4$	.0484	.0221	.0099
$Y4 \& Y5$	.0647	.0294	.0112
$Y4, Y5, Y6$	.0667	.0313	.0120
$Y4, Y5, Y6, Y7$	.0731	.0351	.0133

An examination of the values in the table suggests that including only Y4 in the model as a covariate may be the most beneficial approach. The complete SAS code for finding the estimates using PROC REG is given in Program 5.17. However, the corresponding output is suppressed to save space.

5.6.3   Test of Homogeneity of Regression Coefficients

It is a common practice in most of the growth studies in biological sciences to compare two or more groups by testing the equality of model parameters. First, we consider an example of a linear model and show how to use dummy variables to test the homogeneity of several regressions. The analysis is primarily univariate.

Suppose there are g groups with n_i, $i = 1, \dots, g$ observations in each group. Let the dependent variable be denoted by y and the independent variable be x. Instead of a single independent variable we could have k independent variables. The linear model in this case can be written as

$$y_{ij} = \beta_{i0} + \beta_{i1}x_{ij} + \epsilon_{ij}, \quad j = 1, \dots, n_i, \quad i = 1, \dots, g. \tag{5.18}$$

That is, each group has its own regression line possibly with different intercepts and regression parameters. Assume that ϵ_{ij}, $j = 1, \dots, n_i$, $i = 1, \dots, k$ are all independent $N(0, \sigma^2)$ random variables. In order to test the homogeneity of slopes or homogeneity of intercepts we may first express the above models as a single regression model using dummy variables for identifying the groups. Although PROC GLM directly can be utilized for this analysis we illustrate the approach using dummy variables to facilitate an approach that can be used in the context of nonlinear growth curves discussed in the next section. We thus define

$$D_i = \begin{cases} 1 & \text{if the observation is from group } i, \\ 0 & \text{otherwise;} \end{cases}$$

$i = 1, \ldots, g$. Then Equation 5.18 can be written as

$$y_{ij} = D_1(\beta_{10} + \beta_{11}x_{ij}) + \cdots + D_g(\beta_{g0} + \beta_{g1}x_{ij}) + \epsilon_{ij},$$

or

$$y_{ij} = \beta_{10}D_1 + \beta_{11}D_1x_{ij} + \cdots + \beta_{g0}D_g + \beta_{g1}D_gx_{ij} + \epsilon_{ij}. \tag{5.19}$$

It should be observed that the above model does not have an intercept term. Define $\mathbf{y} = (y_{11}, \ldots, y_{1n_1}, y_{21}, \ldots, y_{2n_2}, \ldots, y_{g1}, \ldots, y_{gn_g})'$, and define ϵ similarly. Also let

$$\mathbf{X} = \begin{bmatrix} 1 & x_{11} & 0 & 0 & \ldots & 0 & 0 \\ \vdots & \vdots & \vdots & \vdots & & \vdots & \vdots \\ 1 & x_{1n_1} & 0 & 0 & \ldots & 0 & 0 \\ 0 & 0 & 1 & x_{21} & \ldots & 0 & 0 \\ \vdots & \vdots & \vdots & \vdots & & \vdots & \vdots \\ 0 & 0 & 1 & x_{2n_2} & \ldots & 0 & 0 \\ \vdots & \vdots & \vdots & \vdots & & \vdots & \vdots \\ 0 & 0 & 0 & 0 & \ldots & 1 & x_{g1} \\ \vdots & \vdots & \vdots & \vdots & & \vdots & \vdots \\ 0 & 0 & 0 & 0 & \ldots & 1 & x_{gn_g} \end{bmatrix}, \quad \boldsymbol{\beta} = \begin{bmatrix} \beta_{10} \\ \beta_{11} \\ \beta_{20} \\ \beta_{21} \\ \vdots \\ \beta_{g0} \\ \beta_{g1} \end{bmatrix}.$$

Then Equation 5.19 can be written in matrix form as

$$\mathbf{y}_{n \times 1} = \mathbf{X}_{n \times m}\boldsymbol{\beta}_{m \times 1} + \epsilon_{n \times 1}, \quad \epsilon \sim N(\mathbf{0}, \sigma^2\mathbf{I}),$$

where $n = \sum_{i=1}^{g} n_i$, and $m = 2g$. Here the matrix $\mathbf{X}$ is of full rank with rank $2g$. The hypotheses $H_0^{(1)} : \beta_{11} = \cdots = \beta_{g1}$ and $H_0^{(2)} : \beta_{10} = \cdots = \beta_{g0}$ can be tested using the standard tests from regression analysis.

EXAMPLE 13   *Homogeneity of Regression, Cabbage Data*   The problem is to compare ascorbic acid content (Y) in cabbage in two genetic lines (LINE) or cultivars planted on three different dates. A completely randomized design with ten experimental units for each combination of planting date and genetic line is used. Thus, $g = 2 \times 3 = 6$, $n_i = 10$ for each of the six treatment groups and hence $n = \Sigma n_i = 60$. As the ascorbic acid content may also depend on the weight (X) of the cabbage, the weight of the cabbage head is taken as a covariate. A model for analyzing these data, taking $\bar{x}$ as the average head weight, is

$$y_{iju} = \mu_{ij} + \beta_{ij}(x_{iju} - \bar{x}) + \epsilon_{iju},$$

$u = 1, \ldots, 10$, $i = 1, 2$, $j = 1, 2, 3$. Here β_{ij} represent the slopes of different regression lines for each of the six groups. It is possible that μ_{ij} themselves may have a model, say of the type $\mu + \alpha_i + \delta_j + \gamma_{ij}$. The problem we consider here is to test for homogeneity of the regression coefficients β_{ij}. That is, we want to test the null hypothesis $H_0^{(1)} : \beta_{ij} = \beta$ for all i and j. The null hypothesis $H_0^{(2)}$ which is the hypothesis of the equality of intercepts can then be tested similarly. It is also possible to test the specific hypotheses on the individual components of μ_{ij}, such as α_i, δ_j or γ_{ij}, $i = 1, 2$; $j = 1, 2, 3$.

Since the main interest here is to test $H_0^{(1)} : \beta_{ij} = \beta$ we consider the alternative, yet equivalent, model

$$y_{iju} = \nu_{ij} + \beta_{ij} x_{iju} + \epsilon_{iju},$$

where $\nu_{ij} = \mu_{ij} + \beta_{ij}\bar{x}$. Of course both models will lead to the same test statistics for $H_0^{(1)}$. The SAS code for performing this test is provided in Program 5.18. Using the two levels (LIN1 and LIN2) of the genetic line variable LINE and three levels (DAT1, DAT2, and DAT3) of the variable DATE we first create six dummy variables D1-D6 to identify the six groups. The respective products of D1-D6 with the covariate X representing interactions are created as the variables XD1-XD6. Then we use the NOINT option in PROC REG to fit the model in Equation 5.19. The TEST statement is used to test the equality of regression parameters β_{ij}. From Output 5.18, the F statistic for testing $H_0 : \beta_{ij} = \beta$ has an observed value of 0.6643. Under $H_0^{(1)}$ it has an F distribution with $(5, 48)$ degrees of freedom thereby giving a p value of 0.6523. The high p value indicates that a common regression parameter can be used in the model. Thus, a simplified model for the analysis of this data set is

$$y_{iju} = \mu_{ij} + \beta(x_{iju} - \bar{x}) + \epsilon_{iju},$$

with $\mu_{ij} = \mu + \alpha_i + \delta_j + \gamma_{ij}$.

```
/* Program 5.18 */

option ls=64 ps=45 nodate nonumber;
title1 'Output 5.18';
data a;
input line $ date $ @;
do i=1 to 5;
input x y@;
output; drop i;
end;
lines;
lin1 dat1 2.5 51 2.2 55 3.1 45 4.3 42 2.5 53
lin1 dat1 4.3 50 3.8 50 4.3 52 1.7 56 3.1 49
lin1 dat2 3.0 65 2.8 52 2.8 41 2.7 51 2.6 41
lin1 dat2 2.8 45 2.6 51 2.6 45 2.6 61 3.5 42
lin1 dat3 2.2 54 1.8 59 1.6 66 2.1 54 3.3 45
lin1 dat3 3.8 49 3.2 49 3.6 55 4.2 49 1.6 68
lin2 dat1 2.0 58 2.4 55 1.9 67 2.8 61 1.7 67
lin2 dat1 3.2 68 2.0 58 2.2 63 2.2 56 2.2 72
lin2 dat2 4.0 52 2.8 70 3.1 57 4.2 58 3.7 47
lin2 dat2 3.0 56 2.2 72 2.3 63 3.8 54 2.0 60
lin2 dat3 1.5 78 1.4 75 1.7 70 1.3 84 1.7 71
lin2 dat3 1.6 72 1.4 62 1.0 68 1.5 66 1.6 72
;
/* Source: Rawlings (1988, p. 219).  Reprinted by permission
of the Wadsworth Publishing Company. */
data cabbage;
set a;
if (line='lin1' and date='dat1') then d1=1;
  else d1=0;
if (line='lin1' and date='dat2') then d2=1;
  else d2=0;
if (line='lin1' and date='dat3') then d3=1;
  else d3=0;
if (line='lin2' and date='dat1') then d4=1;
  else d4=0;
if (line='lin2' and date='dat2') then d5=1;
  else d5=0;
```

```
if (line='lin2' and date='dat3') then d6=1;
  else d6=0;
output;
data cabbage;
set cabbage;
xd1=x*d1;
xd2=x*d2;
xd3=x*d3;
xd4=x*d4;
xd5=x*d5;
xd6=x*d6;
title2 'Homogeneity of Regression';
proc reg data=cabbage;
model y=d1-d6 xd1-xd6/noint;
test xd1=xd2=xd3=xd4=xd5=xd6;
run;
```

Output 5.18

```
                         Output 5.18
                   Homogeneity of Regression


Dependent Variable: Y
Test:    Numerator:     25.5634  DF:    5   F value:  0.6643

         Denominator:  38.48409  DF:   48   Prob>F:   0.6523
```

For testing the equality of the intercepts $(H_0^{(2)})$ for the six groups the following SAS code could be utilized.

```
proc reg;
model y=d1-d6 xd1-xd6/noint;
test d1=d2=d3=d4=d5=d6;
run;
```

5.6.4 Growth as a Nonlinear Regression Model

There may be situations where the linear modeling of growth may not be appropriate. For more realistic applications, especially in fisheries and biological growths, we may need to use nonlinear functions for fitting the growth processes. Although there are several nonlinear models for growth curves, for illustration we will consider only a class of the most celebrated models. These models, called Von Bertalanffy models, are especially useful in fisheries.

The Von Bertalanffy model with additive error in its most general form can be written as

$$y_{ij} = l_{\infty i}[1 - exp\{-k_i(t_{ij} - t_{0i})\}] + \epsilon_{ij}, \tag{5.20}$$

$j = 1, \ldots, n_i$, $i = 1, \ldots, g$, where y_{ij} is the measurement, say, of the length of a fish, on the jth subject (j^{th} fish) from the ith group. Also, t_{ij} is the observed value of a covariate (for example, age of a fish) corresponding to y_{ij}, $l_{\infty i}$ for each group i is the unknown limiting value (as the time variable tends to infinity) of the expectation of y_{ij}, that is, the asymptote of the mean (expected value) as a function of the time variable. For example, $l_{\infty i}$ is the asymptotic length (or the expected length at maturity) of the fish from the ith group. The unknown constant k_i represents the rate at which the asymptotic value $l_{\infty i}$ is achieved, t_{0i} is the unknown hypothetical value of t_{ij} when the value of y_{ij} is zero, and ϵ_{ij} are independently distributed $N(0, \sigma^2)$ error variables.

As in the case of linear regression models, we define the dummy variables

$$D_i = \begin{cases} 1 & \text{if the observation is from group } i, \\ 0 & \text{otherwise;} \end{cases}$$

$i = 1, \ldots, g$. Then the model in Equation 5.20 can be written as

$$y_{ij} = \sum_{u=1}^{g} D_u l_{\infty u}[1 - exp\{-k_u(t_{ij} - t_{0u})\}] + \epsilon_{ij}, \quad j = 1, \ldots, n_i, \ i = 1, \ldots, g.$$

The problem is to compare several of these groups. The likelihood ratio method may be used for such comparisons. The maximum likelihood estimates of the unknown parameters have been obtained using PROC NLIN. See Lakkis and Jones (1992) for a study of this problem.

Rewrite the model Equation 5.20 as

$$y_{ij} = \mu(l_{\infty i}, k_i, t_{0i}, t_{ij}) + \epsilon_{ij},$$

where $\mu(l_{\infty i}, k_i, t_{0i}, t_{ij}) = l_{\infty i}[1 - exp\{-k_i(t_{ij} - t_{0i})\}]$. Then the likelihood function of the parameters given the data $y_{ij}, j = 1, \ldots, n_i, i = 1, \ldots, g$ is

$$f(\boldsymbol{\theta}) = (2\pi\sigma^2)^{-n/2} exp\left\{-\frac{1}{2\sigma^2}\sum_{i=1}^{g}\sum_{j=1}^{n_i}(y_{ij} - \mu(l_{\infty i}, k_i, t_{0i}, t_{ij}))^2\right\}. \quad (5.21)$$

Here $n = \sum_{i=1}^{g} n_i$ and $\boldsymbol{\theta}$ represents all the parameters $l_{\infty 1}, \ldots, l_{\infty g}, k_1, \ldots, k_g, t_{01}, \ldots, t_{0g}$ and σ^2. For a fixed σ^2 maximizing $f(\boldsymbol{\theta})$ in Equation 5.21 with respect to the parameters $l_{\infty i}, k_i, t_{0i}, i = 1, \ldots, g$ is the same as minimizing

$$S(l_{\infty 1}, \ldots, l_{\infty g}, k_1, \ldots, k_g, t_{01}, \ldots, t_{0g}) = \sum_{i=1}^{g}\sum_{j=1}^{n_i}[y_{ij} - \mu(l_{\infty i}, k_i, t_{0i}, t_{ij})]^2$$

with respect to the corresponding parameters and this minimization will need to be done iteratively. Once have we obtained these estimates, the maximum likelihood estimate of σ^2 is given by

$$\hat{\sigma}^2 = \frac{1}{n}S(\hat{l}_{\infty 1}, \ldots, \hat{l}_{\infty g}, \hat{k}_1, \ldots, \hat{k}_g, \hat{t}_{01}, \ldots, \hat{t}_{0g}),$$

where $\hat{l}_{\infty 1}, \ldots, \hat{l}_{\infty g}, \hat{k}_1, \ldots, \hat{k}_g, \hat{t}_{01}, \ldots, \hat{t}_{0g}$ are the estimates obtained by the nonlinear least squares minimization described above.

Consider the general problem of testing a null hypothesis $H_0 : \boldsymbol{\theta} \in \omega$ versus $H_1 : \boldsymbol{\theta} \notin \omega$, where ω is a subset of Ω the parameter space. The likelihood ratio test statistic for this problem is

$$L = \left(\frac{\hat{\sigma}_{\Omega}^2}{\hat{\sigma}_{\omega}^2}\right)^{n/2},$$

where $\hat{\sigma}_{\Omega}^2$ is the maximum likelihood estimate of σ^2 when no restrictions on the parameter space are placed and $\hat{\sigma}_{\omega}^2$ is the maximum likelihood estimate of σ^2 when the linear constraints prescribed by H_0 are put on the parameter space Ω. For large sample size n, the distribution of $-2 \ln L = -n \ln (\hat{\sigma}_{\Omega}^2/(\hat{\sigma}_{\omega}^2))$ is approximated by the chi-square distribution with ν degrees of freedom, where ν is the number of parameters estimated in Ω *minus* the number of parameters estimated in ω (Rao, 1973).

We want to test the following hypotheses:

$H_0^{(1)} : l_{\infty 1} = \cdots = l_{\infty g} (= l_\infty, \ say)$ vs. not all $l_{\infty i}$ are equal

$H_0^{(2)} : k_1 = \cdots = k_g (= k)$ vs. not all k_i are equal

$H_0^{(3)} : t_{01} = \cdots = t_{0g} (= t_0)$ vs. not all t_{0i} are equal

$H_0^{(4)} : l_{\infty 1} = \cdots = l_{\infty g} (= l_\infty)$, $t_{01} = \cdots = t_{0g} (= t_0)$ and $k_1 = \cdots = k_g (= k)$ vs. at least one inequality.

For testing $H_0^{(1)}$, $H_0^{(2)}$, $H_0^{(3)}$, and $H_0^{(4)}$ against the corresponding alternatives, the chi-square approximation of the likelihood ratio test statistics are $-n \, ln(\hat{\sigma}_\Omega^2 / \hat{\sigma}_{\omega_i}^2)$, $i = 1, 2, 3, 4$. Each of these statistics has an approximate chi-square distribution with respective degrees of freedom $(g-1)$, $(g-1)$, $(g-1)$ and $3(g-1)$. The ω_i are the respective subsets of the parameter space Ω defined by the null hypotheses $H_0^{(i)}$, $i = 1, .., 4$. Given the initial estimates $\boldsymbol{\theta} = \boldsymbol{\theta}_0$ and the form of the function $\mu(\boldsymbol{\theta}, x)$, PROC NLIN fits a model of the type $y = \mu(\boldsymbol{\theta}, x) + \epsilon$ given the data (y_i, x_i), $i = 1, \ldots, n$ and obtains the least squares estimates of $\boldsymbol{\theta}$. This procedure can be adapted to the present case of having data on g groups with possibly different growth curves.

EXAMPLE 14   ***Tests of Homogeneity, Fish Growth Data***   This data set from Kimura (1980) and also analyzed by Lakkis and Jones (1992) using SAS programs contains observations on the average lengths at different ages for male and female fish (Pacific hake). The data in Program 5.19 are only a part of a larger data set. As $g = 2$, we introduce two more independent variables D_1 and D_2, respectively, identifying male and female fish. The model to analyze these data is

$$y_{ij} = D_1 l_{\infty 1}[1 - exp\{-k_1(t_{ij} - t_{01})\}]$$

$$+ D_2 l_{\infty 2}[1 - exp\{-k_2(t_{ij} - t_{02})\}] + \epsilon_{ij} \qquad (5.22)$$

$j = 1, \ldots, n_i$, $i = 1, 2$; $n_1 = 11$, $n_2 = 13$, $n = n_1 + n_2 = 24$.

The objective of the study was to fit the separate Von Bertalanffy models to the data on male as well as female Pacific hakes. Interest was also in determining if a single model for the two groups could be considered adequate and if not, if there were certain parameters in the model that could be taken to be the same for the two groups.

The SAS code for calculating the maximum likelihood estimates of the relevant parameters under Ω (no restriction on the parameter space), under ω_1 (parameter space restricted by linear constraint: $l_{\infty 1} = l_{\infty 2} = l_\infty$), ω_2 (parameter space restricted by $k_1 = k_2 = k$), ω_3 (parameter space restricted by $t_{01} = t_{02} = t_0$) and under ω_4 (parameter space restricted by all of the restrictions in ω_1, ω_2, and ω_3 together) is given in Program 5.19. In each case, PROC NLIN requires one to supply the initial estimates of the parameters with the PARMS= option. Following Lakkis and Jones (1992), we first plot for each group (male and female) the length as a function of age, and get the initial estimates for $(t_{0i}, l_{\infty i})$, $i = 1, 2$ by visually examining the graph (not shown here). The initial estimates of k_i are obtained by substituting the corresponding initial estimates for $l_{\infty i}$, t_{0i} and by substituting the average, $\bar{y}_i$ for y_{ij}, the average, $\bar{t}_i$ for t_{ij}, $j = 1, \ldots, n_i$ in Equation 5.20 and then solving for k_i ignoring the error component. The initial estimates under the restricted models are obtained by appropriate simple averaging. For example, under ω_1 the average of initial estimates of $l_{\infty 1}$ and $l_{\infty 2}$ is used as an initial estimate of l_∞ since under $H_0^{(1)}$, we have the restriction $l_{\infty 1} = l_{\infty 2} = l_\infty$.

In the MODEL statement of PROC NLIN, the explicit form of Equation 5.22 (barring error term) is provided. There are many iterative procedures available in the NLIN procedure to fit the model some of which are derivative-based and require the explicit specification of partial derivatives. If no derivatives are provided, the procedure uses the default DUD (Doesn't Use Derivatives), where the derivatives are estimated by the program. The maximum number of iterations are specified in the MAXITER= option. In the output the estimates of each of the parameters and a regression type (ANOVA) summary table are provided. As in the ANOVA, we have partitioned the total sum of squares into sum of squares due to regression and the residual sum of squares. The maximum likelihood estimate of σ^2, say $\hat{\sigma}^2$, is obtained from the residual sum of squares by dividing it by n. Thus, by running the NLIN procedure under no restrictions and under various restrictions specified by $\omega_1 - \omega_4$ we obtain $\hat{\sigma}_\Omega^2$, $\hat{\sigma}_{\omega_1}^2$, ..., $\hat{\sigma}_{\omega_4}^2$. These are used to obtain the test statistics for various hypotheses described above. In the following we present various parameter estimates, test statistics and the p values for the hypotheses $H_0^{(1)}$-$H_0^{(4)}$.

```
/* Program 5.19*/

options ls=64 ps=45 nodate nonumber;
title1 'Output 5.19';
title2 'Nonlinear Growth Curve Analysis';
data fish;
input length d1 d2 age@@;
lines;
15.40 1 0 1.0 26.93 1 0 2.0 42.23 1 0 3.3
44.59 1 0 4.3 47.63 1 0 5.3
49.67 1 0 6.3 50.87 1 0 7.3 52.30 1 0 8.3
54.77 1 0 9.3 56.43 1 0 10.3
55.88 1 0 11.3
15.40 0 1 1.0 28.03 0 1 2.0 41.18 0 1 3.3
46.20 0 1 4.3 48.23 0 1 5.3
50.26 0 1 6.3 51.82 0 1 7.3 54.27 0 1 8.3
56.98 0 1 9.3 58.93 0 1 10.3
59.00 0 1 11.3 60.91 0 1 12.3 61.83 0 1 13.3
;
/* Source: Kimura (1980).  U. S. Fishery Bulletin. */
/* The following code fits von Bertalanffy model
under Omega*/
proc nlin data=fish maxiter=100;
parms l1=55 k1=.276 t1=0 l2=60 k2=.23 t2=0;
model length= l1*d1*(1-exp(-k1*(age-t1)))+
l2*d2*(1-exp(-k2*(age-t2)));
run;
/* The following code fits von Bertalanffy model
under omega 1:l1=l2=l*/
/* The initial estimate of l is taken as the average
of that of l1 and l2*/
proc nlin data=fish maxiter=100;
parms l=57.5 k1=.276 t1=0 k2=.23 t2=0;
model length= l*d1*(1-exp(-k1*(age-t1)))+
l*d2*(1-exp(-k2*(age-t2)));
run;
/* The following code fits von Bertalanffy model
under omega 2:k1=k2=k*/
/* The initial estimate of k is taken as the average
of that of k1 and k2*/
proc nlin data=fish maxiter=100;
parms l1=55 k=.253 t1=0 l2=60 t2=0;
model length= l1*d1*(1-exp(-k*(age-t1)))+
l2*d2*(1-exp(-k*(age-t2)));
run;
/* The following code fits von Bertalanffy model
under omega 3:t1=t2=t*/
/* The initial estimate of t is taken as the average
of that of t1 and t2*/
```

```
proc nlin data=fish maxiter=100;
parms l1=55 k1=.276 t=0 l2=60 k2=.23;
model length= l1*d1*(1-exp(-k1*(age-t)))+
l2*d2*(1-exp(-k2*(age-t)));
run;
/* The following code fits von Bertalanffy model
under omega 4:
l1=l2 ,k1=k2 and t1=t2*/
proc nlin data=fish maxiter=100;
parms l=57.5 k=.253 t=0 ;
model length= l*d1*(1-exp(-k*(age-t)))+
l*d2*(1-exp(-k*(age-t)));
run;
```

Output 5.19

Output 5.19
Nonlinear Growth Curve Analysis

Non-Linear Least Squares Summary Statistics
Dependent Variable LENGTH

Source	DF	Sum of Squares	Mean Square
Regression	6	57120.508434	9520.084739
Residual	18	48.223766	2.679098
Uncorrected Total	24	57168.732200	
(Corrected Total)	23	3989.046050	

Parameter	Estimate	Asymptotic Std. Error	Asymptotic 95 % Confidence Interval	
			Lower	Upper
L1	55.97801266	1.1380114362	53.587155151	58.368870172
K1	0.38558439	0.0409732778	0.299503299	0.471665486
T1	0.17133864	0.1489187227	-0.141525924	0.484203195
L2	61.23333874	1.1709806801	58.773215874	63.693461602
K2	0.29625218	0.0277203755	0.238014214	0.354490140
T2	-0.05726981	0.1691162935	-0.412567606	0.298027995

Non-Linear Least Squares Summary Statistics
Dependent Variable LENGTH

Source	DF	Sum of Squares	Mean Square
Regression	5	57097.130085	11419.426017
Residual	19	71.602115	3.768532
Uncorrected Total	24	57168.732200	
(Corrected Total)	23	3989.046050	

Parameter	Estimate	Asymptotic Std. Error	Asymptotic 95 % Confidence Interval	
			Lower	Upper
L	59.40382630	1.0081829496	57.293690212	61.513962385
K1	0.29675407	0.0270955196	0.240042903	0.353465241
T1	-0.11124358	0.1952230595	-0.519847223	0.297360057
K2	0.33746034	0.0327778212	0.268856063	0.406064619
T2	0.08730393	0.1719128085	-0.272511144	0.447119001

From Output 5.19 (part of which is provided), under no restrictions (Ω) on the parameters of the model in Equation 5.22 the estimates of various parameters are

$$\hat{l}_{\infty 1} = 55.9790, \quad \hat{k}_1 = 0.3855, \quad \hat{t}_{01} = 0.1711$$
$$\hat{l}_{\infty 2} = 61.2352, \quad \hat{k}_2 = 0.2962, \quad \hat{t}_{02} = -0.0575$$

and $n\,\hat{\sigma}_\Omega^2 = 48.2238$ (the residual sum of squares).

Under the restriction $l_{\infty 1} = l_{\infty 2} = l_\infty$ (ω_1), the estimates are $\hat{l}_\infty = 59.4037$, $\hat{k}_1 = 0.2968$, $\hat{t}_{01} = -0.1112$, $\hat{k}_2 = 0.3375$, $\hat{t}_{02} = 0.0873$, and $n\,\hat{\sigma}_{\omega_1}^2 = 71.6021$. The chi-square test statistic for testing $H_0^{(1)}$ is $-24\,ln(\hat{\sigma}_\Omega^2/\hat{\sigma}_{\omega_1}^2) = 9.4865$, which, using the SAS PROBCHI function, yields a p value of 0.0021 based on a chi-square distribution with $g - 1 = 1$ degrees of freedom. Since the p value is small, a rejection of H_{01} occurs and we conclude that male and female fish have different asymptotic lengths upon maturity.

Similarly, under $H_{02} : k_1 = k_2 = k$, $n\,\hat{\sigma}_{\omega_2}^2 = 56.3368$ yielding the chi-square statistic $-24\,ln(\hat{\sigma}_\Omega^2/\hat{\sigma}_{\omega_2}^2) = 3.7319$. The p value based on the chi-square distribution with 1 df is 0.0534. There is some evidence for rejection of $H_0^{(2)}$ although this is only marginal. Output corresponding to $H_0^{(2)}$ is not presented here to save space. For the same reason we have also suppressed the output corresponding to $H_0^{(3)}$ and $H_0^{(4)}$.

Under $H_0^{(3)} : t_{01} = t_{02} = t_0$ $n\,\hat{\sigma}_{\omega_3}^2 = 50.7578$. Hence $-24\,ln(\hat{\sigma}_\Omega^2/\hat{\sigma}_{\omega_3}^2) = 1.2291$ yielding a p value of 0.2676 based on 1 df. We do not reject $H_0^{(3)}$. Finally, with all the three restrictions $n\,\hat{\sigma}_{\omega_4}^2 = 79.7645$ and $-24\,ln(\hat{\sigma}_\Omega^2/\hat{\sigma}_{\omega_4}^2) = 12.0774$ yielding a p value of 0.0071 based on the chi-square distribution with $3(g - 1) = 3$ degrees of freedom. Thus, $H_0^{(4)}$ is rejected.

In summary, the analyses given above suggest that two different growth curves for male and female fish populations are needed with a common length at birth and possibly a common growth rate.

5.7 Crossover Designs

In crossover experiments, treatments are administered in a variety of sequences on various subjects. Thus, each subject may get more than one treatment in the course of the experiment. One reason for doing so is that there is usually more variability across subjects than within subjects. Hence the subjects can be treated as random blocks for the purpose of increasing the precision by controlling experimental error variance. The problem, however, becomes complicated by the fact that the successive measurements on the subjects may be correlated. There is also the possible presence of carryover effect(s) at a given time point, from the treatment(s) applied at previous time point(s). This calls for a careful analysis of the data by incorporating these effects appropriately in the model.

5.7.1   Analysis of Crossover Designs

A linear model for crossover design for the response from the mth subject receiving the ith sequence and the kth treatment in the jth time point is

$$y_{ijkm} = \mu_{ijk} + \eta_{im} + \epsilon_{ijkm}$$

$i = 1, \ldots, s, \quad j = 1, \ldots, p, \quad k = 1, \ldots, a, \quad m = 1, \ldots, n_i,$ where μ_{ijk} is the average response of the kth treatment in sequence i at time point j, η_{im} is the error corresponding to the mth experimental unit in sequence i and ϵ_{ijkm} is the random error associated with time point j of the mth experimental unit in sequence i.

The mean response μ_{ijk} is, for convenience, assumed to be composed of several components as

$$\mu_{ijk} = \mu + \alpha_k + \pi_j + \sum_{r=0}^{j-1} \lambda_{jk_r},$$

where α_k is the treatment effect, π_j the time effect, and λ_{jk_r} is the carryover effect of k_r^{th} treatment administered in one or more times during time points $0, 1, \ldots, j-1$, on the observation at time point j. Initially, at the beginning of any sequence, there is not any carryover effect, so λ_{jk_0} is zero. To simplify the model, we also assume that any carryover effect lasts only up to the next time point, and hence we take $\lambda_{jk_0} = \lambda_{jk_1} = \cdots = \lambda_{jk_{j-2}} = 0$. We also denote, for convenience, $\lambda_{jk_{j-1}}$ by $\lambda_{[k,j-1]}$. In view of this the model becomes

$$y_{ijkm} = \mu + \alpha_k + \pi_j + \lambda_{[k,j-1]} + \eta_{im} + \epsilon_{ijkm}. \tag{5.23}$$

It is further assumed that $\eta_{im} \sim N(0, \sigma_\eta^2)$ are all independent and are independent of $\epsilon_{i'jkm'}$ also for all i, i', j, m, m'. However, the assumption of independence of ϵ_{ijkm} and $\epsilon_{i'j'k'm'}$ may not be realistic. It is so because the observations taken over time on a given subject would most likely be correlated. For convenience the model in Equation 5.23 has been often analyzed as a split plot design, with an assumption of compound symmetric covariance structure for the observation on a given subject. The split plot analysis of Equation 5.23 is straightforward. We will illustrate it by an example adopted from Jones and Kenward (1989, p. 229).

EXAMPLE 15   *Univariate Analysis, Comparison of Drugs*   Two drugs A and B and their combination termed drug C are to be compared for their effectiveness to control hypertension. Each subject was given three drugs in one of the six possible sequences ABC, ACB, BAC, BCA, CAB, and CBA and systolic blood pressure (Y) was measured at the conclusion of each of the three treatments, each of which lasted for four weeks. The data were collected at four different centers. However, we will not consider the variability due to location in our analysis. The data are presented as part of Program 5.20 (the carryover effect for the initial period is defined as zero). The objective of the study is to examine if there were significant differences between drugs, and to examine the period effect and the carryover effect. The model given in Equation 5.23 is the appropriate model and can be fitted using PROC GLM with the following MODEL statement

```
model y= subject period treat carry;
```

where all the variables on the right side are CLASS variables. Since SUBJECT forms a random sample, the variable SUBJECT is declared to be random. This instructs SAS to use the appropriate sums of squares in forming the F ratios for the tests.

```
                    /* Program 5.20 */

     options ls= 64 ps=45 nodate nonumber;
     title1 'Output 5.20';
     title2 'Analysis of Crossover Design';
     data bp;
     infile 'bpress.dat';
     input patient $ y treat $ carry $ period center subject;
     proc glm data = bp ;
     class treat carry period subject;
     model y = subject period treat carry/ ss1 ;
     random subject/test ;
     contrast ' a vs. b' treat 1 -1 0;
     contrast ' a vs. c' treat 1 0 -1;
     contrast ' b vs. c' treat 0 1 -1;
     run;
     proc glm data = bp ;
     class treat carry period subject;
     model y = subject period treat carry/ ss3 ;
     random subject/test ;
     contrast ' a vs. b' treat 1 -1 0;
     contrast ' a vs. c' treat 1 0 -1;
     contrast ' b vs. c' treat 0 1 -1;
     run;
```

Output 5.20

```
                         Output 5.20
                  Analysis of Crossover Design

                 General Linear Models Procedure
```

Dependent Variable: Y

Source	DF	Sum of Squares	F Value	Pr > F
Model	28	40095.9560700	4.95	0.0001
Error	40	11573.9859589		
Corrected Total	68	51669.9420290		

	R-Square	C.V.	Y Mean
	0.776002	9.575829	177.637681

Source	DF	Type I SS	F Value	Pr > F
SUBJECT	22	30646.6086957	4.81	0.0001
PERIOD	2	1395.5942029	2.41	0.1026
TREAT	2	7983.8717062	13.80	0.0001
CARRY	2	69.8814653	0.12	0.8866

Dependent Variable: Y

Contrast	DF	Contrast SS	F Value	Pr > F
a vs. b	1	6303.02403115	21.78	0.0001
a vs. c	1	1471.22264872	5.08	0.0297
b vs. c	1	1491.20262487	5.15	0.0287

The resulting output is presented in Output 5.20, which indicates a significant effect due to the differences in treatments (for Type I SS, the observed value of F(2,40) is 13.80 and the corresponding p value is 0.0001). However, the carryover effect and the period effect are both statistically insignificant. The same conclusions are reached when the analysis is done using Type III SS. Note that because of our convention of using 0 for the carryover effect at the initial period, the sum of squares due to the carryover effect is different from that reported by Jones and Kenward.

In view of the absence of the carryover effect and the period effect, it may be meaningful to compare the treatment means. The estimated difference between the mean effects of A and B, A and C, and B and C are respectively, 25.3971, 12.4140, and -12.9831, which are obtained by first obtaining the least squares solutions using the SOLUTION option in the MODEL statement and then by computing the appropriate differences $\hat{\alpha}_A - \hat{\alpha}_B$, $\hat{\alpha}_A - \hat{\alpha}_C$, and $\hat{\alpha}_B - \hat{\alpha}_C$. Using separate CONTRAST statements, we can perform tests for the significance of treatment differences. All three treatments are significantly different from each other (respective p values are 0.0001, 0.0297 and 0.0287), and A and B are especially markedly different from each other. Treatment C, being the combination of drugs A and B, falls in between the two.

It must be remembered that the course of analysis in crossover designed data depends heavily on what the significant and insignificant effects are. The reason for this is that, depending on the particular design, certain effects may be confounded with each other, and if a sum of squares representing two or more confounded effects is found to be significant, it would require a further careful examination of data to decide which of the many confounded effects may have led to the statistical significance of the particular sum of squares. In fact, if each of the treatment sequences is tried on the groups of subjects, then the group*period interaction is composed of

- the treatment effect,
- carryover effect and treatment*period interaction, and
- other left over effects associated with group*period effects.

Since it is desirable that the effects listed above are all estimated, the choice of an appropriate crossover design becomes of great importance. This topic has been addressed at length in Jones and Kenward (1989) and Ratkowsky, Evans and Alldredge (1993). Later in this section we present the SAS code for generating suitable designs from two useful classes of crossover designs.

In some special cases, it may be possible to perform a straightforward multivariate analysis of crossover data. One such case is the analysis of AB/BA designs, which are two-sequence (namely, AB and BA) designs in two treatments A and B respectively applied to the groups of size, say n_1 and n_2. Since the analysis would follow a familiar pattern described earlier in great detail, it can be best illustrated by an example.

EXAMPLE 16   *Multivariate Analysis, Effect of Onions in Diet*   Dunsmore (1981) presents a case study which was initiated to investigate the effect of including onions in the diet on plasma triglyceride levels. A two-sequence crossover design with two treatments, namely, breakfast without and with onions, to be referred to as A and B respectively, is used. Eight patients were assigned to sequence AB and six to BA. Increases in the plasma triglyceride levels at 1, 2, 3, 4, 5, and 6 hours after breakfast were measured. The analysis here is performed on 10,000 times the logarithm of the plasma triglyceride levels. In the SAS code the respective 12 responses are denoted as Y11, ..., Y16 and Y21, ..., Y26. While Dunsmore analyzed the data using the univariate technique, Grender and Johnson (1992) chose the multivariate route and listed a SAS approach using the 12 by 1 vectors of responses on each of the 14 subjects, which we will briefly present here. Multivariate normality has been assumed for the analysis. There are two groups, namely, AB and BA. These are referred to as Group 1 and 2. Let us denote by μ_{ij}, the 6 by 1 vector of true mean responses on a subject in the ith group and under j^{th} treatment, i, $j = 1, 2$. This is displayed in Table 5.3:

TABLE 5.3 A Two Sequence Crossover Design in Two Treatments

Group	Treatment	Mean	Treatment	Mean
1	A	μ_{11}	B	μ_{12}
2	B	μ_{21}	A	μ_{22}

where $\mu_{ij} = (\mu_{ij1}, \ldots, \mu_{ij6})'$, $i, j = 1, 2$. Also, let $\mu_1' = (\mu_{11}' : \mu_{12}')$ and $\mu_2' = (\mu_{21}' : \mu_{22}')$ and let $\mu_i = \mu + \alpha_i$, $i = 1, 2$ for some μ and α_1, α_2. Then the linear model for these data can be written as

$$\mathbf{Y}_{14 \times 12} = \begin{bmatrix} \mathbf{1}_8 & \mathbf{1}_8 & \mathbf{0} \\ \mathbf{1}_6 & \mathbf{0} & \mathbf{1}_6 \end{bmatrix} \begin{bmatrix} \mu' \\ \alpha_1' \\ \alpha_2' \end{bmatrix} + \mathcal{E}$$

which is in the form of standard linear model

$$\mathbf{Y} = \mathbf{XB} + \mathcal{E}.$$

From Equation 5.23 we have

$$E(y_{ijkm}) = \mu + \alpha_k + \pi_j + \lambda_{[k, j-1]},$$
$$k = 1, 2, \ j = 1, 2, \ i = 1, 2, \ m = 1, \ldots, n_i; \ n_1 = 8, \ n_2 = 6.$$

Thus, we have

$$E(y_{i11m}) = \mu + \alpha_1 + \pi_1,$$
$$E(y_{i21m}) = \mu + \alpha_1 + \pi_2 + \lambda_1,$$
$$E(y_{i12m}) = \mu + \alpha_2 + \pi_1,$$
$$\text{and } E(y_{i22m}) = \mu + \alpha_2 + \pi_2 + \lambda_2.$$

Hence

$$E(y_{111m} + y_{121m}) = 2\mu + (\alpha_1 + \alpha_2) + (\pi_1 + \pi_2) + \lambda_1,$$
$$E(y_{112m} + y_{122m}) = 2\mu + (\alpha_1 + \alpha_2) + (\pi_1 + \pi_2) + \lambda_2,$$
$$E(y_{111m} - y_{121m}) = (\alpha_1 - \alpha_2) + (\pi_1 - \pi_2) - \lambda_1,$$
$$\text{and } E(y_{112m} - y_{122m}) = (\alpha_2 - \alpha_1) + (\pi_1 - \pi_2) - \lambda_2.$$

These sums and differences can be calculated by pre- and post-multiplying $\mathbf{B} = (\mu, \alpha_1, \alpha_2)'$ by certain specific matrices, that is, as $\mathbf{LBM}$, where $\mathbf{L}$ and $\mathbf{M}$ are to be appropriately chosen. As a result, most of the relevant hypotheses can be expressed as $\mathbf{LBM} = \mathbf{0}$.

We adopt the following strategy for the analysis. The appropriate SAS code has been provided by Grender and Johnson (1992), which has been produced here with a few notational changes.

1. First test the null hypothesis that there is no group*time interaction. For this, use

$$\mathbf{L} = \begin{bmatrix} 0 & 1 & -1 \end{bmatrix} \text{ and } \mathbf{M}' = (1, \ 1) \otimes \begin{bmatrix} 1 & -1 & 0 & 0 & 0 & 0 \\ 0 & 1 & -1 & 0 & 0 & 0 \\ 0 & 0 & 1 & -1 & 0 & 0 \\ 0 & 0 & 0 & 1 & -1 & 0 \\ 0 & 0 & 0 & 0 & 1 & -1 \end{bmatrix}$$

$$= \mathbf{M}_1, \text{ (say)}$$

where $\otimes$ stands for Kronecker product (Rao, 1973).

2. Test for the equal carryover effect, that is, $H_0 : \lambda_1 = \lambda_2$. For this, use $\mathbf{L} = (0\ 1\ -1)$ and, say, $\mathbf{M'} = (1\ 1\ 1\ 1\ 1\ 1\ 1\ 1\ 1\ 1\ 1\ 1) = \mathbf{M}_2$.

3. If there are equal carryover effects, then test the hypothesis of no treatment*time interaction. For this, $\mathbf{L} = (0\ 1\ -1)$ and $\mathbf{M'} = \mathbf{M}_1$.

4. If there is no group*time interaction (that is, if H_0 in Step 1 is not rejected), we may test the hypothesis of no period*time interaction. For this $\mathbf{L} = (1\ 0\ 0)$ and $\mathbf{M'} = \mathbf{M}_1$.

5. Test for the time effect by using $\mathbf{L} = (1\ 0\ 0)$ and $\mathbf{M'} = \mathbf{M}_1$.

6. Test for equal treatment effects using $\mathbf{L} = (1\ 0\ 0)$ and $\mathbf{M'} = (1\ 1\ 1\ 1\ 1\ 1\ -1\ -1\ -1\ -1\ -1\ -1) = \mathbf{M}_3$.

7. Test for equal treatment effects using $\mathbf{L} = (1\ 0\ 0)$ and $\mathbf{M'} = \mathbf{M}_3$.

SAS code to analyze Dunsmore's data is presented in Program 5.21. The interpretation of multivariate tests is straightforward and we do not cover it here. The corresponding output has been therefore supressed.

```
/* Program 5.21 */

options ls=64 ps=45 nodate nonumber;
title1 'Output 5.21';
title2 'Multivariate Analysis of Crossover Design';
data onion;
infile 'onion.dat';
input patient meal$ y11 y12 y13 y14 y15 y16 group
patient2 meal2$ y21 y22 y23 y24 y25 y26 group2;
run;
proc glm data = onion ;
class group ;
model y11 y12 y13 y14 y15 y16 y21 y22 y23 y24 y25 y26=
group / nouni;
contrast 'Group*Time  Interaction' group 1 -1 ;
manova
m = y11-y12+y21-y22,
y12-y13+y22-y23,
y13-y14+y23-y24,
y14-y15+y24-y25,
y15-y16+y25-y26;
title3 'Test for Group*Time Interaction';
run;
proc glm data = onion;
class group;
model y11 y12 y13 y14 y15 y16 y21 y22 y23 y24 y25 y26=
group / nouni;
contrast 'Carryover  Effect' group 1 -1 ;
manova
m = y11+y12+y13+y14+y15+y16+y21+y22+y23+y24+y25+y26;
title3 'Test for Equality of Carryover Effects';
run;
proc glm data = onion;
class group ;
model y11 y12 y13 y14 y15 y16 y21 y22 y23 y24 y25 y26=
group / nouni;
contrast 'Teatment*Time  Interaction' group 1 -1 ;
```

```
manova
                      m = y11-y12-y21+y22,
                          y12-y13-y22+y23,
                          y13-y14-y23+y24,
                          y14-y15-y24+y25,
                          y15-y16-y25+y26;
title3 'Test for Treatment*Time Interaction
(Assuming equal Carryover Effects)';
run;
proc glm data = onion ;
class group ;
model y11 y12 y13 y14 y15 y16 y21 y22 y23 y24 y25 y26=
group / nouni;
manova h = intercept  m = y11-y12-y21+y22,
                          y12-y13-y22+y23,
                          y13-y14-y23+y24,
                          y14-y15-y24+y25,
                          y15-y16-y25+y26;
title3 'Test for Period*Time Interaction
(assuming no Group*Time Interaction)';
run;
proc glm data = onion ;
class group ;
model y11 y12 y13 y14 y15 y16 y21 y22 y23 y24 y25 y26=
group / nouni;
contrast 'Treatment  Effect' group 1 -1 ;
manova
m = y11+y12+y13+y14+y15+y16-y21-y22-y23-y24-y25-y26;
title3 'Test for Equality of Treatment Effects';
run;
proc glm data = onion ;
class group ;
model y11 y12 y13 y14 y15 y16 y21 y22 y23 y24 y25 y26=
group / nouni;
manova h = intercept
   m = y11-y12+y21-y22,
       y12-y13+y22-y23,
       y13-y14+y23-y24,
       y14-y15+y24-y25,
       y15-y16+y25-y26;
title3 'Test for Time Effect';
run;
proc glm data = onion ;
class group ;
model y11 y12 y13 y14 y15 y16 y21 y22 y23 y24 y25 y26=
group / nouni;
manova h = intercept
m = y11+y12+y13+y14+y15+y16-y21-y22-y23-y24-y25-y26;
title3 'Test for Equality of Period Effects';
run;
```

5.7.2 Construction of Crossover Designs

There is a vast amount of literature available on the construction of p period crossover designs under various desirable criteria, such as the variance balance or the D-optimality criterion. These criteria have been discussed in detail in Jones and Kenward (1989).

One common crossover design consists of augmenting various mutually orthogonal Latin squares one below the other. These designs are variance balanced in that the variances of the estimates of the treatment differences are minimum. The resulting design is a design in p periods, p treatments that require $p(p-1)$ subjects (provided such a design exists). See Jones and Kenward (1989) for further details.

A complete set of p by p mutually orthogonal Latin squares, if it exists, is equivalent to a resolution III fractional factorial design for $p + 1$, p level variables in p^2 runs that is a $p^{(p+1)-(p-1)}$ fractional factorial design. The problem can be approached by first generating the appropriate fractional factorial layout and then by extracting the elements to arrange $(p-1)$ Latin squares one below the other. In these, the columns denote the period, and rows denote the subject number. Since the transpose of a Latin square is also a Latin square, some caution needs to be exercised about which of the two should be chosen; a simple way to ensure that the correct one of the two is being used is to make sure that all subjects have distinct sequences. The entry at the $(i, j)^{th}$ place indicates the particular treatment administered at the jth time point on the ith subject. The desired $p^{(p+1)-(p-1)}$ fractional factorial design is generated by using the FACTEX procedure. A sample program to generate a 4 period, 4 treatment crossover arrangement requiring 12 subjects is given in Program 5.22. The output is given in Output 5.22. Crossover designs with $a = p$ treatments and $p_1 < p$ periods can be obtained by removing any $(p - p_1)$ columns from the above design.

```
/* Program 5.22 */

options ls=64 ps=45 nodate nonumber;
title1 'Output 5.22';
title2 'Generation of a Latin Square Crossover Design';
title3 ' 4 Treatments, 4 Time Points in 12 Subjects';
proc factex;
factors row col t1-t3/nlev=4;
size design=16;
model resolution=3;
output out=latinsq;
run;
data latinsq1;
set latinsq;
keep t1-t3;
proc iml;
use latinsq1;
read all into z0;
p=4;
latin1 = i(p);
latin2 = i(p);
latin3 = i(p);
do i = 1 to p;
do j = 1 to p;
ij = (j-1)*p +i ;
latin1[i,j] = z0[ij,1];
latin2[i,j] = z0[ij,2];
latin3[i,j] = z0[ij,3];
end;
end;
cross = ((latin1//latin2)//latin3);
print cross;
```

Output 5.22

```
                              Output 5.22
              Generation of a Latin Square Crossover Design
                4 Treatments, 4 Time Points in 12 Subjects


        CROSS
            0            3            1            2
            3            0            2            1
            1            2            0            3
            2            1            3            0
            0            2            3            1
            3            1            0            2
            1            3            2            0
            2            0            1            3
            0            1            2            3
            3            2            1            0
            1            0            3            2
            2            3            0            1
```

Since the design obtained from the complete set of $p \times p$ mutually orthogonal Latin squares requires $p(p-1)$ subjects (and such a design may not even exist, e.g., when $p = 6$), a smaller design may be desired. Williams' designs are possible candidates. These designs are obtained from a $p \times p$ cyclic Latin square (i.e., for which the ith row is $i, i + 1, \ldots, p, 1, \ldots, i - 1$). When p is even, it requires only p subjects. However, when p is odd, the number of subjects required by the Williams' design is $2p$.

Given a cyclic Latin square, we first find its mirror image and interlace the columns of the two. That is, if $A_1, \ldots, A_p$ are columns of the cyclic Latin square and $B_p(= A_1), B_{p-1}(= A_2), \ldots, B_1(= A_p)$ are their respective mirror images (which appear in the order $B_1, \ldots, B_p$), then the resulting arrangement is

$$A_1 B_1 \ A_2 B_2 \ \ldots \ A_p B_p,$$

obtained by interlacing $B_1, \ldots, B_p$ with $A_1, \ldots, A_p$. This gives an arrangement with $2p$ columns. It can be generated by using Program 5.23 and is shown in Output 5.23 for $p = 5$. The desired design in p periods is obtained by taking either the first p or the last p columns of the above arrangement if p is even and by arranging the last p columns below the first p columns if p is odd. By virtue of being generated from a Latin square, all p treatments will appear in the design and every treatment will appear exactly once (if p is even) or twice (if p is odd) in every time period.

```
/* Program 5.23 */

options ls=64 ps=45 nodate nonumber;
title1 'Output 5.23';
title2 'Williams Design: p Treatments in p Time points ';
proc iml;
p=5;
p2=p*2;
pover2 =p/2;
/*Initialize the matrices;
a is the circular p by p Latin Square, b is
its mirror image and the specific;*/
a=i(p);
b=i(p);
w=j(p,p2);
*Create the matrix a;
do i = 1 to p;
```

```
i1=i-1;
do j = 1 to p;
if (i = 1) then  a[i,j] =j;
if (i >1) then do;
if (j < p) then do ;
jmod = j+1;
a[i,j] = a[i1,jmod] ;
end ;
if (j = p) then do ;
jmod = 1 ;
a[i,j] = a[i1,jmod] ;
end ;
end ;
end ;
end ;
* Create b, the mirror image of a;
do i = 1 to p;
do j = 1 to p;
jj=(p+1-j);
b[i,j] = a[i,jj];
end ;
/*Interlace the Circular Latin Square and its
mirror image;*/
do k = 1 to p2;
kby2 =k/2;
if  kby2=floor(kby2) then do;
w[i,k] = b[i,kby2];
end;
if  kby2>floor(kby2) then do;
kk= floor(kby2)+1;
w[i,k] = a[i,kk];
end;
end;
end;
print 'p is equal to ' p;
print 'The following is the Desired Williams Design: ';
print '_____ ';
/*
Design requires p (if p even) or 2p subjects (if p odd).
If p is even take first p or last p columns.
If p is odd slice after first p columns
and augment last p columns below it.
*/
w1 = w*(i(p)//j(p,p,0));
w2 =w*(j(p,p,0)//i(p));
if pover2 > floor(p/2) then do;
williams =w1//w2;
end;
else williams =w1 ;
print williams;
```

Output 5.23

```
                              Output 5.23
                 Williams Design: p Treatments in p Time points

                                           P
                       p is equal to       5

              The following is the Desired Williams Design:
              ---------------------------------------------


         WILLIAMS
                 1        5        2        4        3
                 2        1        3        5        4
                 3        2        4        1        5
                 4        3        5        2        1
                 5        4        1        3        2
                 3        4        2        5        1
                 4        5        3        1        2
                 5        1        4        2        3
                 1        2        5        3        4
                 2        3        1        4        5
```

There are a number of alternative methods available to construct other crossover designs. See, for example, Jones and Kenward (1989) and John (1971). An extensive catalog of many useful designs is given in Ratkowsky, Evans and Alldredge (1993).

5.8 Concluding Remarks

We conclude this chapter with some comments about missing values in repeated measures. One of the very frequent problems in conducting repeated measures experiments is the failure to follow the subject at all time points. As a result, many repeated measures data sets are not balanced. This further complicates the problem in two ways. First, the standard multivariate methods may no longer be applicable. Second, most computer packages including SAS do not deal with missing values in the multivariate data; for example, SAS ignores all the observations on a particular subject if it finds a missing value for any of the dependent variables in the MODEL statement. This not only reduces the sample size substantially but may also result in a sample that is biased due to this implicit self-selection.

A way to alleviate this problem would be the *imputation* of missing values before analyzing the data. There are well-respected approaches, based on the EM algorithm, for imputing the missing values in certain cases of missingness patterns and causes (Little and Rubin, 1987, McLachlan and Krishnan, 1997). Unfortunately, the EM algorithms by definition are very problem specific and often require the identification of appropriate sufficient statistics (for conditioning purposes) even to program the estimation procedure. However, it should be remembered that imputing the missing values and their substitution for further analysis may not necessarily be a desirable choice. This type of analysis may cause the variance terms to be underestimated. The SAS MIXED procedure provides some alternative modeling approaches for data sets of this kind. We will discuss these approaches in Chapter 6.

Analysis of Repeated Measures Using Mixed Models

6.1   Introduction   247
6.2   The Mixed Effects Linear Model   248
6.3   An Overview of the MIXED Procedure   252
6.4   Statistical Tests for Covariance Structures   255
6.5   Models with Only Fixed Effects   265
6.6   Analysis in the Presence of Covariates   274
6.7   A Random Coefficient Model   288
6.8   Multivariate Repeated Measures Data   294
6.9   Concluding Remarks   297

6.1   Introduction

The multivariate and the univariate approaches to analyzing the repeated measures data presented in Chapter 5 represent the extremes of the assumptions made on covariance structures. The former has absolutely no requirements (except that the variance covariance matrix of repeated measures be positive definite) and hence requires one to estimate the maximum possible number of variance and covariance parameters. By contrast, the latter imposes stringent requirements (except when sphericity is required) when the entire covariance structure is governed only by two parameters (in the case of compound symmetry and a few more in the of Huynh-Feldt structure).

Nonetheless, each of these two approaches permits us to use the least squares and the (univariate or multivariate) analysis of variance based approach for data analysis. What is desirable is to assume a covariance structure on repeated measures, which is not as liberal as the multivariate approach but at the same time is not as restrictive and hence not as unrealistic as the univariate approach of Chapter 5. Unfortunately, as soon as any deviation from the two is allowed, the analysis of variance based approach is no longer valid. However, in such a situation one can use an alternative (but not necessarily equivalent) approach based on likelihoods. While the likelihood theory regarding estimation and testing of hypothesis is well established, it must be noted that most of the likelihood based statistical test procedures are asymptotic in nature and hence are only approximate for the finite sample cases.

Some of the useful covariance structures other than compound symmetry for the repeated measure data are the first order autoregressive, unstructured covariance, Toeplitz, and banded Toeplitz. The MIXED procedure provides all of these with several other covariance structures as options. Further, the univariate split plot analysis of repeated measures can be performed by using the MIXED procedure, which makes available several alternative covariance structures including those listed above and in addition to compound symmetry. This kind of analysis under the general mixed effects linear model is the major theme of this chapter. Specifically, this chapter explains how to formulate problems in repeated measures data analysis as problems in the general mixed effects linear model and to utilize the MIXED procedure to solve these problems.

6.2 The Mixed Effects Linear Model

Let $\mathbf{y}_i$ be the $p_i \times 1$ vector of repeated measures on the ith subject. Then consider a mixed effects model described as

$$\mathbf{y}_i = \mathbf{X}_i \boldsymbol{\beta} + \mathbf{Z}_i \boldsymbol{\nu}_i + \boldsymbol{\epsilon}_i, \quad i = 1, \ldots, n, \tag{6.1}$$

where $\mathbf{X}_i$ and $\mathbf{Z}_i$ are the known matrices of orders p_i by q and p_i by r respectively, and $\boldsymbol{\beta}$ is the fixed q by 1 vector of unknown (nonrandom) parameters. The r by 1 vectors $\boldsymbol{\nu}_i$ are random effects with $E(\boldsymbol{\nu}_i) = \mathbf{0}$, and $D(\boldsymbol{\nu}_i) = \sigma^2 \mathbf{G}_1$. Finally $\boldsymbol{\epsilon}_i$ are the p_i by 1 vectors of random errors whose elements are no longer required to be uncorrelated. We assume that $E(\boldsymbol{\epsilon}_i) = \mathbf{0}$, $D(\boldsymbol{\epsilon}_i) = \sigma^2 \mathbf{R}_i$, $cov(\boldsymbol{\nu}_i, \boldsymbol{\nu}_{i'}) = \mathbf{0}$, $cov(\boldsymbol{\epsilon}_i, \boldsymbol{\epsilon}_{i'}) = \mathbf{0}$, $cov(\boldsymbol{\nu}_i, \boldsymbol{\epsilon}_{i'}) = \mathbf{0}$ for all $i \neq i'$, and $cov(\boldsymbol{\nu}_i, \boldsymbol{\epsilon}_i) = \mathbf{0}$. Such assumptions seem to be reasonable in repeated measures data where subjects are assumed to be independent, yet the repeated data on a given subject may be correlated. Note here that $\mathbf{R}_i$ is the appropriate $p_i \times p_i$ submatrix of a $p \times p$ positive definite matrix, where p is the number of time points in the data set where observations have been made. An appropriate covariance structure can be assigned to the data by an appropriate choice of matrices $\mathbf{G}_1$ and $\mathbf{R}_i$. Note that since $\mathbf{y}_i$ is a p_i by 1 vector, $i = 1, \ldots, n$, the model can account for the unbalanced repeated measures data, that is, when data are such that all the subjects have not been observed at all time points.

The n submodels in Equation 6.1 can be stacked one below the other to give a single model

$$\begin{bmatrix} \mathbf{y}_1 \\ \mathbf{y}_2 \\ \vdots \\ \mathbf{y}_n \end{bmatrix} = \begin{bmatrix} \mathbf{X}_1 \\ \mathbf{X}_2 \\ \vdots \\ \mathbf{X}_n \end{bmatrix} \boldsymbol{\beta} + \begin{bmatrix} \mathbf{Z}_1 & \mathbf{0} & \cdots & \mathbf{0} \\ \mathbf{0} & \mathbf{Z}_2 & \cdots & \mathbf{0} \\ \vdots & \vdots & \cdots & \vdots \\ \mathbf{0} & \mathbf{0} & \cdots & \mathbf{Z}_n \end{bmatrix} \begin{bmatrix} \boldsymbol{\nu}_1 \\ \boldsymbol{\nu}_2 \\ \vdots \\ \boldsymbol{\nu}_n \end{bmatrix} + \begin{bmatrix} \boldsymbol{\epsilon}_1 \\ \boldsymbol{\epsilon}_2 \\ \vdots \\ \boldsymbol{\epsilon}_n \end{bmatrix}$$

or

$$\mathbf{y}_{\Sigma p_i \times 1} = \mathbf{X}_{\Sigma p_i \times q} \boldsymbol{\beta}_{q \times 1} + \mathbf{Z}_{\Sigma p_i \times nr} \boldsymbol{\nu}_{nr \times 1} + \boldsymbol{\epsilon}_{\Sigma p_i \times 1}, \tag{6.2}$$

where the definitions of $\mathbf{y}$, $\mathbf{X}$, $\mathbf{Z}$, $\boldsymbol{\nu}$, and $\boldsymbol{\epsilon}$ in terms of the matrices and vectors of submodels are self explanatory. In view of the assumptions made on Equation 6.1, we have $E(\boldsymbol{\nu}) = \mathbf{0}$, $E(\boldsymbol{\epsilon}) = \mathbf{0}$,

$$D(\boldsymbol{\nu}) = \sigma^2 \begin{bmatrix} \mathbf{G}_1 & \mathbf{0} & \cdots & \mathbf{0} \\ \mathbf{0} & \mathbf{G}_1 & \cdots & \mathbf{0} \\ \vdots & \vdots & \cdots & \vdots \\ \mathbf{0} & \mathbf{0} & \cdots & \mathbf{G}_1 \end{bmatrix} = \sigma^2 \mathbf{I}_n \otimes \mathbf{G}_1 = \sigma^2 \mathbf{G}$$

and

$$D(\boldsymbol{\epsilon}) = \sigma^2 \begin{bmatrix} \mathbf{R}_1 & \mathbf{0} & \cdots & \mathbf{0} \\ \mathbf{0} & \mathbf{R}_2 & \cdots & \mathbf{0} \\ \vdots & \vdots & \cdots & \vdots \\ \mathbf{0} & \mathbf{0} & \cdots & \mathbf{R}_n \end{bmatrix} = \sigma^2 \mathbf{R}.$$

The symbol $\otimes$ here stands for the *Kronecker product* (Rao, 1973) defined for two matrices $\mathbf{U}_{s \times t} = (u_{ij})$ and $\mathbf{W}_{l \times m} = (w_{ij})$ as

$$
\mathbf{U} \otimes \mathbf{W} = \begin{bmatrix} u_{11}\mathbf{W} & u_{12}\mathbf{W} & \cdots & u_{1t}\mathbf{W} \\ u_{21}\mathbf{W} & u_{22}\mathbf{W} & \cdots & u_{2t}\mathbf{W} \\ \vdots & \vdots & \cdots & \vdots \\ u_{s1}\mathbf{W} & u_{s2}\mathbf{W} & \cdots & u_{st}\mathbf{W} \end{bmatrix} = (u_{ij}\mathbf{W}).
$$

It follows from Equation 6.2 that

$$
D(\mathbf{y}) = \mathbf{Z}D(\nu)\mathbf{Z}' + D(\epsilon) = \sigma^2[\mathbf{Z}\mathbf{G}\mathbf{Z}' + \mathbf{R}] = \sigma^2\mathbf{V}.
$$

The above representation $D(\mathbf{y}) = \sigma^2\mathbf{V}$ is taken as a convenience in the algorithm of the MIXED procedure. It may be remarked that in many situations, the variance covariance matrix of $\mathbf{y}$ may not be in the above form where the parameter σ^2 has been explicitly factored out. However, with appropriate (but not necessarily unique) modifications in the matrices $\mathbf{G}$ and $\mathbf{R}$, some parameter σ^2 (not necessarily unique) can be factored out. For example, to factor out σ^2, one only needs to divide all elements of $\mathbf{G}$ and $\mathbf{R}$ by σ^2 and use their reparametrized versions for the purpose of defining the appropriate $\mathbf{V}$. Thus, there is no loss of generality in defining the covariance structure as is the case with the MIXED procedure algorithm.

6.2.1   Estimation of Effects When V Is Known

If $\mathbf{G}_1$ and $\mathbf{R}_1, \ldots, \mathbf{R}_n$ are assumed to be known then the Best (minimum mean squared error) Linear Unbiased Estimator (BLUE) using the generalized least squares estimator of β is given by (assuming that it uniquely exists)

$$
\hat{\beta} = [\mathbf{X}'(\mathbf{Z}\mathbf{G}\mathbf{Z}' + \mathbf{R})^{-1}\mathbf{X}]^{-1}\mathbf{X}'(\mathbf{Z}\mathbf{G}\mathbf{Z}' + \mathbf{R})^{-1}\mathbf{y}
$$

$$
= \left[\sum_{i=1}^{n} \mathbf{X}_i'(\mathbf{Z}_i\mathbf{G}_1\mathbf{Z}_i' + \mathbf{R}_i)^{-1}\mathbf{X}_i\right]^{-1} \left[\sum_{i=1}^{n} \mathbf{X}_i'(\mathbf{Z}_i\mathbf{G}_1\mathbf{Z}_i' + \mathbf{R}_i)^{-1}\mathbf{y}_i\right]. \qquad (6.3)
$$

The variance covariance matrix of $\hat{\beta}$ is

$$
\sigma^2[\mathbf{X}'(\mathbf{Z}\mathbf{G}\mathbf{Z}' + \mathbf{R})^{-1}\mathbf{X}]^{-1} = \sigma^2 \left[\sum_{i=1}^{n} \mathbf{X}_i'(\mathbf{Z}_i\mathbf{G}_1\mathbf{Z}_i' + \mathbf{R}_i)^{-1}\mathbf{X}_i\right]^{-1}.
$$

Similarly, the Best Linear Unbiased Predictor (BLUP) of ν is given by $\mathbf{G}\mathbf{Z}'(\mathbf{Z}\mathbf{G}\mathbf{Z}' + \mathbf{R})^{-1}(\mathbf{y} - \mathbf{X}\beta)$. Further an unbiased estimator of σ^2 is obtained as

$$
\hat{\sigma}^2 = \frac{1}{\nu_2}\hat{\epsilon}'\mathbf{V}^{-1}\hat{\epsilon},
$$

where $\hat{\epsilon} = \mathbf{y} - \mathbf{X}(\mathbf{X}'\mathbf{V}^{-1}\mathbf{X})^{-}\mathbf{X}'\mathbf{V}^{-1}\mathbf{y}$ and $\nu_2 = \sum_{i=1}^{n} p_i - Rank(\mathbf{X})$ is the error degrees of freedom. If $\mathbf{X}'(\mathbf{Z}\mathbf{G}\mathbf{Z}' + \mathbf{R})^{-1}\mathbf{X}$ does not admit an inverse, for most estimation problems a generalized inverse would replace the inverse in Equation 6.3 provided estimability of the functions under consideration has been ensured.

The above BLUE of β and the BLUP of ν can also be obtained by solving the system of *mixed model equations* (Henderson, 1984),

$$
\begin{bmatrix} \mathbf{X}'\mathbf{R}^{-1}\mathbf{X} & \mathbf{X}'\mathbf{R}^{-1}\mathbf{Z} \\ \mathbf{Z}'\mathbf{R}^{-1}\mathbf{X} & \mathbf{Z}'\mathbf{R}^{-1}\mathbf{Z} + \mathbf{G}^{-1} \end{bmatrix} \begin{bmatrix} \hat{\beta} \\ \hat{\nu} \end{bmatrix} = \begin{bmatrix} \mathbf{X}'\mathbf{R}^{-1}\mathbf{y} \\ \mathbf{Z}'\mathbf{R}^{-1}\mathbf{y} \end{bmatrix}.
$$

If in addition, multivariate normality is assumed for $\boldsymbol{v}_i$ and $\boldsymbol{\epsilon}_i$, $i = 1, \ldots, n$, then,

$$\mathbf{y} \sim N_{\Sigma p_i}(\mathbf{X}\boldsymbol{\beta}, \sigma^2[\mathbf{Z}\mathbf{G}\mathbf{Z}' + \mathbf{R}]).$$

In this case $\hat{\boldsymbol{\beta}}$ and $\hat{\boldsymbol{v}}$ are also the maximum likelihood estimator and maximum likelihood predictor of $\boldsymbol{\beta}$ and $\boldsymbol{v}$ respectively.

Consider the problem of testing a linear hypothesis of the form $H_0 : \mathbf{L}\boldsymbol{\beta} = \mathbf{0}$, where $\mathbf{L}$ is a full (row) rank matrix. Then the usual test statistic for testing H_0 is

$$F = \frac{\hat{\boldsymbol{\beta}}'\mathbf{L}'(\mathbf{L}(\mathbf{X}'\mathbf{V}^{-1}\mathbf{X})^{-1}\mathbf{L}')^{-1}\mathbf{L}\hat{\boldsymbol{\beta}}}{\hat{\sigma}^2\, Rank(\mathbf{L})},$$

which under the null hypothesis H_0 is distributed as F_{ν_1, ν_2}, where $\nu_1 = Rank(\mathbf{L})$, ν_2 is the error degrees of freedom, and $\mathbf{V} = (\mathbf{Z}\mathbf{G}\mathbf{Z}' + \mathbf{R})$.

6.2.2   Estimation of σ^2 and V

When the matrices $\mathbf{G}$ and/or $\mathbf{R}$ (or $\mathbf{V}$) are unknown, estimation of these matrices can be carried out using the standard likelihood based methods (i.e. ML or REML) under the assumption of joint multivariate normality of $\boldsymbol{v}$ and $\boldsymbol{\epsilon}$. In practice, certain structure on either one or both of these matrices is assumed so that $\mathbf{V}$ is a function of only a few unknown parameters, say $\theta_1, \ldots, \theta_s$. The above methods are iterative in that first for a fixed value of $\mathbf{V}$, an estimator of $\boldsymbol{\beta}$ using the form of the BLUE is obtained. Then the likelihood function of $\mathbf{V}$ is maximized with respect to $\theta_1, \ldots, \theta_s$ to get an estimate of $\mathbf{V}$. These two steps are iterated until a certain user specified convergence criterion is met.

The ML estimators of $\theta_1, \ldots, \theta_s$ and hence of $\mathbf{V}$ (and hence of $\mathbf{G}$ and $\mathbf{R}$) and of σ^2 are obtained by maximizing the logarithm of the normal likelihood function

$$l(\boldsymbol{\theta}) = -\frac{1}{2}ln|\sigma^2\mathbf{V}| - \frac{1}{2\sigma^2}\hat{\boldsymbol{\epsilon}}'\mathbf{V}^{-1}\hat{\boldsymbol{\epsilon}} - \frac{n}{2}ln(2\pi) \tag{6.4}$$

simultaneously with respect to these parameters. The ML estimator of σ^2 expressed in terms of $\hat{\mathbf{V}}$ will be $\hat{\sigma}_n^2 = \hat{\boldsymbol{\epsilon}}'\hat{\mathbf{V}}^{-1}\hat{\boldsymbol{\epsilon}}/n$. The ML estimates of $\theta_1, \ldots, \theta_s$, generally have to be obtained using iterative schemes.

Alternatively, estimators of $\theta_1, \ldots, \theta_s$, and finally of σ^2 can be obtained by maximizing the function:

$$-\frac{1}{2}ln|\mathbf{V}| - \frac{n}{2}ln\hat{\boldsymbol{\epsilon}}'\mathbf{V}^{-1}\hat{\boldsymbol{\epsilon}} - \frac{n}{2}\left[1 + ln\left(\frac{2\pi}{n}\right)\right].$$

which is obtained from the log-likelihood function after factoring and profiling a residual variance $\hat{\sigma}_n^2$.

Similarly, another set of estimators commonly known as the Restricted Maximum Likelihood (REML) estimators is obtained by maximizing the function (after profiling $\hat{\sigma}^2$)

$$-\frac{1}{2}ln|\mathbf{V}| - \frac{1}{2}ln|\mathbf{X}'\mathbf{V}^{-1}\mathbf{X}| - \frac{n-k}{2}ln\hat{\boldsymbol{\epsilon}}'\mathbf{V}^{-1}\hat{\boldsymbol{\epsilon}} - \frac{n-k}{2}\left[1 + ln\left(\frac{2\pi}{n-k}\right)\right],$$

where $k = Rank(\mathbf{X})$. The ML and REML estimators are known to be asymptotically equivalent.

Suppose $\hat{\boldsymbol{\theta}} = (\hat{\theta}_1, \ldots, \hat{\theta}_s)'$ is the ML estimate of $\boldsymbol{\theta} = (\theta_1, \ldots, \theta_s)'$. Let $\mathbf{h}(\boldsymbol{\theta})$ be a certain, possibly vector valued, function of $\boldsymbol{\theta}$. Then the three asymptotic tests to test $H_0 : \mathbf{h}(\boldsymbol{\theta}) = \mathbf{0}$ against the alternative $H_1 : \mathbf{h}(\boldsymbol{\theta}) \neq \mathbf{0}$ are given by

Wald's Statistic: $T_W = n\mathbf{h}(\hat{\boldsymbol{\theta}})'[\mathbf{H}(\hat{\boldsymbol{\theta}})'\mathbf{I}(\hat{\boldsymbol{\theta}})^{-1}\mathbf{H}(\hat{\boldsymbol{\theta}})]^{-1}\mathbf{h}(\hat{\boldsymbol{\theta}})$

Likelihood Ratio Test (LRT) Statistic : $T_L = 2[l(\hat{\boldsymbol{\theta}}) - l(\hat{\boldsymbol{\theta}}_0)]$

Rao's Statistic: $T_R = \frac{1}{n}\mathbf{U}(\hat{\boldsymbol{\theta}}_0)'\mathbf{I}(\hat{\boldsymbol{\theta}}_0)^{-1}\mathbf{U}(\hat{\boldsymbol{\theta}}_0),$

where $\hat{\theta}_0$ is the ML estimator of θ under the null hypothesis H_0, $\mathbf{U}(\theta) = \frac{\partial l}{\partial \theta}$, $\mathbf{H}(\theta) = \frac{\partial \mathbf{h}(\theta)}{\partial \theta}$, and $\mathbf{I}(\theta)$ is the *Fisher information matrix* (Rao, 1973).

Under certain regularity conditions each of the statistics T_W, T_L, and T_R has an asymptotic χ_r^2 distribution under H_0, where $r = Rank(\mathbf{H}(\theta))$. See Rao (1973) and also Sen and Singer (1993) for proofs and more details about these tests. Since REML and ML estimates are asymptotically equivalent one may alternatively use the REML estimates in the above expressions.

Since under certain regularity conditions, the ML estimator $\hat{\theta}$ follows a multivariate normal distribution with the mean vector θ and the variance covariance matrix $\mathbf{I}^{-1}(\theta)$ one can also construct a test for the hypothesis about any component θ_i of θ using the standard normal distribution. This asymptotic test is also known as Wald's test. Using this asymptotic result, approximate confidence intervals can be constructed as well.

6.2.3   Estimation of Effects When V Is Estimated

Suppose $\hat{\mathbf{G}}$ and $\hat{\mathbf{R}}$ are the estimators of $\mathbf{G}$ and $\mathbf{R}$ respectively, obtained by using one of the above two methods. Then the respective estimates of β and ν are obtained by solving the plug-in version of *mixed model equations*,

$$\begin{bmatrix} \mathbf{X}'\hat{\mathbf{R}}^{-1}\mathbf{X} & \mathbf{X}'\hat{\mathbf{R}}^{-1}\mathbf{Z} \\ \mathbf{Z}'\hat{\mathbf{R}}^{-1}\mathbf{X} & \mathbf{Z}'\hat{\mathbf{R}}^{-1}\mathbf{Z} + \hat{\mathbf{G}}^{-1} \end{bmatrix} \begin{bmatrix} \hat{\beta} \\ \hat{\nu} \end{bmatrix} = \begin{bmatrix} \mathbf{X}'\hat{\mathbf{R}}^{-1}\mathbf{y} \\ \mathbf{Z}'\hat{\mathbf{R}}^{-1}\mathbf{y} \end{bmatrix},$$

where the estimators $\hat{\mathbf{G}}$ and $\hat{\mathbf{R}}$ respectively have been used for $\mathbf{G}$ and $\mathbf{R}$ in the mixed model equations stated earlier. Upon solving we obtain $\hat{\beta} = (\mathbf{X}'\hat{\mathbf{V}}^{-1}\mathbf{X})^{-1}\mathbf{X}'\hat{\mathbf{V}}^{-1}\mathbf{y}$ and $\hat{\nu} = \hat{\mathbf{G}}\mathbf{Z}'\hat{\mathbf{V}}^{-1}(\mathbf{y} - \mathbf{X}\hat{\beta})$, where $\hat{\mathbf{V}}$ is obtained by substituting $\hat{\mathbf{G}}$ and $\hat{\mathbf{R}}$ for $\mathbf{G}$ and $\mathbf{R}$ respectively in $\mathbf{V}$. Note that $\hat{\beta}$ is an estimator of the best linear unbiased estimator (BLUE) $(\mathbf{X}'\mathbf{V}^{-1}\mathbf{X})^{-}\mathbf{X}'\mathbf{V}^{-1}\mathbf{y}$ of β and $\hat{\nu}$ is an estimator of the best linear unbiased predictor (BLUP) $\mathbf{G}\mathbf{Z}'\mathbf{V}^{-1}(\mathbf{y} - \mathbf{X}(\mathbf{X}'\mathbf{V}^{-1}\mathbf{X})^{-}\mathbf{X}'\mathbf{V}^{-1}\mathbf{y})$ of the random effects vector ν.

For simplicity of presentation, let us denote the estimate of σ^2 by $\hat{\sigma}^2$, whatever the method may have been used for the estimation. The estimated variances and covariance matrices of these estimators are: $\hat{D}(\hat{\beta}) = \hat{\sigma}^2\mathbf{C}_{11} = \hat{\sigma}^2(\mathbf{X}'\hat{\mathbf{V}}^{-1}\mathbf{X})^{-}$, $\hat{cov}(\hat{\beta}, \hat{\nu}) = \hat{\sigma}^2\mathbf{C}_{21} = -\hat{\sigma}^2\hat{\mathbf{G}}\mathbf{Z}'\hat{\mathbf{V}}^{-1}\mathbf{X}\mathbf{C}_{11}$, and $\hat{D}(\hat{\nu}) = \hat{\sigma}^2\mathbf{C}_{22} = \hat{\sigma}^2((\mathbf{Z}'\hat{\mathbf{R}}^{-1}\mathbf{Z} + \hat{\mathbf{G}}^{-1})^{-1} - \mathbf{C}_{21}\mathbf{X}'\hat{\mathbf{V}}^{-1}\mathbf{Z}\mathbf{G})$. It may however be cautioned that

$$\hat{\sigma}^2 \begin{bmatrix} \mathbf{C}_{11} & \mathbf{C}'_{21} \\ \mathbf{C}_{21} & \mathbf{C}_{22} \end{bmatrix}$$

usually underestimates $D(\hat{\beta}', \hat{\nu}')'$, the true variance covariance matrix of $(\hat{\beta}', \hat{\nu}')'$.

6.2.4   Tests for Fixed Effect Parameters

Consider the problem of testing a linear hypothesis of the form $H_0 : \mathbf{L}\beta = \mathbf{0}$, where $\mathbf{L}$ is a full rank matrix. A suggested test statistic for H_0 is

$$F = \frac{\hat{\beta}'\mathbf{L}'(\mathbf{L}\mathbf{C}_{11}\mathbf{L}')^{-1}\mathbf{L}\hat{\beta}}{\hat{\sigma}^2 \; Rank(\mathbf{L})}.$$

The exact distribution of F is complicated due to many facts. For example, $\hat{\beta}$ is only an approximate version of BLUE since $\mathbf{G}_1$ and $\mathbf{R}_1, \ldots, \mathbf{R}_n$ are unknown and hence their estimates have been used in their expressions. The matrix $\hat{\sigma}^2\mathbf{C}_{11}$ is also an estimated version of the variance covariance matrix of $\hat{\beta}$. Further, the distribution of F also depends on the type of unbalancedness that exists in the data. However for large samples the test statistic F will have an approximate F distribution with numerator degrees of freedom $\nu_1 = Rank(\mathbf{L})$ and denominator degrees of freedom ν_2 appropriately estimated.

A brief description of the MIXED procedure follows. This procedure implements the likelihood based approach described above and hence is useful in the repeated measures context. More specific details of this procedure will be provided in later sections as the need arises.

6.3   An Overview of the MIXED Procedure

PROC MIXED, available in SAS/STAT, can be used to analyze data under the model stated earlier in Equation 6.1. To describe the choices available in PROC MIXED, we will follow the notations used above. More details can be found in *SAS/STAT Software: Changes and Enhancements through Release 6.12.*

Consider the following mixed effects model defined earlier in Equation 6.1,

$$\mathbf{y} = \mathbf{X}\boldsymbol{\beta} + \mathbf{Z}\boldsymbol{v} + \boldsymbol{\epsilon}. \tag{6.5}$$

The first statement that invokes this procedure is

```
proc mixed;
```

The inferential approach in the MIXED procedure is predominantly likelihood based. The multivariate normality assumptions as stated in the previous section are needed. The ML and REML estimation procedures have been implemented for the estimation of all parameters and the prediction of random effects. The test procedures for the purpose of hypothesis testing rely heavily on likelihood based functions. Examples of such tests are LRT and Wald's tests. Additionally, another estimation procedure suggested by C. R. Rao (1972) known as MIVQUE0 is also available. No normality assumptions need to be made for this method. Consequently, no statistical tests based on MIVQUE0 are available.

6.3.1   Structures for G and R

Recall that $\mathbf{G}$ is the variance covariance matrix of random effects and $\mathbf{R}$ is the variance covariance matrix of error vectors corresponding to repeated measures on the same subject. Various choices of structures for $\mathbf{G}$ and $\mathbf{R}$ are available in the MIXED procedure. Accordingly, structures for $\mathbf{G}$ are selected using the TYPE= option in the RANDOM statement and those for $\mathbf{R}$ are selected using the TYPE= option in the REPEATED statement of PROC MIXED. The MIXED procedure also has the ability to allow the Kronecker product (named as a *direct product* (notation: @) in SAS/IML documentation) covariance structure for $\mathbf{R}$. See Example 10 for an illustration.

6.3.2   Estimation of G and R

As stated earlier estimation of the covariance parameters is carried out using one of the three methods, namely the maximum likelihood (ML), restricted maximum likelihood (REML), and minimum variance quadratic unbiased estimation (MIVQUE0). The first two are iterative. For these we need the joint multivariate normality of the error vector $\boldsymbol{\epsilon}$ and the random effects $\boldsymbol{v}$.

The minimum variance quadratic unbiased estimator, MIVQUE0 (developed using the formulas given in Rao (1972), is non-iterative and no multivariate normality assumption is needed. It is usually used as an initial estimator in the iterative process of the ML or REML method. For certain designs with balanced data this estimator coincides with the REML estimator.

Using the METHOD = ML, REML, or MIVQUE0 option of the PROC MIXED statement, one of the three methods of estimating the covariance parameters can be adopted. For example, the syntax for using the REML method is

```
proc mixed method=reml;
```

It must be remarked that this specification also implements the same method of estimation for the fixed effect parameters in the case of ML and REML. For MIVQUE0, the fixed effects are obtained by generalized least squares where the estimate of **G** and **R** have been used in place of true values.

6.3.3   Selection of Appropriate Structure for G and R

Given numerous choices of structures for **G** and **R**, one of the problems a practitioner faces is the selection of appropriate structure. Under the heading "Model Fitting Information," PROC MIXED prints out certain useful statistics that are helpful in selecting an appropriate covariance structure for either **G** or for **R** or for both. Two of these which are used often are Akaike's Information Criterion (AIC) and Schwarz's Bayesian Criterion (BIC).

Akaike's Information Criterion (AIC) is defined as

$$AIC = l(\hat{\theta}) - q,$$

where $l(\theta)$ is the log-likelihood function as given in Equation 6.4 (or the restricted log-likelihood function) and $l(\hat{\theta})$ is the maximum log-likelihood function (or the restricted maximum log-likelihood function) and q is the number of the estimated covariance parameters. The structure expressed in terms of θ with the largest AIC is preferred.

Schwarz's Bayesian Criterion (BIC) is defined as

$$l(\hat{\theta}) - \frac{1}{2}q \, log(n^*),$$

where $n^* = n$ for ML and $(n - k)$ for REML. Similar to AIC interpretation, a model with a larger value of BIC is preferred.

Keselman, Algina, Kowalchuk and Wolfinger (1998) indicate through extensive simulation studies that AIC performs better than BIC in trying to identify the true models. However, both criteria frequently fail to identify the correct covariance structure. These authors have also speculated that the poor performance of BIC may be due to the fact that in PROC MIXED the penalty criterion for BIC is a function of n, the total number of observations, rather than the number of subjects. Further, in SAS Version 7, the number of subjects rather than the number of observations are used in the penalty criterion for BIC.

In the context of selecting a covariance structure for **R**, LRT on covariance structure can be performed to decide if the particular covariance structure is deemed adequate. This approach will be discussed in detail in the next section.

6.3.4   Inference for Covariance Parameters

The estimates, standard errors (SEs) of the estimates, and the asymptotic tests using the standard normal distribution (Wald's test) for each of the covariance parameters are produced when the COVTEST option is specified in the PROC MIXED statement. The standard errors and (Wald's) tests are determined from the general theory of the maximum likelihood estimates which states that the vector of ML estimates of a vector parameter, under certain regularity conditions, is consistent and has a multivariate normal distribution with the inverse of the Fisher's information matrix as its variance covariance matrix. The tests provided on covariance parameters are for two-sided alternative hypotheses. Thus, care should be exercised in interpreting the p value since in certain cases it is more meaningful to test a particular hypothesis under a one sided alternative (e. g., when the parameters are interpreted as the variance components).

PROC MIXED also provides confidence intervals for all the unknown parameters in the variance covariance matrix. The 95% confidence intervals for these parameters can be obtained using the CL option in the PROC MIXED statement. The default 95% for the confidence level can be changed, if needed, using the ALPHA= option of the PROC MIXED statement. For the parameters which have a natural lower bound constraint of zero (for example, the variance components and the diagonal elements of the variance covariance matrix), the confidence intervals are provided using the Satterthwaite approximation. For all the other parameters, the confidence limits are obtained using the corresponding Wald's statistics.

6.3.5   Inference for Fixed and Random Effects Parameters

As indicated earlier, a linear hypothesis of the form $H_0 : \mathbf{L}\boldsymbol{\beta} = \mathbf{0}$, where $\mathbf{L}$ is a full rank matrix is tested using the approximate F test described in the previous section. The test statistic

$$F = \frac{\hat{\boldsymbol{\beta}}' \mathbf{L}' (\mathbf{L} \mathbf{C}_{11} \mathbf{L}')^{-1} \mathbf{L} \hat{\boldsymbol{\beta}}}{\hat{\sigma}^2 \; Rank(\mathbf{L})}$$

under the null hypothesis H_0 is approximately distributed as an F with the degrees of freedom ν_1 and ν_2, where $\nu_1 = Rank(\mathbf{L})$ and ν_2 is the degrees of freedom of the error sum of squares. However, different estimates of ν_2, to improve the approximation, can be used in practice. The MIXED procedure allows one to specify predetermined degrees of freedom using the DDF= option in the MODEL statement. The procedure also provides several built-in choices for ν_2 using the DDFM= option in the MODEL statement. For example, the DDFM=RESIDUAL option conducts all the tests using the error sum of squares degrees of freedom, which is $n - Rank(X)$. The DDF=SATTERTH option conducts a general Satterthwaite approximation for obtaining ν_2. The default sums of squares used are of Type III. The Type I sums of squares can also be utilized using the HTYPE=1 option in the MODEL statement. Approximate p values of the tests are reported using a certain standard estimate of ν_2.

As an alternative to the above F statistic one can use the log-likelihood ratio test statistic or the chi-square statistic associated with that. The degrees of freedom of the chi-square distribution are determined by taking the difference between the number of parameters in the full model and that in the reduced (under the null hypothesis) model. This chi-square test can be requested using the CHISQ option in the MODEL statement and METHOD=ML option in the PROC MIXED statement. Since the REML method produces estimators that are not the maximum likelihood estimators, whenever chi-square tests are requested the ML and not the REML method must be used for estimating the parameters.

The estimates of the fixed effects are obtained using the LSMEANS statement of PROC MIXED. The multiple comparison tests of these effects using one of the standard methods (for example Tukey's method) can be carried out using the ADJUST= option in the LSMEANS statement. Estimation and testing of the hypotheses of certain specific contrasts of the parameters can be carried out using the ESTIMATE and CONTRAST statements in PROC MIXED.

In the following sections, we provide several applications of this approach in conjunction with PROC MIXED for a variety of models. It is not possible to address all the aspects and options of this very general procedure. For different applications and detailed description, we refer the reader to Littell, Milliken, Stroup, and Wolfinger (1996) and The MIXED Procedure Chapter in *SAS/STAT Software: Changes and Enhancements through Release 6.12*.

6.4 Statistical Tests for Covariance Structures

In Chapter 5, we encountered the spherical or compound symmetric covariance structures as a required prerequisite for the univariate analysis of variance methods to be valid for the repeated measures data. It was suggested that one should first test if such a covariance structure can be assumed using the likelihood ratio tests. In many real world applications the hypothesis of spherical or compound symmetric covariance structure will be rejected. While the univariate analysis of variance approach is then invalidated, one may still pursue the (univariate-like) analysis using the asymptotic results based on the maximum likelihood theory. However, in such problems, first one should also determine the appropriate covariance structure so as to enable one to estimate the unknown parameters, including the ones represented in the covariance structure. As earlier, the likelihood ratio tests can be used to test such hypotheses. However, unlike the cases discussed in Chapter 5, it may not be possible to obtain a closed form test statistic when testing for the particular covariance structure. Fortunately, in the data analysis problems, such an issue is only secondary as long as the evaluation of the test statistics is computationally feasible and its theoretical properties are known.

Let $\mathbf{y}_{p \times 1}$ be a $p-$variate normally distributed random vector with mean $\boldsymbol{\mu}$ and the variance covariance matrix $\boldsymbol{\Sigma}$. Suppose we have our data as a random sample $\mathbf{y}_1, \ldots, \mathbf{y}_n$ from this population. The appropriate likelihood ratio test statistic for the null hypothesis,

$$H_0 : \boldsymbol{\Sigma} \text{ has a given covariance structure}$$

is given by,

$$L = \frac{\max\limits_{H_0} g(\boldsymbol{\Sigma}|data)}{\max\limits_{\text{unrestricted}} g(\boldsymbol{\Sigma}|data)},$$

where $g(\boldsymbol{\Sigma}|data)$ is same as the $f(\mathbf{y}_1, \ldots, \mathbf{y}_n)$, the joint density function of $\mathbf{y}_1, \ldots, \mathbf{y}_n$ and for optimization purposes, it is treated as a function of $\boldsymbol{\Sigma}$ for given data. Of course, any other unknown parameters, that may appear in the function $g(.|data)$ will be replaced by their maximum likelihood estimates.

It may be noted that in certain problems and for the calculation of the denominator of L, the matrix $\boldsymbol{\Sigma}$ may not be completely unrestricted and from the very context of the problem at hand, it may still have some covariance structure. The likelihood ratio test can still be performed so long as this restricted parameter space itself contains as a subset, the parameter space restricted by H_0. A simple example of such a situation is when we want to test if a variance covariance matrix additionally has a compound symmetric structure while it is a priori assumed that this matrix already has a circular covariance structure — a structure which contains all compound symmetric variance covariance matrices as a further specialization. In Example 3, we consider yet another context where such a situation arises.

From the likelihood theory it follows that for large n, and under the null hypothesis, $-2 \ln L$ approximately follows a chi-square distribution with f degrees of freedom, where the quantity f is computed as

f = *number of unknown parameters in unstructured* $\boldsymbol{\Sigma}$ $-$ *number of unknown parameters in* $\boldsymbol{\Sigma}$ *under the covariance structure specified by* H_0.

For a better approximation, often a correction factor b known as the Bartlett's correction is multiplied to $-2 \ln L$ and hence $-2 b \ln L$ is used as the modified test statistic. However, we will not concern ourselves with this modification in this chapter.

Often, instead of maximizing the likelihood, only *a part of the likelihood which is invariant of the fixed effects* is maximized. This results in what is known as the restricted maximum likelihood (REML) estimate of the variance covariance matrix. Using these estimates, another test statistic L^R instead of L defined above can be devised. Corresponding to L^R, the quantity $-2 ln L^R$ also follows a chi-square distribution and in fact, it is asymptotically equivalent to the test based on $-2 ln L$. Since in PROC MIXED the default estimation is REML, the quantities reported under the default correspond to $-2 ln L^R$ rather than $-2 ln L$.

Using SAS, a number of covariance structure choices can be adopted and hence tests for these can be performed using the MIXED procedure. Here we give a list of a few selected covariance structures. The complete list can be found in *SAS/STAT Software: Changes and Enhancements through Release 6.12*. Corresponding SAS options to be used in the REPEATED statement in PROC MIXED are indicated within the parentheses.

1. $\boldsymbol{\Sigma} = \sigma^2 \mathbf{I}$ (VC)
2. $\boldsymbol{\Sigma} = \sigma_1^2 \mathbf{11}' + \sigma_2^2 \mathbf{I}$ (CS)
3. $\boldsymbol{\Sigma}$ Unstructured (UN)
4. $\boldsymbol{\Sigma} = diag(\sigma_1^2, \ldots, \sigma_p^2)$: Banded main diagonal (UN(1))
5.

$$\boldsymbol{\Sigma} = \sigma^2 \begin{bmatrix} 1 & \rho & \cdots & \rho^{p-1} \\ \rho & 1 & \cdots & \rho^{p-2} \\ \vdots & & & \\ \rho^{p-1} & \rho^{p-2} & \cdots & 1 \end{bmatrix} : \text{ Autoregressive of order 1 (AR(1))}$$

6.

$$\boldsymbol{\Sigma} = \begin{bmatrix} \sigma_0 & \sigma_1 & \cdots & \sigma_{p-1} \\ \sigma_1 & \sigma_0 & \cdots & \sigma_{p-2} \\ \vdots & & & \\ \sigma_{p-1} & \sigma_{p-2} & \cdots & \sigma_0 \end{bmatrix} : \text{ Toeplitz (TOEP)}$$

7.

$$\boldsymbol{\Sigma} = \begin{bmatrix} \sigma_0 & \sigma_1 & 0 & 0 & \cdots & 0 \\ \sigma_1 & \sigma_0 & \sigma_1 & 0 & \cdots & 0 \\ \vdots & & & & & \\ 0 & \cdots & \cdots & 0 & \sigma_1 & \sigma_0 \end{bmatrix} : \text{ Two Bands Toeplitz (TOEP(2))}$$

8. $\boldsymbol{\Sigma} = \sigma^2 (\rho_{ij}^{d_{ij}})$, $\rho_{ii} = 1$: Spatial Power or Markov (SP(POW)(c))
9. $\boldsymbol{\Sigma} = (\sigma_{ij})$, $\sigma_{ij} = \frac{\sigma_{ii} + \sigma_{jj}}{2} - \lambda$, if $i \neq j$: Huynh $-$ Feldt (HF)

Linear Structures In addition, one can test the null hypothesis (as well as perform the estimation) for any variance covariance matrix $\boldsymbol{\Sigma}$, which can be represented as a linear combination of known matrices. Such a structure is called a *linear covariance structure* and is given by

$$\boldsymbol{\Sigma} = c_1 \mathbf{A}_1 + \cdots + c_k \mathbf{A}_k,$$

where $\mathbf{A}_1, \ldots, \mathbf{A}_k$ are known matrices but the scalars $c_1, \ldots, c_k$ are all unknown and are functionally unrelated to each other. For example, the matrix

$$\boldsymbol{\Sigma}_1 = \begin{bmatrix} \sigma_1^2 + \sigma_2^2 & \sigma_1^2 & \sigma_1^2 \\ \sigma_1^2 & \sigma_1^2 + \sigma_2^2 & \sigma_1^2 \\ \sigma_1^2 & \sigma_1^2 & \sigma_1^2 + \sigma_2^2 \end{bmatrix} = \sigma_1^2 \begin{bmatrix} 1 & 1 & 1 \\ 1 & 1 & 1 \\ 1 & 1 & 1 \end{bmatrix} + \sigma_2^2 \begin{bmatrix} 1 & 0 & 0 \\ 0 & 1 & 0 \\ 0 & 0 & 1 \end{bmatrix}$$

does have the above structure but the matrix

$$\boldsymbol{\Sigma}_2 = \sigma^2 \begin{bmatrix} 1 & \rho & \rho^2 \\ \rho & 1 & \rho \\ \rho^2 & \rho & 1 \end{bmatrix} = \sigma^2 \begin{bmatrix} 1 & 0 & 0 \\ 0 & 1 & 0 \\ 0 & 0 & 1 \end{bmatrix} + \sigma^2 \rho \begin{bmatrix} 0 & 1 & 0 \\ 1 & 0 & 1 \\ 0 & 1 & 0 \end{bmatrix} + \sigma^2 \rho^2 \begin{bmatrix} 0 & 0 & 1 \\ 0 & 0 & 0 \\ 1 & 0 & 0 \end{bmatrix}$$

does not because the three parameters $c_1 = \sigma^2$, $c_2 = \sigma^2 \rho$ and $c_3 = \sigma^2 \rho^2$ are essentially dependent on only two quantities namely, σ^2 and ρ and therefore are functionally related. The SAS option for linear structure is TYPE =LIN(k) which is specified as part of the REPEATED statement.

It may be remembered that presently we are dealing with the fixed effects models of the type $\mathbf{y}_i = \boldsymbol{\mu} + \boldsymbol{\epsilon}_i$ only and hence the variance covariance matrix of $\mathbf{y}_i$ is the same as the variance covariance matrix of $\boldsymbol{\epsilon}_i$. If the p by 1 vector $\boldsymbol{\mu}$ corresponds to fewer than p parameters and is linearly expressed as $\boldsymbol{\mu} = \mathbf{X}\boldsymbol{\beta}$ for some design matrix $\mathbf{X}$ then the MODEL statement can be used to specify such details for fixed effects. In the REPEATED statement of the MIXED procedure, one can specify the covariance structure of $\boldsymbol{\epsilon}_i$ and hence in the present case, for $\mathbf{y}_i$. This will not be the case if the model also has some other random effects.

From the expression of L, it is clear that,

$$-2 \ln L = -2 \left[\ln \max_{H_0} g\left(\boldsymbol{\Sigma}|data\right) - \ln \max_{unrestricted} g(\boldsymbol{\Sigma}|data) \right]$$

$$= [-2 \ln g(\hat{\boldsymbol{\Sigma}}_{H_0}|data)] - [-2 \ln g(\hat{\boldsymbol{\Sigma}}_{unrestricted}|data)],$$

where $\hat{\boldsymbol{\Sigma}}_{H_0}$ and $\hat{\boldsymbol{\Sigma}}_{unrestricted}$ are the maximum likelihood estimators of $\boldsymbol{\Sigma}$ under H_0 and without any restrictions on $\boldsymbol{\Sigma}$, respectively. The two terms in the above expression can be obtained by executing the MIXED procedure twice, for the appropriate models and under appropriate covariance structures, which are specified as the TYPE = option in the REPEATED statements. The option TYPE =UN will result in the value of $-2 \ln g(\hat{\boldsymbol{\Sigma}}_{unrestricted}|data)$. The appropriate option to obtain $-2 \ln g(\hat{\boldsymbol{\Sigma}}_{H_0}|data)$ depends on what H_0 is.

EXAMPLE 1   ***Testing Covariance Structure, Glucose Data***   Six volunteers were observed for the blood glucose levels over a period of time after eating a certain test meal. Measurements were taken 15 minutes before the meal, immediately before (0 hours) and then at every half an hour for the next two hours. Later, hourly readings were taken for the next four hours. The measurements from time point 0 onward are denoted by $y_1, \ldots, y_9$. The time points are represented by the variable TIMEPT in the program. The experiment was repeated six times with the meals taken at different times of the day, thus giving the six treatments (represented by GROUP in the program). Any carryover effect of treatments was assumed to be nonexistent. Data were obtained from Crowder and Hand (1990, p. 14); we have replaced the two missing values in the data set by certain estimates.

To do any meaningful analysis, we must first determine the appropriate covariance structure for the errors. To begin with, an AR(1) structure seems a potential choice but the observations are not made at equal time intervals. In view of this, a Markov covariance structure given by $\boldsymbol{\Sigma} = (\sigma_{ij})$, $\sigma_{ij} = \sigma^2 \rho^{d_{ij}}$, where $d_{ij} = |t_i - t_j|$ is the time elapsed between i^{th} and j^{th} repeated measures, can be used. We will therefore test the null hypothesis that the error covariance structure is of the Markov type. As there are no other random effects (except of course subjects on which repeated measures are taken) in the present context, the test on the covariance structure for error is the same as that for the dependent variable. Consequently, we will test for the Markov covariance structure after correcting for the fixed effects present in the model, namely, the GROUP, TIMEPT and their interaction. In SAS, the Markov structure is referred to as the *Spatial power structure* and is specified by the option TYPE = SP(POW) (V_BLE) where V_BLE defines the variable taking values as time points $t_1, \ldots, t_p$.

A SAS program for this test is presented in Program 6.1. First of all, the data on
$y_1, \ldots, y_9$ at the nine time points should be arranged on a single variable y, at various
levels of the variable TIMEPT. This is done using the first part of the code presented in
Program 6.1. Output 6.1 follows.

```
/* Program 6.1 */

options ls =64 ps=45 nodate nonumber;
title1 'Output 6.1';
title2 'Test for Markov Covariance Structure: Glucose Data';
data a;
infile 'glucose.dat' obs=36;
input id group$ before y1-y9;
run;
data c;
set a;
array t{9} y1-y9;
subj+1;
do timept=1 to 9;
y=t{timept};
if timept = 1 then realtime = 0;
if timept = 2 then realtime = .5;
if timept = 3 then realtime = 1;
if timept = 4 then realtime = 1.5;
if timept = 5 then realtime = 2;
if timept = 6 then realtime = 3;
if timept = 7 then realtime = 4;
if timept = 8 then realtime = 5;
if timept = 9 then realtime = 6;
output;
end;
drop y1-y9;
run;
proc mixed method=ml;
class subj  group timept;
model y= group timept group*timept;
repeated/type=un subject=subj;
title3 'Unstructured Covariance';
run;
proc mixed method=ml;
class subj group timept;
model y=group timept group*timept;
repeated/type=sp(pow)(realtime) subject=subj;
title3 'Markov Covariance';
run;
```

Output 6.1

```
                      Output 6.1
     Test for Markov Covariance Structure: Glucose Data
                   The MIXED Procedure


                 Unstructured Covariance
               Model Fitting Information for Y


           Description                      Value

           Observations                  324.0000
           Log Likelihood                -288.269
           Akaike's Information Criterion -333.269
```

```
Schwarz's Bayesian Criterion        -418.336
-2 Log Likelihood                    576.5388
Null Model LRT Chi-Square            211.7630
Null Model LRT DF                     44.0000
Null Model LRT P-Value                 0.0000


                    Markov Covariance
            Model Fitting Information for Y

        Description                     Value

        Observations                 324.0000
        Log Likelihood              -385.092
        Akaike's Information Criterion -387.092
        Schwarz's Bayesian Criterion -390.873
        -2 Log Likelihood            770.1850
        Null Model LRT Chi-Square     18.1168
        Null Model LRT DF              1.0000
        Null Model LRT P-Value         0.0000
```

Note that the variable REALTIME which is different from the variable TIMEPT, takes values corresponding to the actual time elapsed. This information is needed in specifying the covariance structure.

To compute the likelihood ratio L, the two likelihood functions are to be maximized separately. This is done using the option METHOD = ML in the PROC MIXED statement in the corresponding two executions. As explained earlier, the corresponding fixed effects are to be specified in the MODEL statement. The options TYPE = UN and TYPE = SP(POW)(REALTIME) are used for specifying the appropriate covariance structures used in the calculations of the denominator and the numerator of L. The results of these two runs of PROC MIXED are reported in Output 6.1.

Since

$$-2ln\ L = \left[-2ln \max_{H_0} g(\mathbf{\Sigma}|data)\right] - \left[-2ln \max_{unrestricted} g(\mathbf{\Sigma}|data)\right],$$

rather than working with the individual maximized likelihoods, it suffices to only know *-2 times their natural logarithm*. Fortunately, these are reported as *-2 Log Likelihood* in the output. Thus the observed value of the chi-square test statistic is $\chi^2 = -2lnL = 770.1850 - 576.5388 = 193.6462$. This test statistic follows a chi-square distribution with degrees of freedom f = *Null Model LRT df (Under H_0) - Null Model LRT df (Unrestricted)* = 44 - 1 = 43.

This observed value of the test statistics is highly significant at any reasonable level of significance (such as 5% or 1%) and hence the null hypothesis of Markov covariance structure can be rejected.

Testing for Linear Structures   As discussed earlier, a variance covariance matrix $\mathbf{\Sigma}$ is said to have a linear structure if it can be written as, $\mathbf{\Sigma} = c_1\mathbf{A}_1 + \cdots + c_k\mathbf{A}_k$, for some k and for some known matrices $\mathbf{A}_1, \ldots, \mathbf{A}_k$. The parameters $c_1, \ldots, c_k$ are unknown and assumed to be *functionally* not related to each other, or can be reparametrized to do so.

There are a number of models and consequently a number of situations, where the variance covariance matrix has a linear structure. The hypothesis about such covariance structures can also be tested using the MIXED procedure. For this the option TYPE = LIN(k), where k is the number of parameters in the linear structure can be used in the REPEATED statement. However, the known matrices $\mathbf{A}_i$ corresponding to parameters $c_i, i = 1, \ldots, k$

are to be specified separately in a data set. We will illustrate this using the cork data discussed extensively in the earlier chapters.

EXAMPLE 2 *Cork Data, Testing Circular Structure for Measurements on a Tree* We have already provided PROC IML code to test the circulant covariance structure in Chapter 5. The approach here provides an alternative (although not universally applicable due to convergence problems) approach. Due to the very nature of cork data, where the measurements on the trees were taken along four directions namely, North, East, South and West, the assumption of circular covariance structure seems to be a viable one. The variance covariance matrix,

$$
\Sigma = \begin{bmatrix} \sigma_0 & \sigma_1 & \sigma_2 & \sigma_1 \\ \sigma_1 & \sigma_0 & \sigma_1 & \sigma_2 \\ \sigma_2 & \sigma_1 & \sigma_0 & \sigma_1 \\ \sigma_1 & \sigma_2 & \sigma_1 & \sigma_0 \end{bmatrix}
$$

can be written either as

$$
\Sigma = \sigma_0 \begin{bmatrix} 1 & 0 & 0 & 0 \\ 0 & 1 & 0 & 0 \\ 0 & 0 & 1 & 0 \\ 0 & 0 & 0 & 1 \end{bmatrix} + \sigma_1 \begin{bmatrix} 0 & 1 & 0 & 1 \\ 1 & 0 & 1 & 0 \\ 0 & 1 & 0 & 1 \\ 1 & 0 & 1 & 0 \end{bmatrix} + \sigma_2 \begin{bmatrix} 0 & 0 & 1 & 0 \\ 0 & 0 & 0 & 1 \\ 1 & 0 & 0 & 0 \\ 0 & 1 & 0 & 0 \end{bmatrix}
$$

$$
= c_1 \mathbf{A}_1 + c_2 \mathbf{A}_2 + c_3 \mathbf{A}_3
$$

or as

$$
\Sigma = \sigma_2 \begin{bmatrix} 1 & 1 & 1 & 1 \\ 1 & 1 & 1 & 1 \\ 1 & 1 & 1 & 1 \\ 1 & 1 & 1 & 1 \end{bmatrix} + (\sigma_1 - \sigma_2) \begin{bmatrix} 1 & 1 & 0 & 1 \\ 1 & 1 & 1 & 0 \\ 0 & 1 & 1 & 1 \\ 1 & 0 & 1 & 1 \end{bmatrix} + (\sigma_0 - \sigma_1) \begin{bmatrix} 1 & 0 & 0 & 0 \\ 0 & 1 & 0 & 0 \\ 0 & 0 & 1 & 0 \\ 0 & 0 & 0 & 1 \end{bmatrix}
$$

$$
= c_1^* \mathbf{A}_1^* + c_2^* \mathbf{A}_2^* + c_3^* \mathbf{A}_3^*
$$

among certain other choices. We will however use the first representation. Clearly, $k = 3$ and $\mathbf{A}_1$, $\mathbf{A}_2$ and $\mathbf{A}_3$ are known. The fact that these matrices are to be retrieved from another data set titled CIRC has been specified in the option LDATA = CIRC in the REPEATED statement. The PARMS statement is optional within the code but has been used to provide some initial guesses of the parameters σ_0, σ_1 and σ_2 for the iterative process. Such a specification is helpful and highly recommended to attain the convergence because often during the iterations the parameters may be estimated by quantities which may violate the required positive definiteness of the matrix Σ, thereby making the likelihood function unbounded and growing indefinitely. It may be pointed out that for the present example, the convergence is not obtained unless the PARMS statement with appropriate initial values is added.

Returning to the specification of matrices $\mathbf{A}_1$, $\mathbf{A}_2$ and $\mathbf{A}_3$, these are specified row by row in the data set CIRC. The variable PARM taking values 1, 2 and 3, specifies which parameter — first, second or third, that is, c_1, c_2 or c_3 — does the particular matrix correspond to. Accordingly, in the output, these parameters are referred to as LIN(1), LIN(2) and LIN(3) respectively.

The program for this testing problem is shown in Program 6.2. The corresponding output appears in Output 6.2. The chi-square test statistic $\chi^2 = -2lnL = 811.1686 - 791.4164 = 19.7522$ at degrees of freedom $9 - 2 = 7$ is significant at any reasonable level of significance. We therefore reject the hypothesis of circulant structure.

```
/* Program 6.2 */

options ls =64 ps=45 nodate nonumber;
title1 'Output 6.2';
```

```
title2 'Test for Linear Structure: Circulant for Cork Data';
data cork;
infile 'cork.dat';
input n e w s;
data cork1;
set cork;
array t{4} n e s w;
tree+1;
do dir = 1 to 4;
dir1 = dir;
y = t{dir};
output;
end;
drop n e s w;
run;
data circ;
input parm row col1-col4;
datalines;
1 1 1 0 0 0
1 2 0 1 0 0
1 3 0 0 1 0
1 4 0 0 0 1
2 1 0 1 0 1
2 2 1 0 1 0
2 3 0 1 0 1
2 4 1 0 1 0
3 1 0 0 1 0
3 2 0 0 0 1
3 3 1 0 0 0
3 4 0 1 0 0
;
proc mixed data =cork1 method = ml;
class tree dir;
model y =dir;
repeated/ type =un subject = tree;
title3 'Unstructured Covariance';
run;
proc mixed data = cork1 method = ml;
class tree dir;
model y = dir;
repeated/type = lin(3) ldata = circ subject = tree;
parms (190) (30) (70);
title3 'Circulant';
run;
```

Output 6.2

Output 6.2
Test for Linear Structure: Circulant for Cork Data
The MIXED Procedure

Unstructured Covariance
Model Fitting Information for Y

Description	Value
Observations	112.0000
Log Likelihood	-395.708
Akaike's Information Criterion	-405.708
Schwarz's Bayesian Criterion	-419.301
-2 Log Likelihood	791.4164

```
Null Model LRT Chi-Square        150.0319
Null Model LRT DF                  9.0000
Null Model LRT P-Value             0.0000


                      Circulant
            Model Fitting Information for Y

Description                         Value

Observations                     112.0000
Log Likelihood                   -405.584
Akaike's Information Criterion   -408.584
Schwarz's Bayesian Criterion     -412.662
-2 Log Likelihood                 811.1686
PARMS Model LRT Chi-Square         78.0542
PARMS Model LRT DF                  2.0000
PARMS Model LRT P-Value             0.0000
```

We have already emphasized the usefulness of specifying the initial guess(es) of the parameters in the PARMS statement. In fact, in our experience, we have encountered many data sets where despite our best efforts by specifying the good initial guesses, we could not succeed in obtaining the convergence. Fortunately, in the case of circulant covariance structure, we do have an alternative namely, that given in Program 5.4.

Prespecified Known Variance Covariance Matrix   In many situations, it is often of interest to test the hypothesis that the population variance covariance matrix is equal to a given known positive definite matrix. This can be done using a combination of TYPE = LIN(k) and NOITER options along with PARMS statement. The following example illustrates the approach.

EXAMPLE 3   *Quality Control for Car Door Panels, Warpage Data*   During the manufacturing process in the automotive industry, an undesirable amount of warpage on car door panels is often observed. During a manufacturing process, warpage measurements are taken at three different locations on the same car door panel. The three observations can be treated as repeated measures. The two sources of variation during the process are the variations within and between the car door panels denoted respectively as σ_1^2 and σ_2^2. Thus the variance covariance matrix of three measurement on the same panel will have a compound symmetric structure given by,

$$\Sigma = \begin{bmatrix} \sigma_1^2 + \sigma_2^2 & \sigma_1^2 & \sigma_1^2 \\ \sigma_1^2 & \sigma_1^2 + \sigma_2^2 & \sigma_1^2 \\ \sigma_1^2 & \sigma_1^2 & \sigma_1^2 + \sigma_2^2 \end{bmatrix} = \sigma_1^2 \mathbf{1}\mathbf{1}' + \sigma_2^2 \mathbf{I}$$

For control chart purposes, it is believed that $\sigma_1^2 = 1.0$ and $\sigma_2^2 = 0.1$. Thus $\Sigma = \Sigma_0 = \mathbf{1}\mathbf{1}' + (0.1)\mathbf{I}$ is a known compound symmetric matrix. Periodically, checks are to be made to see if the process is stable and that the variability of the process has not changed. Thus, it is appropriate to test the hypothesis

$$\Sigma = \begin{bmatrix} 1.1 & 1 & 1 \\ 1 & 1.1 & 1 \\ 1 & 1 & 1.1 \end{bmatrix} = 0.1 \begin{bmatrix} 1 & 0 & 0 \\ 0 & 1 & 0 \\ 0 & 0 & 1 \end{bmatrix} + 1.0 \begin{bmatrix} 1 & 1 & 1 \\ 1 & 1 & 1 \\ 1 & 1 & 1 \end{bmatrix} = \Sigma_0.$$

Data on fifteen panels were collected for all three measurements and are reported as part of Program 6.3. Clearly, $k = 2$. Thus, we use the option TYPE = LIN(2) in the REPEATED statement. As in the previous example, the matrices $\mathbf{A}_1 = \mathbf{1}\mathbf{1}'$ and $\mathbf{A}_2 = \mathbf{I}$ are specified in the data set MATRIX. We use the PARMS statement with initial guesses for c_1 and c_2 as 0.1 and 1.0 respectively but allow no iterations for maximum likelihood estimation using the option NOITER. Consequently, no iterations will take place and the matrix Σ will remain as $\Sigma_0 = c_1 \mathbf{A}_1 + c_2 \mathbf{A}_2$ with $c_1 = 0.1$ and $c_2 = 1.0$ for the computation of likelihood.

However, in this particular context, we may also note that regardless of any shifts in the process variances, the compound symmetric structure of Σ will still be intact. Thus, to maximize the denominator of L, one should still use the compound symmetric structure $\Sigma = c_1 \mathbf{A}_1 + c_2 \mathbf{A}_2$, with c_1 and c_2 unknown. Therefore, we must use the TYPE =CS option rather than TYPE =UN in this case as has been done in Program 6.3. Output 6.3 follows.

```
/* Program 6.3 */

options ls =64 ps=45 nodate nonumber;
title1 'Output 6.3';
title2 'Test for \Sigma =\Sigma_0: Warpage Data';
data warpage;
input y1 y2 y3;
lines;
3.3 3.7 3.8
4.1 4.3 4.9
2.1 1.9 2.0
1.8 1.4 2.2
3.9 3.9 3.7
2.5 2.7 2.5
4.4 3.9 4.0
3.3 3.7 4.0
4.4 3.8 4.2
1.1 1.7 1.5
1.8 1.4 1.8
4.4 4.8 4.3
3.4 3.4 3.9
2.9 2.5 2.4
3.6 3.1 3.8
;
data warpage1;
set warpage;
array t{3} y1-y3;
panel+1;
do location = 1  to 3;
y = t{location};
output;
end;
drop y1-y3;
run;
data matrix;
input parm row col1-col3;
datalines;
1 1 1 1 1
1 2 1 1 1
1 3 1 1 1
2 1 1 0 0
2 2 0 1 0
2 3 0 0 1
;
proc mixed data =warpage1 method = ml;
class panel location ;
model y =location ;
repeated/ type =cs subject = panel;
title3 'Compound Symmetry';
```

```
run;
proc mixed data =warpage1 method = ml;
class panel location ;
model y =location ;
parms (1) (.1)/noiter;
repeated/ type =lin(2) ldata =matrix subject = panel;
title3 'Known Fixed \Sigma_0';
run;
```

Output 6.3

```
                          Output 6.3
              Test for \Sigma =\Sigma_0: Warpage Data
                       The MIXED Procedure


                     Compound Symmetry
              Model Fitting Information for Y


        Description                         Value


        Observations                        45.0000
        Log Likelihood                     -32.9845
        Akaike's Information Criterion     -34.9845
        Schwarz's Bayesian Criterion      -36.7911
        -2 Log Likelihood                   65.9690
        Null Model LRT Chi-Square           65.2826
        Null Model LRT DF                    1.0000



                    Known Fixed \Sigma_0
              Model Fitting Information for Y


        Description                         Value


        Observations                        45.0000
        Log Likelihood                     -33.6813
        Akaike's Information Criterion     -35.6813
        Schwarz's Bayesian Criterion      -37.4880
        -2 Log Likelihood                   67.3627
            Test for \Sigma =\Sigma_0: Warpage Data
```

From Output 6.3, it is evident that $-2 ln L = 67.3627 - 65.9690 = 1.3937$. This follows a chi-square distribution with degrees of freedom equal to the difference between the *Null Model LRT DF* under two scenarios. Since no parameters have been estimated when Σ has been specified as Σ_0, the degrees of freedom for our χ^2 test statistics obtained from the likelihood ratio is 1. Since the observed value of 1.3937 is quite small and smaller than the χ^2 cut off point at any reasonable level of significance, we do not reject the null hypothesis in this case and claim that the two components of the process variability are unchanged.

Finally a few comments on choosing a covariance structure among several competing ones are in order. It is possible that for two different covariance structures, when LR tests are applied, they deem both covariance structures as acceptable. Which of the two should one choose? While no conclusive well defined preference policy can be laid out, certain measures can still be used to evaluate such preferences. These are Akaike's and Schwartz's information criteria defined earlier and popularly referred to as AIC and BIC respectively. The larger value of these criteria indicates a higher degree of preference. It may however be remarked that recent simulation studies by Keselman, Algina, Kowalchuk and Wolfinger (1998) indicate that both of these methods often fail to identify the true covariance structure, although AIC performs a little better than BIC in this respect.

Assuming no random effects, the tests presented above require that the fixed effect part of the model remain the same while maximizing the two likelihoods. However, estimating the appropriate fixed effect part of the model is itself a step in modeling and thus the fixed effects will seldom be known a priori. Diggle, Liang and Zeger (1995) suggest that to determine the appropriate covariance structure, one may use the saturated model. Once an appropriate covariance structure has been determined, the modeling of fixed effects and its reduction to a more parsimoneous model can be done under the selected covariance structure.

As an ad-hoc procedure, the above two stage procedure can also be applied when the model contains some random effects as well. In this case, for the selection of an appropriate covariance structure, residuals instead of data on response variables should be used. These residuals may be obtained by initially fitting the saturated fixed and random effect parts of the model under the spherical covariance structure. Once these are obtained, two separate runs (under H_0 and unrestricted respectively) of MIXED procedure as detailed earlier will enable one to test the hypothesis on the error covariance structure.

6.5   Models with Only Fixed Effects

Consider the following fixed effects model

$$\mathbf{y}_{\Sigma p_i \times 1} = \mathbf{X}_{\Sigma p_i \times q}\boldsymbol{\beta}_{q \times 1} + \boldsymbol{\epsilon}_{\Sigma p_i \times 1}, \tag{6.6}$$

with $D(\boldsymbol{\epsilon}) = \sigma^2\mathbf{R} = \sigma^2 diag(\mathbf{R}_1,\ldots,\mathbf{R}_n) = \sigma^2\mathbf{V}$, which is a special case of the model in Equation 6.2 with $\boldsymbol{v} = \mathbf{0}$. We have considered this type of model in Chapter 5 but the variance covariance matrix of the error was assumed to be one of the two extremes, namely, the compound symmetry and completely unstructured. In the following two subsections, we describe two examples of the above fixed effects model in the context of repeated measures. The model for the first example differs from the model used for that data in Chapter 5, only in its error covariance structure. As we will see in the following analysis, such a difference, however, affects the entire analysis including the method of estimation and the approach taken for the hypothesis testing.

6.5.1   Repeated Measures with AR(1) Structure

The split plot model given in Equation 5.11, viz.,

$$y_{iju} = \mu + \alpha_i + \beta_j + (\alpha\beta)_{ij} + \delta_{iu} + \epsilon_{iju},$$

has two random effects, namely, δ_{iu} and ϵ_{iju}, $i = 1,\ldots,k$, $j = 1,\ldots,p, u = 1,\ldots,n_i$, corresponding to the whole plot and subplot errors. Under the assumptions stated there, this in turn leads to the variance covariance matrix of the repeated observations on the uth subject of the ith group, that is of $\mathbf{y}_{iu} = (y_{i1u}, y_{i2u},\ldots,y_{ipu})'$ as

$$\Sigma_{subject} = \begin{bmatrix} \sigma_\delta^2 + \sigma^2 & \sigma_\delta^2 & \cdots & \sigma_\delta^2 \\ \sigma_\delta^2 & \sigma_\delta^2 + \delta^2 & \cdots & \sigma_\delta^2 \\ \vdots & \vdots & \cdots & \vdots \\ \sigma_\delta^2 & \sigma_\delta^2 & \cdots & \sigma_\delta^2 + \sigma^2 \end{bmatrix} = \frac{1}{\sigma_\delta^2 + \sigma^2}\begin{bmatrix} 1 & \rho & \cdots & \rho \\ \rho & 1 & \cdots & \rho \\ \vdots & \vdots & \cdots & \vdots \\ \rho & \rho & \cdots & 1 \end{bmatrix},$$

where $\rho = \sigma_\delta^2/(\sigma_\delta^2 + \sigma^2)$. The above matrix has a compound symmetry covariance structure, a structure which was convenient for the analysis as discussed in Chapter 5. However since $(y_{i1u}, y_{i2u}, \ldots, y_{ipu})'$ are the measurements over time on the same subject, a more realistic covariance structure may be that of the first order autoregressive process, viz.,

$$\Sigma_{subject} = \sigma^2 \begin{bmatrix} 1 & \rho & \rho^2 & \cdots & \rho^{p-1} \\ \rho & 1 & \rho & \cdots & \rho^{p-2} \\ \vdots & \vdots & \vdots & \cdots & \vdots \\ \rho^{p-1} & \rho^{p-2} & \rho^{p-3} & \cdots & 1 \end{bmatrix}$$

or possibly some other suitable structures such as the Toeplitz. The unstructured covariance can also be used. Thus instead of the split plot model with compound symmmetric structure, the model,

$$y_{iju} = \mu + \alpha_i + \beta_j + (\alpha\beta)_{ij} + \epsilon_{iju}, \tag{6.7}$$

where ϵ_{iju} now represents the combined random error of the whole as well as the subplot with an assumed suitable covariance structure, such as the first order autoregressive (AR(1)) structure can be viewed as more accommodating. Of course the model in Equation 5.11 is a special case of this model when ϵ_{iju} can be expressed additively in terms of two identifiable components δ_{iu} and ϵ_{iju}^* representing the whole plot and sub-plot errors respectively. We note that the model in Equation 6.7 can be expressed in matrix notation as the model in Equation 6.6 with

$$\sigma^2 \mathbf{V} = \sigma^2 \mathbf{R} = diag(\Sigma_{subject}, \ldots, \Sigma_{subject}).$$

Now we will illustrate the analysis of the model in Equation 6.6 under an autoregressive error of order one, commonly known as AR(1) using the heart rate data discussed in Chapter 5. Without presenting any details or corresponding output, it may be mentioned that LRT to test the AR(1) covariance structure for this data set supports this assumption. The chi-square test statistic corresponding to LRT in this case is $\chi^2 = 6.8954$ on 8 degrees of freedom which is less than the corresponding 5% cutoff point of 15.51.

EXAMPLE 4   ***Heart Rate Data (continued)***   These data have been previously analyzed using both the multivariate and univariate methods discussed in Chapter 5. To reanalyze the data under AR(1) covariance structure and the general linear model set up, we first need to arrange all repeated measures as the values of a single dependent variable observed at various levels of the longitudinal variable TIME. This is done using the following code, included in Program 6.4.

```
data split;
set heart;
array t{4} y1-y4;
subject +1;
do time =1 to 4;
y=t{time};
output;
end;
drop y1-y4;
run;
```

The SET statement creates a new data set named SPLIT by reading observations from the data set HEART. The new variable Y is defined as taking values Y1, Y2, Y3, and Y4 corresponding to the values 1, 2, 3, and 4 of the variable TIME respectively. This is done by first defining a 4 by 1 array T containing data on Y1, Y2, Y3, and Y4 and then transferring the values in the array T to the variable Y within a DO loop which goes through 4 iterations (TIME = 1 to 4) for every value of SUBJECT.

Under the model given in Equation 6.7 with AR(1) covariance structure for $\epsilon_{iu} = (\epsilon_{iju}, \ldots, \epsilon_{ipu})'$, let α_i represent the treatment (drug) effect, β_j the time effect, and $(\alpha\beta)_{ij}$ the DRUG*TIME interaction. The error subvectors ϵ_{iu}, $i = 1, \ldots, 3$, $u = 1, \ldots, 8$, individually follow independent and identical AR(1) processes. The statements given in Program 6.4 fit the model and perform the subsequent analysis. The results appear in Output 6.4.

```
/* Program 6.4 */

options ls=64 ps=45 nodate nonumber;
title1 ' Output 6.4';
title2 'Analysis of Heart Rate Data';
data heart;
infile 'heart.dat';
input drug $ y1 y2 y3 y4;

proc glm data=heart;
class drug;
model y1-y4=drug/nouni;
repeated time 4;
run;
data split;
set heart;
array t{4} y1-y4;
subject+1;
do time=1 to 4;
y=t{time};
output;
end;
drop y1-y4;
run;
* AR(1) Covariance Structure;
proc mixed data = split covtest method = reml;
class drug subject time;
model y =  drug time time*drug;
repeated /type = ar(1) subject = subject r ;
title3 'AR(1) Covariance Structure';
run;
*Compound Symmetry Structure;
proc mixed data = split covtest method = reml;
class drug subject time;
model y =  drug time time*drug;
repeated /type = cs subject = subject r ;
title3 'Compound Symmetry Structure';
*Unstructured Covariance;
proc mixed data = split covtest method = reml;
class drug subject time;
model y =  drug time time*drug;
repeated /type = un subject = subject r ;
title3 'Unstructured Covariance';
run;
```

In general, the MODEL statement in the MIXED procedure is similar to the MODEL statement in the GLM procedure, except for the fact that only fixed effects are to be listed on the right-hand side of the MODEL statement. However, a very different set of options is available for the analysis using PROC MIXED. For example, the option METHOD=REML specifies that the restricted maximum likelihood estimation procedure should be used. The option COVTEST requests Wald's test for the parameters of variance covariance matrix.

For the first set of statements in Program 6.4 these parameters are the error variance σ^2 and the autocorrelation ρ.

The covariance structure of the error is specified in the REPEATED statement. The SUBJECT = option specifies the independent random blocks and accordingly it defines the way the block diagonal matrix for all the errors is created. For example in our heart rate data for each of the twenty-four independent subjects there are four repeated measures. Hence the variance covariance matrix of the error vector ϵ is a diagonal matrix of twenty four blocks each of size 4 by 4, namely $diag(\Sigma_{subject}, \ldots, \Sigma_{subject})$ where the 4 by 4 matrix $\Sigma_{subject}$ has an AR(1) covariance structure,

$$\Sigma_{subject} = \sigma^2 \begin{bmatrix} 1 & \rho & \rho^2 & \rho^3 \\ \rho & 1 & \rho & \rho^2 \\ \rho^2 & \rho & 1 & \rho \\ \rho^3 & \rho^2 & \rho & 1 \end{bmatrix},$$

and SAS provides an estimate of $\Sigma_{subject}$ using the estimation procedure indicated in the METHOD = option.

Output 6.4

```
                        Output 6.4
                Analysis of Heart Rate Data

                   The MIXED Procedure
                 AR(1) Covariance Structure

                 Class Level Information

     Class     Levels  Values

     DRUG          3   ax23 bww9 control
     SUBJECT      24   1 2 3 4 5 6 7 8 9 10 11 12 13
                       14 15 16 17 18 19 20 21 22 23
                       24
     TIME          4   1 2 3 4


            REML Estimation Iteration History

     Iteration  Evaluations    Objective     Criterion

             0            1  403.36154087
             1            2  330.03350930  0.00002071
             2            1  330.03005861  0.00000000

             Convergence criteria met.


               R Matrix for SUBJECT 1

     Row         COL1          COL2          COL3          COL4

       1   32.28611617   26.52424845   21.79065925   17.90183919
       2   26.52424845   32.28611617   26.52424845   21.79065925
       3   21.79065925   26.52424845   32.28611617   26.52424845
       4   17.90183919   21.79065925   26.52424845   32.28611617
```

Covariance Parameter Estimates (REML)

Cov Parm	Subject	Estimate	Std Error	Z	Pr > |Z|
AR(1)	SUBJECT	0.82153729	0.04957930	16.57	0.0001
Residual		32.28611617	7.85162497	4.11	0.0001

Model Fitting Information for Y

Description	Value
Observations	96.0000
Res Log Likelihood	-242.206
Akaike's Information Criterion	-244.206
Schwarz's Bayesian Criterion	-246.637
-2 Res Log Likelihood	484.4117
Null Model LRT Chi-Square	73.3315
Null Model LRT DF	1.0000
Null Model LRT P-Value	0.0000

AR(1) Covariance Structure
Tests of Fixed Effects

Source	NDF	DDF	Type III F	Pr > F
DRUG	2	21	6.39	0.0068
TIME	3	63	15.85	0.0001
DRUG*TIME	6	63	13.34	0.0001

Compound Symmetry Structure
Tests of Fixed Effects

Source	NDF	DDF	Type III F	Pr > F
DRUG	2	21	5.92	0.0092
TIME	3	63	12.68	0.0001
DRUG*TIME	6	63	12.00	0.0001

Unstructured Covariance
Tests of Fixed Effects

Source	NDF	DDF	Type III F	Pr > F
DRUG	2	21	5.92	0.0092
TIME	3	21	15.82	0.0001
DRUG*TIME	6	21	21.91	0.0001

As shown in Output 6.4, for this data set the restricted maximum likelihood (REML) procedure converged in just two iterations. The REML estimate of $\Sigma_{subject}$ is $\hat{\Sigma}_{subject}$ (identified as the R matrix in SAS output) and is given by

$$\hat{\Sigma}_{subject} = \begin{bmatrix} 32.2861 & 26.5242 & 21.7907 & 17.9018 \\ 26.5242 & 32.2861 & 26.5242 & 21.7907 \\ 21.7907 & 26.5242 & 32.2861 & 26.5242 \\ 17.9018 & 21.7907 & 26.5242 & 32.2861 \end{bmatrix}.$$

The estimate of σ^2 is $\hat{\sigma}^2 = 32.2861$ and that of the AR(1) correlation parameter ρ is $\hat{\rho} = 0.8215$. The respective asymptotic standard errors are $\hat{se}(\hat{\sigma}^2) = 7.8516$ and $\hat{se}(\hat{\rho}) = 0.0496$.

The tests for fixed effects, namely the variables DRUG and TIME, and TIME*DRUG, are shown next. The approximate F tests indicate that the interaction as well as DRUG and TIME effects are significant. For example, the TIME*DRUG interaction effect has the F statistic value 13.34 with 6 and 63 degrees of freedom and a p value 0.0001.

Suppose the chi-square test is used (program-output not shown) by specifying the CHISQ option in the MODEL statement, and the maximum likelihood estimation is implemented by using the option METHOD=ML in PROC MIXED statement. Then the chi-square statistic for testing TIME*DRUG interaction effect will have a chi-square statistic value 91.49 with 6 degrees of freedom and a p value 0.0001. The conclusions here are consistent with those observed using the mutivariate and univariate methods discussed in Chapter 5.

The univariate split-plot analysis under the model given in Equation 5.11 can be performed by adopting the option TYPE=CS in the REPEATED statement. The general (unstructured) covariance structure can also be adopted by using the option TYPE=UN. These options are also used in Program 6.4. The corresponding tests on fixed effects are presented in Output 6.4 for a comparison. A multivariate approach can also be taken as in Chapter 5 using PROC GLM. Of course, the tests used by PROC MIXED are different from multivariate tests. For example, for testing the TIME*DRUG interaction, the exact (for this example) F test corresponding to Wilks' Λ has the F statistic value 12.7376 with 6 and 38 degrees of freedom and a p value 0.0001, whereas the approximate F test in MIXED procedure has an F statistic value of 21.91 with 6 and 21 degrees of freedom and a p value 0.0001. The conclusions are however the same.

A few brief comments about the preference for the covariance structure are in order. To choose a covariance structure among the three used here, we may look at AIC and BIC values produced in the output of PROC MIXED. These are reported in the following table.

TABLE 6.1 Values of Various Information Criteria: Heart Data

Covariance	AIC	BIC
$AR(1)$	-244.206	-246.637
CS	-245.917	-248.348
UN	-248.758	-260.912

Since the values of AIC and BIC are both maximum for the AR(1) structure, this structure seems to be most appropriate among the choices considered in the above table.

6.5.2   Unbalanced and Unequally Spaced Data

Unbalanced and unequally spaced data occur in practice due to many factors. These data are especially common in clinical experiments where patients reschedule their appointments and/or drop out. Additionally, many consumer preference surveys, where two or more groups of consumers are asked to try out the products over time, and then report their preferences will also yield such data. The techniques illustrated here are useful when there is a reason to believe that the dropouts are fairly random. In cases, when there is some nonrandom assignable cause for the dropouts, the techniques given here are not applicable. See Little and Rubin (1987) for details on appropriate techniques for such data sets.

EXAMPLE 5   *Fitting Markov Structure, Audiology Data*   The data are the percentage of correct scores on a sentence hearing test administered to two groups of subjects wearing two different cochlear implant types denoted by A and B respectively. There are 19 subjects in group A and 16 subjects in group B. The hearing tests are administered 1, 9, 18, and 30 months after the implantation of the devices. The objective of the study is *(i)* to determine if there is any difference between the two cochlear implants and also *(ii)* to determine the average improvement curves as functions of the length of time since implantation. The raw data have several missing values and are observed at unequally spaced time points.

Suppose we decide to fit two different quadratic functions for different groups, as functions of time since implantation, for the scores on the hearing tests. For the uth individual in the ith group, $i = 1, 2$, we consider the following model relating the improvement as a function of TIME,

$$y_{i(time)u} = \beta_0 + \beta_{0i} + \beta_1 time + \beta_{1i} time + \beta_2 time^2 + \beta_{2i} time^2 + \epsilon_{iu}$$

$time = t_1, t_2, t_3, t_4$; $i = 1, 2$; $u = 1, \ldots, n_i$, $n_1 = 19$, $n_2 = 16$. We assume that ϵ_{iu} are all independently distributed as $N(0, \sigma^2)$, β_{0i} for $i = 1, 2$; $u = 1, \ldots, n_i$. The coefficients β_{1i} and β_{2i}, $i = 1, 2$ allow the curves for the two groups to be different in their linear and quadratic time components. Thus corresponding terms in Program 6.5 represent the linear and quadratic interactions of TIME with the group effect GP. We also assume that the variance covariance matrix of the p_{iu} repeated measurements, collected on a given subject over time since the implantation of the hearing device, is given by

$$\sigma^2 \mathbf{R}_{iu} = \sigma^2 \begin{bmatrix} 1 & \rho^{t_2-t_1} & \rho^{t_3-t_1} & \rho^{t_4-t_1} \\ \rho^{t_2-t_1} & 1 & \rho^{t_3-t_2} & \rho^{t_4-t_2} \\ \rho^{t_3-t_1} & \rho^{t_3-t_2} & 1 & \rho^{t_4-t_3} \\ \rho^{t_4-t_1} & \rho^{t_4-t_2} & \rho^{t_4-t_3} & 1 \end{bmatrix},$$

where for our data $t_1 = 1$, $t_2 = 9$, $t_3 = 18$, and $t_4 = 30$. The above covariance structure is often referred to as the Markov covariance structure and is especially useful in modeling spatial correlations. Since this covariance structure involves the powers of the parameter ρ, it is also referred to as the spatial power covariance structure. To fit this covariance structure the appropriate TYPE = option in the REPEATED statement is SP(POW)(TIME1), where TIME1 is the variable taking values as the actual time points for which the data were observed (in our example these are 1, 9, 18, and 30 respectively). Program 6.5 is used to analyze the audiology data. Output 6.5 follows.

```
/* Program 6.5 */

options ls=64 ps=45 nodate nonumber;
title1 'Output 6.5';
data aud;
infile 'audiology.dat';
input gp$ y1-y4;
data aud_n;
set aud;
array t{4} y1-y4;
subject+1;
do i=1 to 4;
if (i=1) then time=1;
if (i=2) then time=9;
if (i=3) then time=18;
if (i=4) then time=30;
time1=time;
y=t{i};
output;
end;
drop i y1-y4;
run;
```

```
title2 'Fit Different Quadratic Curves for Groups A and B';
proc mixed data=aud_n method=reml covtest;
class gp subject;
model y= gp time time*gp time*time time*time*gp/htype=1;
repeated/type=sp(pow)(time1) subject=subject r;
run;
title2 'Common Quadratic Term for Groups A and B';
proc mixed data=aud_n method=reml covtest;
class gp subject;
model y= gp time time*gp time*time;
repeated/type=sp(pow)(time1) subject=subject r;
run;
title2 'Common Linear and Quadratic Terms for Groups A and B';
proc mixed data=aud_n method=reml covtest;
class gp subject;
model y= gp time time*time/s;
repeated/type=sp(pow)(time1) subject=subject r;
run;
title2 'Common Quadratic Curve for Groups A and B';
proc mixed data=aud_n method=reml covtest;
class gp subject;
model y= time time*time/s;
repeated/type=sp(pow)(time1) subject=subject r;
run;
```

Some explanation is needed about the MODEL statements used in Program 6.5. Since for each of the two groups (GP) the two models will have different coefficients, we introduce a CLASS variable GP and incorporate that in the model along with linear and quadratic components of the interaction with TIME. These are denoted by GP, TIME*GP, and TIME*TIME*GP respectively. If any of these are found to be statistically not significant the corresponding terms can perhaps be dropped in the process of finalizing the model. Further, the curves for the two groups will be deemed parallel if both of the interactions TIME*GP and TIME*TIME*GP are zero. Additionally, if a common quadratic curve can be fitted it will amount to saying that the two curves are identical and the GP effect is also absent.

The acceptance of the null hypothesis

$$H_0^{(1)} : \beta_{2i} = 0, \; i = 1, \, 2$$

indicates that a common quadratic term can be fit for the two groups. Since the polynomial growth curves are fit in a sequence, TYPE I sums of squares can be utilized to test this hypothesis. The SAS code for testing this hypothesis is provided in the first MODEL statement in Program 6.5. From Output 6.5, we see that the approximate p value computed for the F statistic using the TYPE I sum of squares is 0.8508. Under H_0, F follows an F-distribution with $(1, 71)$ degrees of freedom. The observed value of F = 0.8508 is not significant at any reasonable level of significance. Hence we do not reject $H_0^{(1)}$. Thus a common quadratic term can be used for the two groups.

Given that the two groups have common quadratic terms, acceptance of the null hypothesis

$$H_0^{(2)} : \beta_{1i} = 0, \; i = 1, \, 2$$

implies that the two groups have parallel quadratic improvement curves. That is, these two growth curves are possibly different only in their intercept terms. The intercept for the two groups are respectively $\beta_0 + \beta_{01}$ and $\beta_0 + \beta_{02}$. The second MODEL statement in Program 6.5 tests this hypothesis which has been formally expressed as $H_0^{(2)}$. From Output 6.5, the p value for test is 0.6649. Thus $H_0^{(2)}$ is not rejected.

Finally, not rejecting the hypothesis

$$H_0^{(3)} : \beta_{0i} = 0, \ \ i = 1, \ 2$$

implies that a common quadratic improvement curve fits both the groups. The third
MODEL statement in Program 6.5 tests $H_0^{(3)}$. Since the p value for testing $H_0^{(3)}$, given
that the quadratic curves are parallel, is 0.1514 this hypothesis is not rejected as well.

Output 6.5

```
                              Output 6.5
              Fit Different Quadratic Terms for Groups A and B
                         Tests of Fixed Effects

            Source          NDF   DDF   Type I F   Pr > F

            GP               1    33      2.23     0.1453
            TIME             1    71     64.20     0.0001
            TIME*GP          1    71      0.20     0.6533
            TIME*TIME        1    71     33.43     0.0001
            TIME*TIME*GP     1    71      0.04     0.8508


                Common Quadratic Term for Groups A and B
                         Tests of Fixed Effects

            Source          NDF   DDF   Type III F  Pr > F

            GP               1    33      1.32     0.2581
            TIME             1    72     85.48     0.0001
            TIME*GP          1    72      0.19     0.6649
            TIME*TIME        1    72     33.85     0.0001


            Common Linear and Quadratic Terms for Groups A and B
                         Solution for Fixed Effects

        Effect     GP     Estimate      Std Error      DF      t

        INTERCEPT        15.36207906   5.81203232     33    2.64
        GP         a     11.05462289   7.52633962     33    1.47
        GP         b      0.00000000        .          .     .
        TIME              2.87736050   0.30970055     73    9.29
        TIME*TIME        -0.05748638   0.00983687     73   -5.84


                      Solution for Fixed Effects

                      Pr > |t|

                        0.0125
                        0.1514
                          .
                        0.0001
                        0.0001
```

```
                    Tests of Fixed Effects

          Source      NDF   DDF   Type III F   Pr > F

          GP           1     33         2.16   0.1514
          TIME         1     73        86.32   0.0001
          TIME*TIME    1     73        34.15   0.0001


          Common Quadratic Curve for Groups A and B
                   Solution for Fixed Effects

Effect          Estimate     Std Error     DF        t   Pr > |t|

INTERCEPT     21.32088661    4.20808393    34     5.07    0.0001
TIME           2.88426715    0.31011367    73     9.30    0.0001
TIME*TIME     -0.05773884    0.00984910    73    -5.86    0.0001

                    Tests of Fixed Effects

          Source      NDF   DDF   Type III F   Pr > F

          TIME         1     73        86.50   0.0001
          TIME*TIME    1     73        34.37   0.0001
```

Having eliminated the possibility of any differences between the two model we fit a common quadratic curve for the two groups using the last MODEL statement of Program 6.5. As seen from the bottom part of Output 6.5, the common quadratic curve $\hat{y} = \hat{\beta}_0 + \hat{\beta}_1 \times time + \hat{\beta}_2 \times time^2$ fits well to the data, where $\hat{\beta}_0 = 21.3209$, $\hat{\beta}_1 = 2.8843$ and $\hat{\beta}_2 = -0.0577$. Since $\hat{\beta}_1$ is positive and $\hat{\beta}_2$ is negative, this curve which, representing the effectiveness of implantation over a period of time, increases initially, stabilizes, and slowly decreases after a period of time.

6.6 Analysis in the Presence of Covariates

In many biological, medical and industrial applications, we often come across repeated measures data along with covariates that influence the response variable. It is important to account for the effects of such covariates. In clinical trials, the baseline measurements can be thought of as the useful covariates for analyzing response patterns at successive visits. For example, in therapies for the treatment of chronic stable angina, treadmill walking time (a covariate) is recorded just before the administration of a dose, and then at some post-dose times. The effectiveness of the visit doses is evaluated relative to the corresponding baseline walking times (Patel, 1986). Here the covariate information available for each person remains unchanged during the trial.

In Chapter 5 (Section 5.5) two types of covariates, viz. fixed over time and varying over time, were discussed and the analyses of repeated measures data with the covariates fixed over time were shown using both the multivariate and univariate approaches. These approaches make the assumption of either no covariance structure whatsoever (for multivariate approach) or the simplistic assumption of compound symmetry (for univariate approach). Both of these approaches fail when the correlation structure is not one of the two extremes or when the data are unbalanced along the time axis. We will now explore the analysis of covariance problem in the presence of various other covariance structures.

6.6.1   Covariates Fixed Over Time

Let y_{iju} represent the observed value of the response variable on the uth subject from the ith treatment group at the jth occasion and x_{iu} be an observed value of a covariate which may depend on the subject but does not depend on time. As a first step towards the analysis consider the model

$$y_{iju} = \mu + \alpha_i + \beta_j + (\alpha\beta)_{ij} + \lambda x_{iu} + \lambda_j x_{iu} + \delta_i x_{iu} + \eta_{ij} x_{iu} + \epsilon_{iju},$$

$$j = 1, \ldots, p_{iu}, \ u = 1, \ldots, n_i, \ i = 1, \ldots, k$$

where $\epsilon_{iu} = (\epsilon_{i1u}, \ldots, \epsilon_{ip_{iu}u})'$ are independently distributed as p_{iu}-variate normal with mean vector zero and variance covariance matrix $\sigma^2 \mathbf{R}_{iu}$. If needed, these variance covariance matrices can be allowed to be different for different treatment groups, but their apparent dependence on u (that is, $\mathbf{R}_{iu}$ depending on the uth subject) is only due to the fact that the dimensions of $\mathbf{R}_{iu}$, may be different for different subjects. Let us assume for the moment that they do not differ for different groups. Then the matrices $\mathbf{R}_{iu}$ themselves will not be completely different, and will just be submatrices of a matrix, say $\mathbf{R}_{max}$, which corresponds to the (possibly hypothetical) subject with repeated measures at all time points. For example, suppose corresponding to this possibly hypothetical subject the error vector has a compound symmetric structure. That is, $\mathbf{R}_{max} = (1 - \rho)\mathbf{I} + \rho\mathbf{11}'$. Then any matrix $\mathbf{R}_{iu}$ will be a submatrix of this and will (still) depend on only two parameters.

The model considered above may appear rather general and complex. However, to account for various effects we do need to consider this model, at least in the initial stages of model building. We will first provide a brief explanation of the terms in the model. The term α_i represents the effect of the treatment group, β_j the effect of time, $(\alpha\beta)_{ij}$ the interaction effect between the variables, the treatment and time. The term λx_{iu} is considered to represent the common slope λ of the line relating covariate to the response variable. The terms $\lambda_j x_{iu}$ and $\delta_i x_{iu}$ respectively represent different slopes at different occasions and different slopes at different treatment groups respectively. Finally the term $\eta_{ij} x_{iu}$ represents the interaction effects between the variables representing the treatment group, the time or occasion, and the covariate. In other words, the statistical significance of the term η_{ij} implies that the slopes of the lines relating x_{iu} to y_{iju} are different for different levels of the treatment and time.

Noting $D(\mathbf{y}_{iu}) = \sigma^2 \mathbf{R}_{iu}$, where $\mathbf{y}_{iu} = (y_{i1u}, \ldots, y_{ip_{iu}u})'$, we can express the above model as

$$\mathbf{y} = \mathbf{X}\boldsymbol{\beta} + \boldsymbol{\epsilon},$$

where $\mathbf{y} = (\mathbf{y}'_{11}, \ldots, \mathbf{y}'_{1n_1}, \ldots, \mathbf{y}'_{k1}, \ldots, \mathbf{y}'_{kn_k})'$, and $\boldsymbol{\epsilon}$ is similarly defined, $\boldsymbol{\beta} = (\mu, \alpha_1, \ldots, \alpha_k, \beta_1, \ldots..)'$ is the vector of the parameters in the model, $\mathbf{X}$ is the appropriately chosen design matrix, and $\boldsymbol{\epsilon} \sim N_n(\mathbf{0}, \sigma^2 \mathbf{R})$, where $\mathbf{R} = diag(\mathbf{R}_{11}, \ldots, \mathbf{R}_{kn_k})$, and $n = \sum_{i=1}^{k} n_i$. This model is a special case of the general linear mixed effects model when there are no random effects. Thus, the testing of linear hypotheses of the form $H_0 : \mathbf{L}\boldsymbol{\beta} = \mathbf{0}$ for specific choices of matrix $\mathbf{L}$ can be easily carried out by appealing to the likelihood theory described earlier. If appropriate, a compound symmetric or any other alternative structures can be selected for $\mathbf{R}_{iu}$. To illustrate the analysis under this model we once again analyze the diabetic patients study data considered in Chapter 5, but with the aid of PROC MIXED.

EXAMPLE 6   *Subject-specific Covariates, Diabetic Patients Study Data*   Three groups of diabetic patients, without complications (DINOCOM), with hypertension (DIHYPER), and with postural hypotension (DIHYPOT) respectively and a control (CONTROL) group of healthy subjects were asked to perform a small physical task at time zero. A particular response was observed at times $-30, -1, 1, 2, 3, 4, 5, 6, 8, 10, 12$, and 15 minutes. The responses at ten time points starting from 1 onward are denoted by Y1 through Y10 respectively. The corresponding subject specific covariates representing pre-performance responses (at times -30 and -1) are denoted by X1 and X2 and are used as covariates. The objective is to assess the group differences after correcting for the effects of covariates. We analyze these data using Program 6.6 under compound symmetry and AR(1) covariance structures. The corresponding output is presented in Output 6.6.

```
/* Program 6.6 */

options ls=64 ps=45 nodate nonumber;
title1 'Output 6.6';
data task;
infile 'task.dat';
input group$ x1 x2 y1-y10;
data a;
set task;
array t{10} y1-y10;
subject+1;
do time=1 to 10;
y=t{time};
output;
end;
drop y1-y10;
run;
proc mixed data=a method=reml covtest;
classes group time subject;
model y=group time time*group x1 x2 time*x1 time*x2 group*x1
        group*x2 time*group*x1 time*group*x2;
repeated /type=cs subject=subject r;
title2 'Analysis Under CS Covariance Structure';
run;
proc mixed data=a method=reml covtest;
classes group time subject;
model y=group time time*group x1 x2 time*x1 time*x2 group*x1
        group*x2 time*group*x1 time*group*x2;
repeated /type=ar(1) subject=subject r;
title2 'Analysis Under AR(1) Covariance Structure';
run;
```

Output 6.6

```
                           Output 6.6
                Analysis Under CS Covariance Structure
                   Model Fitting Information for Y


            Description                       Value

            Observations                    247.0000
            Res Log Likelihood             -365.546
            Akaike's Information Criterion -367.546
            Schwarz's Bayesian Criterion  -370.390
            -2 Res Log Likelihood          731.0924
            Null Model LRT Chi-Square        32.5465
            Null Model LRT DF                 1.0000
            Null Model LRT P-Value            0.0000
```

Tests of Fixed Effects

Source	NDF	DDF	Type III F	Pr > F
GROUP	3	14	2.58	0.0953
TIME	9	113	2.08	0.0373
GROUP*TIME	27	113	5.51	0.0001
X1	1	14	3.89	0.0685
X2	1	14	240.25	0.0001
X1*TIME	9	113	1.53	0.1473
X2*TIME	9	113	13.87	0.0001
X1*GROUP	3	14	1.58	0.2390
X2*GROUP	3	14	14.16	0.0002
X1*GROUP*TIME	27	113	1.80	0.0178
X2*GROUP*TIME	27	113	20.91	0.0001

Analysis Under AR(1) Covariance Structure
Model Fitting Information for Y

Description	Value
Observations	247.0000
Res Log Likelihood	-350.808
Akaike's Information Criterion	-352.808
Schwarz's Bayesian Criterion	-355.653
-2 Res Log Likelihood	701.6169
Null Model LRT Chi-Square	62.0220
Null Model LRT DF	1.0000
Null Model LRT P-Value	0.0000

Tests of Fixed Effects

Source	NDF	DDF	Type III F	Pr > F
GROUP	3	14	2.95	0.0692
TIME	9	113	1.08	0.3859
GROUP*TIME	27	113	3.52	0.0001
X1	1	14	4.75	0.0469
X2	1	14	274.13	0.0001
X1*TIME	9	113	1.61	0.1218
X2*TIME	9	113	8.51	0.0001
X1*GROUP	3	14	1.62	0.2290
X2*GROUP	3	14	15.57	0.0001
X1*GROUP*TIME	27	113	1.05	0.4175
X2*GROUP*TIME	27	113	10.64	0.0001

From Output 6.6, we observe that under the TYPE=CS option, many interactions are found to be statistically significant. Specifically, at 5% level of significance, the interaction effects of X1*GROUP*TIME, X2*GROUP*TIME, X2*GROUP, X2*TIME, GROUP*TIME and the effects of TIME and X2 are significant.

Suppose instead of the compound symmetry covariance structure we use the autoregressive (AR(1)) structure for the error covariance. When the analysis is performed using the TYPE=AR(1) option, the TIME, X1*TIME, and X1*GROUP*TIME effects were found to be statistically not significant. Additionally the model appears to be more parsimonious in the sense that fewer significant effects are observed. This reveals that AR(1) structure was perhaps more appropriate and that in the analysis under CS structure many interactions with TIME were declared as statistically significant possibly due to misspecified covariance structure. The fact that AR(1) is more suitable is also evident from comparatively larger AIC($= -352.808$) and BIC($= -355.653$) values.

Assuming that the AR(1) structure is a reasonable choice for the variance covariance matrix of the error, we may conclude that the covariate X1 does not play any significant role in the model. Since the interaction between the covariate X2 and GROUP*TIME is significant (p value $= 0.0001$) we suggest including all the terms in the model except of course those involving the covariate X1.

6.6.2   Time Varying Covariates

In many repeated measures studies, the covariates themselves may vary over time and so in addition to being subject specific, they may also be time specific. In the medical field we often come across repeated measures data with covariates that influence the response variable at every time period where the measurements are made. For example, the effectiveness of a drug in the treatment of arteriosclerosis is probably influenced by many factors such as diet, exercise and smoking (Patel, 1986) which can be viewed as the covariates. In this situation the value of a covariate is being measured at each time point along with the response variable leading to what are termed as the time varying covariates.

Assume for simplicity that there is only one covariate, represented by x. The data in this case can be represented by (y_{iju}, x_{iju}), $j = 1, \ldots, p$ (time periods), $u = 1, \ldots, n_i$, $i = 1, \ldots, k$ (treatment groups). We also assume that there are no missing data for any time point. Then the following multivariate approach has been suggested by Patel (1986).

The multivariate approach   Let $\mathbf{y}_{iu}$ be a $p \times 1$ vector of responses and $\mathbf{x}_{iu}$ be $p \times 1$ vector of the corresponding values of the covariate, taken over p occasions on the jth individual. Define $\mathbf{Y}_{n \times p} = (\mathbf{y}_{11} : \ldots : \mathbf{y}_{1n_1} : \mathbf{y}_{21} : \ldots : \mathbf{y}_{2n_2} : \ldots : \mathbf{y}_{kn_k})'$, $n = \sum_{i=1}^{k} n_i$ and $\mathbf{X}_{n \times p} = (\mathbf{x}_{11}, : \ldots : \mathbf{x}_{1n_1} : \mathbf{x}_{21} : \ldots : \mathbf{x}_{2n_2} : \ldots : \mathbf{x}_{kn_k})'$. Then consider the following model similar (but not the same) as that in Equation 4.7,

$$\mathbf{Y}_{n \times p} = \mathbf{A}_{n \times m} \boldsymbol{\xi}_{m \times p} + \mathbf{X}_{n \times p} \boldsymbol{\Gamma}_{p \times p} + \mathcal{E}_{n \times p}. \tag{6.8}$$

Here $\mathbf{A}$ is a design matrix, $\boldsymbol{\xi}$ is a matrix of unknown parameters, $\boldsymbol{\Gamma}$ is a diagonal matrix with the unknown diagonal elements, $\gamma_1, \ldots, \gamma_p$, representing the slopes of the line relating the covariate and the response variable at the time points $1, \ldots, p$ respectively. The matrix $\mathcal{E}$ is an $n \times p$ error matrix, rows of which are assumed to be independently distributed with a common multivariate normal distribution that has a zero mean vector and a $p \times p$ covariance matrix $\mathbf{V}$. Note that the model (6.8) is still different from the usual multivariate analysis of covariance model discussed in Chapter 4 in the sense that its parameter matrix $\boldsymbol{\Gamma}$ is known (to be zero) except at the diagonal entries. This case of a partially known and partially unknown parameter matrix makes it harder to handle the analysis of this model in a routine MANOVA setup.

Patel (1986) provided an iterative algorithm describing the computation of the maximum likelihood estimators of the unknown parameters and for the likelihood ratio test for any general linear hypothesis of the form $H_0 : \mathbf{L}_{b \times m} \boldsymbol{\xi}_{m \times p} \mathbf{M}_{p \times c} = \mathbf{0}$. It can be shown that the likelihood function under model (6.8) is

$$L = (2\pi)^{-\frac{1}{2}np} |\mathbf{V}|^{-\frac{1}{2}n} exp \left\{ -\frac{1}{2} tr(\mathbf{V}^{-1} \mathcal{R}) \right\}, \tag{6.9}$$

where $\mathcal{R} = (\mathbf{Y} - \mathbf{A}\boldsymbol{\xi} - \mathbf{X}\boldsymbol{\Gamma})'(\mathbf{Y} - \mathbf{A}\boldsymbol{\xi} - \mathbf{X}\boldsymbol{\Gamma})$.

The maximum likelihood estimators of $\boldsymbol{\xi}$ and $\boldsymbol{\Gamma}$ are obtained by minimizing $\phi = ln\,|\mathcal{R}|$. The reason for this is that, if $\boldsymbol{\xi}$ and $\boldsymbol{\Gamma}$ are known then the MLE of $\mathbf{V}$ is $n^{-1}\mathcal{R}$. Substituting this, the log-likelihood function reduces to $ln\,|\mathcal{R}|$ apart from a constant. Hence the computation procedure iterates between computing $\hat{\mathbf{V}}$ as $\hat{\mathbf{V}} = n^{-1}\hat{\mathcal{R}}$ and estimating $\hat{\mathcal{R}}$ by minimizing $ln\,|\mathcal{R}|$ with respect to $\boldsymbol{\xi}$ and $\boldsymbol{\Gamma}$. Thus,

$$\hat{\mathcal{R}} = \mathcal{R}|_{\boldsymbol{\xi}=\hat{\boldsymbol{\xi}},\boldsymbol{\Gamma}=\hat{\boldsymbol{\Gamma}}}, \tag{6.10}$$

where $\hat{\boldsymbol{\xi}}$ and $\hat{\boldsymbol{\Gamma}}$ are the estimated values of $\boldsymbol{\xi}$ and $\boldsymbol{\Gamma}$ at the particular iteration. It can be shown that $\boldsymbol{\xi}$ and $\boldsymbol{\Gamma}$ are respectively estimated from the equations

$$\hat{\boldsymbol{\xi}} = (\mathbf{A}'\mathbf{A})^{-1}\mathbf{A}'(\mathbf{Y} - \mathbf{X}\hat{\boldsymbol{\Gamma}}), \tag{6.11}$$

and

$$diag\{\mathcal{R}^{-1}\mathbf{X}'(\mathbf{Y} - \mathbf{A}\hat{\boldsymbol{\xi}} - \mathbf{X}\hat{\boldsymbol{\Gamma}})\} = \mathbf{0}$$

which together imply,

$$diag\{\hat{\mathcal{R}}^{-1}\mathbf{X}'(\mathbf{I} - \mathbf{A}(\mathbf{A}'\mathbf{A})^{-1}\mathbf{A}')\mathbf{Y}\} = diag\{\hat{\mathcal{R}}^{-1}\mathbf{X}'\mathbf{X}\hat{\boldsymbol{\Gamma}}\}. \tag{6.12}$$

Based on this fact, Patel (1986) suggests the following iterative steps to solve the ML equations:

1. Initially take $\hat{\mathcal{R}} = \mathbf{I}$ and compute $\hat{\boldsymbol{\Gamma}}$ from (6.12).
2. Compute $\hat{\boldsymbol{\xi}}$ from (6.11).
3. Compute $\hat{\mathcal{R}}$ using $\hat{\boldsymbol{\xi}}$ and $\hat{\boldsymbol{\Gamma}}$ from (6.10). Then compute $\hat{\phi} = ln\,|\hat{\mathcal{R}}|$.
4. Compute a revised estimate $\hat{\boldsymbol{\Gamma}}$ from (6.12) using $\hat{\mathcal{R}}$ obtained in Step 3.
5. Repeat Steps 2, 3 and 4 until the convergence has been obtained up to a desired degree of accuracy under a suitable convergence criterion.

This iterative scheme was initially implemented using the IML procedure by S. Rao (1995) for a data set provided by Dr. Barbara Hargrave of Old Dominion University. A cosmetically improved version is presented here as Program 6.7.

Suppose it is of interest to test for the significance of the covariate X. For this, we also need to estimate the parameters of the reduced model under $H_0 : \boldsymbol{\Gamma} = \mathbf{0}$. Under H_0 we have the usual multivariate linear model, $\mathbf{Y}_{n \times p} = \mathbf{A}_{n \times m}\boldsymbol{\xi}_{m \times p} + \mathcal{E}_{n \times p}$. Hence the likelihood ratio test for testing the significance of the covariate can be easily constructed as described earlier.

To test a general linear hypothesis $H_0 : \mathbf{L}\boldsymbol{\xi}\mathbf{M} = \mathbf{0}$, Patel (1986) provides a similar algorithm for computing the ML estimates under the null hypothesis $H_0 : \mathbf{L}\boldsymbol{\xi}\mathbf{M} = \mathbf{0}$. As before, using the maximum likelihood estimates under no restriction and those under the null hypothesis, the likelihood ratio test for H_0 can be obtained.

Under no restrictions on the parameters, the maximum of the likelihood can be shown to be

$$max\,L = (2\pi)^{-\frac{1}{2}np}|\hat{\mathbf{V}}|^{-\frac{1}{2}n}exp\left(-\frac{1}{2}np\right), \tag{6.13}$$

where $\hat{\mathbf{V}} = \hat{\mathcal{R}}/n$ is the maximum likelihood estimator of $\mathbf{V}$. Similarly, the maximum likelihood estimator of $\mathbf{V}$ under H_0, is $\tilde{\mathbf{V}}$, where $\tilde{\mathbf{V}} = \tilde{\mathcal{R}}/n$. The maximum of L under $H_0 : \mathbf{L}\boldsymbol{\xi}\mathbf{M} = \mathbf{0}$ is

$$max\,L_{H_0} = (2\pi)^{-\frac{1}{2}np}|\tilde{\mathbf{V}}|^{-\frac{1}{2}n}exp\left(-\frac{1}{2}np\right). \tag{6.14}$$

Thus, by taking the ratio of the quantities in Equations 6.14 and 6.13, the likelihood ratio test for testing $H_0 : \mathbf{L}\xi\mathbf{M} = \mathbf{0}$ rejects the null hypothesis if

$$\lambda = |\hat{\mathbf{V}}\tilde{\mathbf{V}}^{-1}|^{\frac{1}{2}n} \le C_\alpha, \tag{6.15}$$

where C_α is a constant satisfying $Pr(\lambda \le C_\alpha|H_0) = \alpha$. Using the standard likelihood theory, under H_0, the quantity $-2\ln\lambda$ asymptotically follows a chi-square distribution with $mp - bc$ degrees of freedom.

Patel's approach has been illustrated through the following example.

EXAMPLE 7   *Time Varying Covariates, Sheep Data*   The effects of phenylephrine induced increase in arterial pressure on the secretion of atrial natriuretic peptide (ANP) in the ovine fetus have been studied by Hargrave and Castle (1995). A set of 16 chronically cannulated fetal sheep was divided into two groups, the young and old. Arterial pressure was increased by infusing phenylephrine to the fetus from each of the two groups. Systematic mean arterial pressure (MAP), plasma ANP concentrations and plasma renin activity (PRA) were measured at three time points (5 min, 15 min, and 30 min) after infusion. PRA is used as the response variable and MAP is an accompanying time varying covariate. In our case, $k = 2$, $p = 3$, $n_1 = 6$ and $n_2 = 10$ and one covariate. The data set (SHEEP DATA) and the program appear in Program 6.7.

```
/* Program 6.7 */

option ls=64 ps=45 nodate nonumber;
title1 'Output 6.7';
title2 'Multivariate Analysis of Time Varying
Covariates Data';
proc iml;
y={66        59       47      -16      -29      6.9      -16
   29        5.8     -40      -80       13     -1.6       60
   16       -17,
  -12        59       6.7      2.6     -53      -29       2.6
  -40       -29      -60      170      -45      4.1        6.6
  -73       -38,
   60        46       6.7    -18.4     -61      -35      -18.4
  -38        5.8     -58       80      -25     -7.1        0.2
  -41       -50};
a={1         1        1        1        1        1        0
   0         0        0        0        0        0         0
   0         0,
   0         0        0        0        0        0         1
   1         1        1        1        1        1         1
   1         1};
x1={135      21       55       22      -22      7.6       55
   875       23       86      -9.6     -16       46        31
    15       13,
   545       97       78      -28      155      169       158
   158       56      121       2.4      0.2     104        81
    85       54,
   840      224      210      -33      629      754       133
   133      177       82       76      -16       25       442
   830      205};

/* Enter the parameters for each problem */
n=16;
p=3;
eps=10;
```

```
* Computation of gamma hat;
z=block(y, x1);
g=i(n)-(a'*inv(a*a')*a);
* Initial value of Vhat;
nvhat=i(p);
do iteratio=1 to 50 while (eps > 0.0001);
vhat=nvhat;
pihat=log(det(vhat));
vhatinv=inv(vhat);
w=vhatinv#(y*g*x1');
T=vhatinv#(x1*g*x1');
onep=j(1,p,1);
gamma=onep*W*inv(T);

* Computation of xihat;
irp=repeat(i(p), 1, 2);
icn=i(n) // -i(n);
* Compute ygx= Y-Sum(Gamma*X);
gone=onep || gamma;
irpp=irp*diag(gone);
ygx=irpp*z*icn;
xihat= ygx*a'*inv(a*a');

* Compute new Vhat and new pihat;
nvhat=(ygx-xihat*a)*(ygx-xihat*a)';
npihat=log(det(nvhat));
eps=abs(pihat-npihat);
end;
* keep vhat for the computation of lambda;
vhat=nvhat;
print vhat;

/* Compute Vhat under the null hypothesis of
no covariate effect (we called this value sighat).
We have used PRINTE option of MANOVA statement of
PROC GLM for getting this value. These statements
are commented out in the program */
/*
yt=y';
at=a';
var1=({y1 y2 y3});
var2=({a1 a2});
create ndata1 from yt [colname=var1];
append from yt;
close ndata1;
create ndata2 from at [colname=var2];
append from at;
close ndata2;
data ndata;
merge ndata1 ndata2;
proc glm data=ndata;
model y1 y2 y3=a1 a2/noint nouni;
manova/printe;
*/
sighat={21554.744333   -8727.857667    3420.5183333,
        -8727.857667   49988.409333    29065.183333,
         3420.5183333   29065.183333    25186.393333};
* Test for testing no covariate effect;
lambda1=(det(vhat*inv(sighat)))**(n/2);
```

```
print lambda1;
chi1=-2*log(lambda1);
print chi1;
pval1=1-probchi(chi1,3);
print pval1;

* Test of hypothesis: No time effect, algorithm;
/* Give L and M matrices */
L=i(2);
M={ 1 -1 0, 1 0 -1};
eps=10;
/* Use nvhat and gamma from the previous step as initial
        values. So initial value of ygx=Y-Sum(Gamma*X) is also from
        the previous step */
do iter=1 to 50 while (eps > 0.0001);
pihat=log(det(nvhat));
delta=2*inv(m*nvhat*m')*(m*ygx*a'*inv(a*a')*l)*
inv(l'*inv(a*a')*l);
* Compute xi-tilde;
xtilde=(ygx*a'-0.5*nvhat*m'*delta*l')*inv(a*a');
vtilde=(ygx-xtilde*a)*(ygx-xtilde*a)';
vtnv=inv(vtilde);
W=vtnv#(y*g*x1');
T=vtnv#(x1*g*x1');
onep=j(1,p,1);
gmt=onep*W*inv(T);
/* Update ygx=Y-Sum(gmt*X) in preparation for the next
iteration irp and icn are defined earlier */
icn=i(n) // -i(n);
gtn=onep || gmt;
irpp=irp*diag(gtn);
ygx=irpp*z*icn;
npihat=log(det(vtilde));
eps=abs(pihat-npihat);
nvhat=vtilde;
end;
print vtilde;
lambda=(det(vhat*inv(vtilde)))**(n/2);
print lambda;
chi=-2*log(lambda);
print chi;
pval2=1-probchi(chi,2);
print pval2;
quit;
```

In Program 6.7 we first test for the significance of the covariate in the model. The unrestricted ML estimates of $\mathcal{R} = n\mathbf{V}$ is found to be

$$\hat{\mathcal{R}} = \begin{bmatrix} 19807.802 & -4186.271 & 4663.295 \\ -4186.271 & 48285.522 & 30775.957 \\ 4663.295 & 30775.957 & 24974.933 \end{bmatrix}.$$

Under the null hypothesis of no covariate effect, the ML estimate $\hat{\mathcal{R}}$ of $\mathcal{R} = n\mathbf{V}$ is the same as that obtained from the standard multivariate regression model. This estimate is obtained by using PROC GLM in Program 6.7 and is denoted as SIGHAT in the program. The complete output however is not shown. The results for the hypothesis test extracted from the output are summarized below:

λ	df	Approx. χ^2	p value
.0053	3	10.4793	.0149

Since the p value is smaller than 0.05 we conclude that the *covariate effect* is statistically not significant at a 5% level of significance.

In Program 6.7 we also provide PROC IML code for computing the likelihood ratio for the null hypothesis of no time effect. The corresponding null hypothesis

$$H_0 : \xi_{11} = \xi_{12} = \xi_{13} = 0$$

$$\xi_{21} = \xi_{22} = \xi_{23} = 0,$$

or

$$H_0 : \xi_{11} - \xi_{12} = 0, \quad \xi_{11} - \xi_{13} = 0$$

$$\xi_{21} - \xi_{22} = 0, \quad \xi_{21} - \xi_{23} = 0.$$

This can be written as $H_0 : \mathbf{L}\xi\mathbf{M} = \mathbf{0}$ with $\mathbf{L} = \mathbf{I}_{2\times2}$ and

$$\mathbf{M} = \begin{bmatrix} 1 & 1 \\ 1 & 1 \\ 0 & -1 \end{bmatrix}.$$

The ML estimate $\tilde{\mathcal{R}}$ of $\mathcal{R} = n\mathbf{V}$ under this null hypothesis is given by

$$\tilde{\mathcal{R}} = \begin{bmatrix} 20212.801 & -4704.252 & 5694.256 \\ -4704.252 & 48330.394 & 30810.573 \\ 5694.256 & 30810.573 & 28866.521 \end{bmatrix}.$$

Accordingly from Equation 6.15 the value of λ is 0.0025. Thus, $-2\,ln\,\lambda = 11.9920$, which is statistically significant with the corresponding p value of 0.0025. The program can be appropriately modified to test other relevant linear hypotheses of interest with the appropriate choices of $\mathbf{L}$ and $\mathbf{M}$. We summarize the results below for three such hypotheses of interest.

H_0 on	λ	df	Approx. χ^2	p value
Group	0.1042	2	4.5225	0.1042
Time	0.0025	2	11.9920	0.0025
Time*Group	0.5993	2	1.0241	0.5993

Using a univariate approach, Verbyla (1988) points out that model in Equation 6.8 can be expressed as Zellner's seemingly unrelated regression (SUR) model and a two-stage estimation procedure can be utilized. This method and its variations are discussed in detail using IML procedure by Timm and Mieczkowski (1997). We will not discuss these here.

A General Linear Model Approach The multivariate approach described above cannot handle unbalanced data and ignores any observations with missing values. An alternative univariate approach which can handle the unbalanced data will be described below (Rao, 1995).

Let $1 \times p$ vector $\mathbf{y}_i'$ be the ith row of $\mathbf{Y}$, $i = 1, \ldots, n$ in the model given in Equation 6.8 assuming only one treatment group (that is, $k = 1$) for illustration. Similarly consider other vectors and matrices of model Equation 6.8. Then tentatively assuming no missing values, we can write the model from Equation 6.8 as,

$$\mathbf{y}_{i_{1\times p}}' = \mathbf{a}_i'[\xi_1 : \ldots : \xi_p] + [x_{1i}, \ldots, x_{pi}]diag(\gamma_1, \ldots, \gamma_p) + \epsilon_i' \tag{6.16}$$

$$= [\mathbf{a}_i'\xi_1 : \ldots : \mathbf{a}_i'\xi_p] + [\gamma_1, \ldots, \gamma_p]diag(x_{i1}, \ldots, x_{ip}) + \epsilon_i',$$

where $\mathbf{a}_i'$ is the ith row of matrix $\mathbf{A}$ and other quantities are similarly defined. Transposing both sides, we obtain

$$
\mathbf{y}_i = \begin{bmatrix} \mathbf{a}_i' & 0 & . & . & . & 0 \\ 0 & \mathbf{a}_i' & . & . & . & 0 \\ . & . & . & & & . \\ . & . & & . & & . \\ 0 & 0 & . & . & . & \mathbf{a}_i' \end{bmatrix} \begin{bmatrix} \boldsymbol{\xi}_1 \\ . \\ . \\ . \\ \boldsymbol{\xi}_p \end{bmatrix} + \begin{bmatrix} x_{1i} & 0 & . & . & . & 0 \\ 0 & x_{2i} & . & . & . & 0 \\ . & . & . & & & . \\ . & . & & . & & . \\ 0 & 0 & . & . & . & x_{pi} \end{bmatrix} \begin{bmatrix} \gamma_1 \\ \gamma_2 \\ . \\ . \\ \gamma_p \end{bmatrix} + \boldsymbol{\epsilon}_i
$$

$$
= \mathbf{A}_{i\,p\times mp}\,\boldsymbol{\eta}_{mp\times 1} + \mathbf{D}_{i\,p\times p}\,\boldsymbol{\gamma}_{p\times 1} + \boldsymbol{\epsilon}_i,
$$

where $\mathbf{A}_i = diag(\mathbf{a}_i', \ldots, \mathbf{a}_i')$, $\mathbf{D}_i = diag(x_{1i}, \ldots, x_{pi})$, $\boldsymbol{\eta}' = [\boldsymbol{\xi}_1', .., \boldsymbol{\xi}_p']$, and $\boldsymbol{\gamma}' = (\gamma_1, \ldots, \gamma_p)$.

The above equation can again be rewritten in the form

$$
\mathbf{y}_i = [\mathbf{A}_i : \mathbf{D}_i] \begin{bmatrix} \boldsymbol{\eta} \\ \boldsymbol{\gamma} \end{bmatrix} + \boldsymbol{\epsilon}_i
$$

or

$$
\mathbf{y}_{i\,p\times 1} = \mathbf{B}_{i\,p\times (m+1)p}\,\boldsymbol{\beta}_{(m+1)p\times 1} + \boldsymbol{\epsilon}_i,
$$

where $\mathbf{B}_i = [\mathbf{A}_i : \mathbf{D}_i]$, and $\boldsymbol{\beta} = \begin{bmatrix} \boldsymbol{\eta} \\ \boldsymbol{\gamma} \end{bmatrix}$.

Let $\mathbf{y} = (\mathbf{y}_1', \ldots, \mathbf{y}_n')'$ and $\boldsymbol{\epsilon} = (\boldsymbol{\epsilon}_1', \ldots, \boldsymbol{\epsilon}_n')'$. Then

$$
\mathbf{y} = \begin{bmatrix} \mathbf{B}_1 \\ . \\ . \\ . \\ \mathbf{B}_p \end{bmatrix} \boldsymbol{\beta} + \boldsymbol{\epsilon}
$$

or

$$
\mathbf{y} = \mathbf{B}\boldsymbol{\beta} + \boldsymbol{\epsilon},
$$

which is a special case of linear model in the Equation 6.1 with no random effects and $D(\boldsymbol{\epsilon}) = \sigma^2\mathbf{R} = \sigma^2\mathbf{I}_n \otimes \mathbf{V}(\boldsymbol{\theta})$, where $\mathbf{V}(\boldsymbol{\theta})$ is a $p \times p$ variance covariance matrix of the repeated measures depending on a vector of parameters $\boldsymbol{\theta}$, and $\mathbf{B} = (\mathbf{B}_1', \ldots, \mathbf{B}_p')'$. It may be noted that although we have presented here the model with no missing covariates, the only change, were there any missing covariate values at any time points for a particular subject, will be to discard the particular values of the dependent variable (corresponding to the particular time points only and, only for the particular subject). This model, being a special case of a general mixed effects linear model, can be analyzed using the MIXED procedure. For illustration, we consider the following example.

EXAMPLE 7   **Sheep Data (continued)**   To illustrate, we again consider the SHEEP DATA where $k = 2$, $p = 3$, $n_1 = 6$ and $n_2 = 4$ and there is one covariate, namely MAP. The SAS code to analyze these data under a general linear model using PROC MIXED is given in Program 6.8 and the corresponding output is presented in Output 6.8.

```
/* Program 6.8 */

options ls=64 ps=45 nodate nonumber;
title1 'Output 6.8';
title2 'Analysis of Time Varying Covariates Data';
title3 'Analysis with Independent Error for Subjects';
data one;
infile 'sheep.dat';
input group$ sheep time pra anp map;
```

```
data one;
set one;
y=pra;
x=anp;
z=map;
run;
proc mixed data=one method=reml;
classes group time sheep;
model y=group time group*time time*z group*z group*time*z z;
repeated /type=simple subject=sheep r;
run;
title3 'Analysis with Compound Symmetry Error for Subjects';
proc mixed data=one method=reml;
classes group time sheep;
model y=group time group*time time*z group*z group*time*z z;
repeated /type=cs subject=sheep r;
run;
```

Output 6.8

```
                            Output 6.8
                Analysis of Time Varying Covariates Data
                Analysis with Independent Error for Subjects



                          The MIXED Procedure

                        Class Level Information

           Class     Levels   Values

           GROUP          2   old young
           TIME           3   5 15 30
           SHEEP         10   24 49 502 505 528 599 617 618
                              717 722



                    REML Estimation Iteration History

        Iteration  Evaluations      Objective      Criterion

               0            1   195.78315742
               1            1   195.78315742   0.00000000

                       Convergence criteria met.



                        R Matrix for SHEEP 24

             Row          COL1          COL2          COL3

              1   1091.5163360
              2                  1091.5163360
              3                                1091.5163360



           Covariance Parameter Estimates (REML)

             Cov Parm   Subject      Estimate

             DIAG       SHEEP     1091.5163360
```

```
                Analysis with Independent Error for Subjects

                    Model Fitting Information for Y

            Description                          Value

            Observations                         30.0000
            Res Log Likelihood                  -114.432
            Akaike's Information Criterion      -115.432
            Schwarz's Bayesian Criterion       -115.878
            -2 Res Log Likelihood               228.8649
            Null Model LRT Chi-Square             0.0000
            Null Model LRT DF                     0.0000
            Null Model LRT P-Value                1.0000


                       Tests of Fixed Effects

            Source          NDF   DDF   Type III F   Pr > F

            GROUP            1     8       4.84       0.0590
            TIME             2     10      0.89       0.4394
            GROUP*TIME       2     10      0.28       0.7612
            Z*TIME           2     10      0.84       0.4596
            Z*GROUP          1     10      0.26       0.6231
            Z*GROUP*TIME     2     10      1.63       0.2444
            Z                1     10      2.30       0.1600



        Analysis with Compound Symmetry Error for Subjects

                       The MIXED Procedure

                     Class Level Information

        Class      Levels   Values

        GROUP         2      old young
        TIME          3      5 15 30
        SHEEP        10      24 49 502 505 528 599 617 618
                            717 722


                 REML Estimation Iteration History

        Iteration   Evaluations      Objective       Criterion

              0          1        195.78315742
              1          2        182.64116970      0.00035155
              2          1        182.60566734      0.00001167
              3          1        182.60457671      0.00000002
              4          1        182.60457534      0.00000000

                   Convergence criteria met.
```

R Matrix for SHEEP 24

Row	COL1	COL2	COL3
1	1067.2157907	865.09810940	865.09810940
2	865.09810940	1067.2157907	865.09810940
3	865.09810940	865.09810940	1067.2157907

Covariance Parameter Estimates (REML)

Cov Parm	Subject	Estimate
CS	SHEEP	865.09810940
Residual		202.11768133

Analysis with Compound Symmetry Error for Subjects
Model Fitting Information for Y

Description	Value
Observations	30.0000
Res Log Likelihood	-107.843
Akaike's Information Criterion	-109.843
Schwarz's Bayesian Criterion	-110.734
-2 Res Log Likelihood	215.6864
Null Model LRT Chi-Square	13.1786
Null Model LRT DF	1.0000
Null Model LRT P-Value	0.0003

Tests of Fixed Effects

Source	NDF	DDF	Type III F	Pr > F
GROUP	1	8	4.72	0.0615
TIME	2	10	6.08	0.0187
GROUP*TIME	2	10	0.38	0.6907
Z*TIME	2	10	3.16	0.0862
Z*GROUP	1	10	1.87	0.2014
Z*GROUP*TIME	2	10	2.67	0.1179
Z	1	10	0.06	0.8110

We assume the compound symmetric error structure and, thus, perform the analysis with the TYPE = CS option in the REPEATED statement. From the output the estimated compound symmetry correlation coefficient is equal to $\hat{\rho} = 865.0981/1067.2158 = 0.8106$. Further, from the bottom part of the output (under 'Tests of Fixed Effects') we conclude that only the TIME factor is statistically significant at a 5% level of significance. This agrees with our earlier conclusion arrived at by using the multivariate approach, except that the covariate effect was also significant in the multivariate approach.

It may be emphasized that if time varying covariates are present then the estimates and the tests under the options TYPE=CS and TYPE=SIMPLE will be different (compare the two sets of outputs given above corresponding to these options). This is in stark contrast from the situation of subject specific covariates. This is so because in the former situation, the *generalized* least squares estimators, with the compound symmetry covariance structure for the error, are not the same as the ordinary least squares estimators.

6.7   A Random Coefficient Model

Another approach to the modeling of repeated measures data is to use an appropriate random coefficient model. In this approach, one or more regression parameters are assumed to be a random sample from a population of regression coefficients. These models are useful whenever the regression model for fitting the repeated data on a subject can be assumed to be a random variation of a population regression model. This approach can be especially useful due to its capacity to handle unequally spaced and/or unbalanced growth curve type data.

Let $\mathbf{y}_{iu}$ be the $p_{iu} \times 1$ vector of repeated measures on the uth subject of the ith treatment group. Then consider a mixed effects model described as

$$\mathbf{y}_{iu} = \mathbf{X}_{iu}\boldsymbol{\beta} + \mathbf{Z}_{iu}\boldsymbol{\nu}_{iu} + \boldsymbol{\epsilon}_{iu}, \ \ u = 1, \ldots, n_i, \ \ i = 1, \ldots, k, \qquad (6.17)$$

where $\mathbf{X}_{iu}$ and $\mathbf{Z}_{iu}$ are the known matrices of orders p_{iu} by q and p_{iu} by r respectively, and $\boldsymbol{\beta}$ is the fixed q by 1 vector of unknown nonrandom regression coefficients. The r by 1 vectors $\boldsymbol{\nu}_{iu}$ are random effects with $E(\boldsymbol{\nu}_{iu}) = \mathbf{0}$, and $D(\boldsymbol{\nu}_{iu}) = \sigma^2 \mathbf{G}_1$. Also $\boldsymbol{\epsilon}_{iu}$ are the p_{iu} by 1 vectors of random errors with $E(\boldsymbol{\epsilon}_{iu}) = \mathbf{0}$, $D(\boldsymbol{\epsilon}_{iu}) = \sigma^2 \mathbf{R}_{iu}$. The usual further assumption that the various variables are uncorrelated is also made.

When both $\mathbf{X}_{iu}$ and $\mathbf{Z}_{iu}$ correspond to quantitative variables and $\mathbf{Z}_{iu}$ is a submatrix of $\mathbf{X}_{iu}$, the model in Equation 6.17 is referred as a random coefficient model. The fact that $\mathbf{Z}_{iu}$ is a submatrix of $\mathbf{X}_{iu}$ distinguishes this situation from the general mixed model (Equation 6.1) where no such assumption need be made. A situation where $\mathbf{Z}_{iu}$ will be a submatrix of $\mathbf{X}_{iu}$ can be described formally as follows. Suppose for a certain random regression coefficient, say γ_{iul}, $E(\gamma_{iul}) = \beta_l$, so that it can be written as

$$\gamma_{iul} = E(\gamma_{iul}) + \nu_{iul} = \beta_l + \nu_{iul},$$

where ν_{iul} is random with $E(\nu_{iul}) = 0$. Thus any random regression coefficient ν_{iul} with $E(\nu_{iul}) = 0$ has its fixed effects counterpart, namely β_l. Therefore, when the model is expressed in the matrix form as in Equation 6.17, the columns corresponding to β_l and ν_{iul} in the matrices $\mathbf{X}_{iu}$ and $\mathbf{Z}_{iu}$ respectively are identical, thereby making $\mathbf{Z}_{iu}$ a submatrix of $\mathbf{X}_{iu}$.

The random coefficient models provide ample flexibility to deal with the repeated measures data. Within-subject variability is conveniently dealt with by modeling it through the random errors $\boldsymbol{\epsilon}_{iu}$ and the random slope coefficients $\boldsymbol{\nu}_{iu}$ for changes in repeated measures specific to the uth subject in the ith treatment group. The correlation structures such as compound symmetry and autoregressive or unstructured covariances can be assumed for $\mathbf{G}_1$ and $\mathbf{R}_{iu}$, $u = 1, \ldots, n_i$, $i = 1, \ldots, k$. The development of an appropriate model and corresponding analysis can best be illustrated through an example.

EXAMPLE 8   ***Random Coefficients, A Pharmaceutical Stability Study***   This example is adopted from *SAS/STAT Software: Changes and Enhancements through Release 6.12*, pp. 684–685. The pharmaceutical stability data (used with permission from Glaxo Wellcome Inc.) presents replicate assay results as the observed responses for the shelf life of various drugs (in months). The response variable is potency of the drug relative to the percentage claim on the label. There are three batches of products which may differ in initial potency represented by intercepts and in degradation rates represented by the slope parameters. Since the batches are taken randomly, these intercepts and slope parameters are assumed to be the random coefficients. Note that p_{iu}, the number of repeated measurements on each subject, are not all equal. The model can be expressed as

$$y_{iju} = \gamma_{0i} + \gamma_{1i} age_{iu} + \epsilon_{iju}, \ \ i = 1, 2, 3; \ \ j = 1, \ldots, p_{iu}; \ \ u = 1, \ldots, 6,$$

where γ_{0i} and γ_{1i} respectively are random intercepts and slopes normally distributed with mean β_0 and β_1 respectively. We write $\gamma_{0i} = \beta_0 + v_{0i}$ and $\gamma_{1i} = \beta_1 + v_{1i}$. Then v_{0i} and v_{1i} are normally distributed with zero means. Thus the above model can be reexpressed as

$$y_{iju} = \beta_0 + \beta_1 age_{iu} + v_{0i} + v_{1i} age_{iu} + \epsilon_{iju},$$

where $\epsilon_{iju} \sim N(0, \sigma^2)$ are independent. The coefficients $(v_{0i}, v_{1i})'$ are assumed to be independently jointly distributed as bivariate normal with zero mean vector and variance covariance matrix $\sigma^2 \mathbf{G}_1$. The independence of $(v_{0i}, v_{1i})'$ and $\epsilon_{i'ju}$ is also assumed for all i, i', j, u.

Let $\boldsymbol{\beta} = (\beta_0, \beta_1)'$ and $\boldsymbol{v}_i = (v_{0i}, v_{1i})'$. For our example, we can provide an interpretation of the above model as follows. Since there are two random coefficients, namely the batch intercept and the batch slope in vector $\boldsymbol{v}_i$ both of which may not have zero means, we write their effects as $\boldsymbol{\beta} + \boldsymbol{v}_1, \boldsymbol{\beta} + \boldsymbol{v}_2, \ldots$. In the interpretation, $\boldsymbol{\beta}$ is the fixed effect part, representing the mean initial potency and mean degradation rate, and the vectors $\boldsymbol{v}_1, \boldsymbol{v}_2, \ldots$ are all 2 by 1 vectors with their variance-covariance matrix $\sigma^2 \mathbf{G}_1$, where $\mathbf{G}_1$ a 2 by 2 matrix, which we assume to be unstructured. The three unknown parameters of $\mathbf{G}_1$, namely $g_{11}, g_{12} = g_{21}$, and g_{22}, need to be estimated, along with several other parameters. For the error vector $\boldsymbol{\epsilon}_{iu} = (\epsilon_{i1u}, \ldots, \epsilon_{ip_{iu}u})'$, we assume the spherical covariance structure. We will use the restricted maximum likelihood (REML) procedure for the estimation of parameters of the variance covariance matrix.

The subject effect is represented by three batches and on each batch, data are collected at 0, 1, 3, 6, 9, and 12 months. As mentioned earlier, the age of the drug in months represents a fixed effect factor as well, along with an intercept in the model which can be interpreted as the mean initial potency.

To analyze this data using PROC MIXED, we essentially need to spell out these facts in compact form within short SAS code. Specifically, we must indicate that BATCH plays the role of SUBJECT (SUBJECT=BATCH); that INTERCEPT and AGE are random effects (RANDOM=INT AGE); and we specify only the fixed effects part of the model, which includes an intercept, (which need not be specified as SAS adds the intercept by default) and the variable AGE. The SAS code is given as Program 6.9 and the output is presented as Output 6.9.

```
/* Program 6.9 */

options ls = 64 ps = 45 nodate nonumber;
title1 'Output 6.9';
title2 'A Pharmaceutical Stability Study';
data rc;
input batch age@;
do i=1 to 6;
input y@;
output;
end;
cards;
1 0 101.2 103.3 103.3 102.1 104.4 102.4
1 1 98.8 99.4 99.7 99.5 . .
1 3 98.4 99.0 97.3 99.8 . .
1 6 101.5 100.2 101.7 102.7 . .
1 9 96.3 97.2 97.2 96.3 . .
1 12 97.3 97.9 96.8 97.7 97.7 96.7
2 0 102.6 102.7 102.4 102.1 102.9 102.6
2 1 99.1 99.0 99.9 100.6 . .
2 3 105.7 103.3 103.4 104.0 . .
2 6 101.3 101.5 100.9 101.4 . .
2 9 94.1 96.5 97.2 95.6 . .
2 12 93.1 92.8 95.4 92.5 92.2 93.0
```

```
3 0 105.1 103.9 106.1 104.1 103.7 104.6
3 1 102.2 102.0 100.8 99.8 . .
3 3 101.2 101.8 100.8 102.6 . .
3 6 101.1 102.0 100.1 100.2 . .
3 9 100.9 99.5 102.5 100.8 . .
3 12 97.8 98.3 96.9 98.4 96.9 96.5
;
/*Source: Obenchain (1990). Data Courtesy of R. L. Obenchain*/

proc mixed data=rc;
class batch;
model y=age/s;
random int age/type=un sub=batch s;
run;
```

Output 6.9

Output 6.9
A Pharmaceutical Stability Study
Covariance Parameter Estimates (REML)

Cov Parm	Subject	Estimate
UN(1,1)	BATCH	0.97292750
UN(2,1)	BATCH	-0.10192674
UN(2,2)	BATCH	0.03649300
Residual		3.30229533

Solution for Fixed Effects

Effect	Estimate	Std Error	DF	t	Pr > |t|
INTERCEPT	102.70159884	0.64480457	2	159.28	0.0001
AGE	-0.52417636	0.11845227	2	-4.43	0.0475

Solution for Random Effects

Effect	BATCH	Estimate	SE Pred	DF	t
INTERCEPT	1	-0.99744294	0.68336297	78	-1.46
AGE	1	0.12668799	0.12362914	78	1.02
INTERCEPT	2	0.38582987	0.68336297	78	0.56
AGE	2	-0.20397070	0.12362914	78	-1.65
INTERCEPT	3	0.61161307	0.68336297	78	0.90
AGE	3	0.07728271	0.12362914	78	0.63

Solution for Random Effects

Pr > |t|

```
0.1484
0.3087
0.5740
0.1030
0.3735
0.5337
```

Tests of Fixed Effects

```
   Source        NDF   DDF   Type III F   Pr > F

   AGE            1     2      19.58      0.0475
```

The output shows that the REML estimate of the matrix $\mathbf{G}_1$ is

$$\hat{\mathbf{G}}_1 = \begin{bmatrix} 0.9729 & -0.1019 \\ -0.1019 & 0.0365 \end{bmatrix}$$

and therefore, the REML estimator of $\mathbf{G}$ is

$$\hat{\mathbf{G}} = \begin{bmatrix} \hat{\mathbf{G}}_1 & \mathbf{0} & \mathbf{0} \\ \mathbf{0} & \hat{\mathbf{G}}_1 & \mathbf{0} \\ \mathbf{0} & \mathbf{0} & \hat{\mathbf{G}}_1 \end{bmatrix}$$

which is a 6 by 6 block diagonal matrix. Since the covariance structure for error is assumed to be spherical ($\sigma^2 \mathbf{I}$), we do not need to specify this default choice in the SAS code. A REPEATED statement would be needed if any other covariance structure for error were to be specified. Under assumed sphericity, the estimated error variance is $\hat{\sigma}^2 = 3.3023$.

The effects of the intercept and slope are in part fixed and in part random. These are represented as $\boldsymbol{\beta} + \boldsymbol{v} = (\beta_0 + v_0, \beta_1 + v_1)'$, where β_0 and β_1 are the fixed parameters and v_0 and v_1 are the random coefficients, in each case respectively for the INTERCEPT and slope for AGE. The estimate of $\boldsymbol{\beta}$ and the predicted values of the random effects $\boldsymbol{v}$ are presented in two separate tables in Output 6.9. Specifically,

$$\hat{\boldsymbol{\beta}} = \begin{bmatrix} 102.7016 \\ -0.5242 \end{bmatrix} \text{ and } \hat{\boldsymbol{v}} = \begin{cases} (-0.9974, \ 0.1267)' & \text{for Batch } 1 \\ (0.3858, \ -0.2040)' & \text{for Batch } 2 \\ (0.6116, \ 0.0773)' & \text{for Batch } 3. \end{cases}$$

Also presented are the corresponding standard errors, the prediction errors, and corresponding tests for significance. It may be remarked that since v_i's are random rather than fixed parameters, hypothesis testing on them may not be meaningful.

EXAMPLE 9   *Modeling Linear Growth, Ramus Heights Data*   To further illustrate the use of the random coefficients models, we will analyze the ramus heights data of Elston and Grizzle (1962) where the heights of the ramus bone (in mm) for 20 boys were measured at 8, $8\frac{1}{2}$, 9, and $9\frac{1}{2}$ years of age. We may want to model the ramus height, say y_t, at age t as a polynomial growth function of their ages. Since these boys are a sample from a hypothetical population of all boys, the modeled growth curve can be thought of as the common growth curve for the population. However, due to many genetic and environmental variations, each boy would have his own individual growth curve which can be thought of as a random variation of the population growth curve.

For a given boy, we consider the model

$$y_t = \beta_0 + \beta_1 age_t + \epsilon_t, \tag{6.18}$$

where $\epsilon_t \sim N(0, \sigma^2)$ are independent and the values of β_0 and β_1 are specific to the specific boy. In other words, β_0 and β_1 are random coefficients. It is possible that only one of the two coefficients may be random. For example, if β_0 is fixed and β_1 is random then all the boys will have the common intercept of the population but different rate of growth. The difference is modeled as $\beta_1 = \beta_{1F} + \beta_{1R}$, where β_{1F} is the fixed common slope for the population and β_{1R} is the random part representing the amount of change from the common slope for the specific individual (boy). We assume $E(\beta_{1R}) = 0$ and $var(\beta_{1R}) = \sigma^2_{\beta_1}$. Similarly, if β_1 is fixed then all the boys will have the common rate of growth of the population but will have different intercepts. The appropriate assumptions to accommodate this case are $\beta_0 = \beta_{0F} + \beta_{0R}$, where β_{0F} is the fixed common intercept for the population and β_{0R} is the random part representing the amount of change from the common intercept for the specific individual (boy). We assume $E(\beta_{0R}) = 0$ and $var(\beta_{0R}) = \sigma^2_{\beta_0}$. If both β_0 and β_1 are random then the above assumptions and interpretations hold for both coefficients. We will illustrate the case when only β_1 is random.

The SAS code is presented as Program 6.10. For the sake of completeness we have also added the code for the other two cases in the same program (but they have been commented out).

If only β_1 is random but β_0 is fixed then the model (6.18) results in

$$y_t = \beta_0 + \beta_{1F}age_t + \beta_{1R}age_t + \epsilon_t. \tag{6.19}$$

Thus $\beta_0 + \beta_{1F}age_t$ represents the fixed part of the model and β_{1R} is a normally distributed random coefficient with zero mean, and variance $\sigma^2_{\beta_1}$. Thus, $\sigma^2 G_1$ is a 1 by 1 matrix, namely $(\sigma^2_{\beta_1})$. The errors ϵ_t are assumed to be independent and have a common variance, say, σ^2. That is, the spherical structure for the variance covariance matrix of the errors is assumed. Since $\sigma^2 G_1 = \sigma^2_{\beta_1} I$, or equivalently, $G_1 = \frac{\sigma^2_{\beta_1}}{\sigma^2} I = \delta I$, the option TYPE=SIM for the covariance structure of G_1 can be used (in fact, this is a default option). Thus the appropriate MODEL and RANDOM statements are given by

```
model y=age/s;
random age/type=sim subject=boy;
```

where the option S in the MODEL statement requests that the solution for the fixed effects be printed. The option SUBJECT = BOY indicates that the observations on a given BOY constitute the vector of repeated measures. We have chosen to use the METHOD=REML option as the choice of estimation procedure. The complete code is given in Program 6.10. The corresponding output appears as Output 6.10.

```
/* Program 6.10 */

options ls = 64 ps=45 nodate nonumber;
title1 'Output 6.10';
title2 'Analysis of Ramus Data';
data ramus;
input boy y1 y2 y3 y4;
y=y1;
age=8;
output;
y=y2;
age=8.5;
output;
y=y3;
age=9;
output;
y=y4;
age=9.5;
output;
lines;
1 47.8 48.8 49.  49.7
2 46.4 47.3 47.7 48.4
3 46.3 46.8 47.8 48.5
4 45.1 45.3 46.1 47.2
5 47.6 48.5 48.9 49.3
6 52.5 53.2 53.3 53.7
7 51.2 53. 54.3 54.5
```

```
8 49.8 50. 50.3 52.7
9 48.1 50.8 52.3 54.4
10 45. 47. 47.3 48.3
11 51.2 51.4 51.6 51.9
12 48.5 49.2 53. 55.5
13 52.1 52.8 53.7 55.
14 48.2 48.9 49.3 49.8
15 49.6 50.4 51.2 51.8
16 50.7 51.7 52.7 53.3
17 47.2 47.7 48.4 49.5
18 53.3 54.6 55.1 55.3
19 46.2 47.5 48.1 48.4
20 46.3 47.6 51.0 51.8
;
/* Source: Elston and Grizzle (1962).  Reproduced by
permission of the International Biometric Society. */
proc mixed data=ramus covtest;
class boy;
model y=age/s;
random age /type = simple subject = boy;
title3 'Only slope is random';
run;
/*
proc mixed data=ramus covtest;
class boy;
model y=age/s;
random int /type = simple subject = boy;
title3 'Only intercept is random';
run;
proc mixed data=ramus covtest;
class boy;
model y=age/s;
random int age /type = simple subject = boy;
title3 'Both intercept and slope are random';
run;
*/
```

Output 6.10

```
                        Output 6.10
                   Analysis of Ramus Data
                   Only slope is random

          Covariance Parameter Estimates (REML)
```

Cov Parm	Subject	Estimate	Std Error	Z	Pr > |Z|
AGE	BOY	0.07944986	0.02647746	3.00	0.0027
Residual		0.66113890	0.12172554	5.43	0.0001

```
                Solution for Fixed Effects
```

Effect	Estimate	Std Error	DF	t	Pr > |t|
INTERCEPT	33.77000000	1.42583383	59	23.68	0.0001
AGE	1.86300000	0.17440771	19	10.68	0.0001

Tests of Fixed Effects

Source	NDF	DDF	Type III F	Pr > F
AGE	1	19	114.10	0.0001

From Output 6.10 we find the estimates of variance components as $\hat{\sigma}^2 = 0.6611$, $\hat{\delta} = 0.0794$, and $\hat{\sigma}^2_{\beta_1} = \hat{\sigma}^2\hat{\delta} = 0.0525$. It is appropriate to test the null hypothesis $H_0 : \sigma^2_{\beta_1} = 0$ against the one-sided alternative. Testing $H_0 : \sigma^2_{\beta_1} = 0$ is equivalent to testing $H_0 : \delta = 0$. Wald's test given in the output tests this null hypothesis against the two-sided alternative. The p value under the one-sided alternative can be computed as one-half of the reported p value. For our case, it is $0.0027/2 \approx 0.0014$, which is statistically significant. Thus we can claim $\sigma^2_{\beta_1}$ to be nonzero. Since $E(y_t) = \beta_0 + \beta_1 Fage_t$, an estimate of average ramus height at time t is given by

$$\hat{E}(y_t) = \hat{\beta}_0 + \hat{\beta}_1 Fage_t. = 33.7700 + 1.8630 age_t.$$

Suppose instead of fitting Equation 6.19, we decide to fit Equation 6.18 with a 4 by 1 vector of ϵ_t for each subject having a compound symmetric (CS) structure. In this case, there are no random effects in the model given in Equation 6.18, but the covariance structure within each subject would need to be specified using the REPEATED statement

```
repeated/type=cs subject=boy r;
```

where SUBJECT=BOY indicates that repeated measures which are assumed to be independent are taken on boys. Option R in the REPEATED statement prints a typical diagonal block of the block diagonal matrix $\mathbf{R}$. However, we will not discuss this model further here.

6.8 Multivariate Repeated Measures Data

When the data are collected on a set of q variables and on each of these variables at p different occasions, we have a set of *multivariate repeated measures* data. Analysis of such data is further complicated by the existence of correlation among the measurements on different variables in addition to the correlation among measurements taken at different occasions. Several approaches to analyze these data exist in the literature. We will briefly describe one of these here.

Let y_{ijul}, $l = 1, \ldots, q$; $j = 1, \ldots, p$; $u = 1, \ldots, n_i$; $i = 1, \ldots, k$, be the observation on the lth variable for the uth individual in the ith treatment group at the jth occasion and let $\mathbf{y}_{iu} = (y_{i1u1}, \ldots, y_{ipu1}, y_{i1u2}, \ldots, y_{ipuq})'$. Then $\mathbf{y}_{iu}$ is pq by 1 random vector of responses corresponding to the uth individual in the ith group. Let $D(\mathbf{y}_{iu}) = \mathbf{\Omega}$, for $u = 1, \ldots, n_i$; $i = 1, \ldots, k$ and $n = \sum_{i=1}^{k} n_i$. By defining

$$\mathbf{Y}_{n \times pq} = \begin{bmatrix} \mathbf{y}'_{11} \\ \mathbf{y}'_{12} \\ \vdots \\ \mathbf{y}'_{kn_k} \end{bmatrix},$$

and following the approach described in Chapter 3, a multivariate linear model with an appropriate $\mathbf{X}$ matrix consisting of design (assumed to be fixed) and/or regression variables can be written as $\mathbf{Y} = \mathbf{XB} + \mathcal{E}$. The rows of error matrix $\mathcal{E}$ are assumed to be independent and distributed as pq-variate normal with a zero mean vector and the variance covariance matrix $\boldsymbol{\Omega}$. Thus, following the approaches presented in Chapters 3, 4, and 5 any linear hypothesis about the elements of $\mathbf{B}$ can be formulated in the form of a general linear hypothesis $H_0 : \mathbf{LBM} = \mathbf{0}$, for known and full (row and column respectively) rank matrices $\mathbf{L}$ and $\mathbf{M}$. Using Wilks' Λ or any other standard multivariate tests, such a hypothesis can be tested.

This approach for analyzing the multivariate repeated measures data, known as the *doubly multivariate model* (DMM) analysis, is commonly adopted in practice. Since this approach has been discussed extensively in previous chapters we will not discuss this further here. However, see Timm (1980) for a detailed description of the theory and Timm and Mieczkowski (1997) for SAS applications of these methods.

An alternative approach (Naik and Rao, 1997) is to begin by making certain assumptions on the covariance structure of $\boldsymbol{\Omega}$. We assume

$$D(\mathbf{y}_{iu}) = \boldsymbol{\Omega}_{pq \times pq} = \boldsymbol{\Sigma}_{q \times q} \otimes \mathbf{V}_{p \times p},$$

where $\mathbf{V}$ and $\boldsymbol{\Sigma}$ respectively are p by p and q by q positive definite matrices and $\otimes$ stands for the Kronecker product. In view of the defined arrangement of the elements in $\mathbf{y}_{iu}$ the matrix $\mathbf{V}$ represents the variance covariance matrix of repeated measures for a given response variable. This is assumed to be same for all response variables. The matrix $\boldsymbol{\Sigma}$ represents the variance covariance matrix between the measurements on all response variables at a given time point. It is assumed that this does not depend on the particular time point and is the same for all time points.

This assumed structure of $\boldsymbol{\Omega}$ has certain advantages over the general covariance structure. First, the variance covariance matrix of the repeated measures may have a simpler structure such as compound symmetric or AR(1). It is easier to accommodate different structures for the variance covariance matrix of repeated measures $\mathbf{V}$ in this formulation. Secondly, the number of unknown parameters of the variance covariance matrix $\boldsymbol{\Omega}$, in this set up, viz, $[q(q+1)/2 + p(p+1)/2]$, is smaller as compared to that in the general covariance structure, viz, $[pq(pq+1)/2]$. Further, this structure enables us to handle unbalanced multivariate repeated measures data more easily. Statistical analysis of multivariate repeated measures data assuming the above structure can be performed by using PROC MIXED. We illustrate this and the subsequent analysis in the following example.

EXAMPLE 10   *Multivariate Repeated Measures Data, Data from a Dental Study*   Data used in Program 6.11 were collected by T. Zullo of the School of Dental Medicine at the University of Pittsburgh and have been considered in Timm (1980, Table 7.2). The study is concerned with determining the relative effectiveness of two orthopedic adjustments of the mandible. Nine subjects were assigned to each of two orthopedic treatments called activator treatments. The measurements were made on $p = 3$ characteristics to assess the changes in the vertical position of the mandible at $t = 3$ time points of activator treatment. The problem is to compare the two treatments, and study the time effect and the interaction between time and treatment.

As discussed earlier, here the assumed covariance structure for observations on any subject is of the form

$$D(\mathbf{y}_{iu}) = \boldsymbol{\Omega} = \boldsymbol{\Sigma} \otimes \mathbf{V}.$$

We will assume $\boldsymbol{\Sigma}$ to be unstructured (UN) and $\mathbf{V}$ to have an AR(1) covariance structure. This can be specified using the option TYPE=UN @ AR(1) in the REPEATED statement. In this specification, the Kronkecker product $\otimes$ is specified by @. The complete SAS program is presented as Program 6.11.

```
/* Program 6.11 */

option ls=64 ps=45 nodate nonumber;
title1 'Output 6.11';
data a;
```

```
input y1-y9;
cards;
   117     59      10.5    117.5   59      16.5    118.5   60      16.5
   109     60      30.5    110.5   61.5    30.5    111     61.5    30.5
   117     60      23.5    120     61.5    23.5    120.5   62      23.5
   112     67.5    33      126     70.5    32      127     71.5    32.5
   116     61.5    24.5    118.5   62.5    24.5    119.5   63.5    24.5
   123     65.5    22      126     61.5    22      127     67.5    22
   130.5   68.5    33      132     69.5    32.5    134.5   71      32
   126.5   69      20      128.5   71      20      130.5   73      20
   113     58      25      116.5   59      25      118     60.5    24.5
   128     67      24      129     67.5    24      131.5   69      24
   116.5   63.5    28.5    120     65      29.5    121.5   66      29.5
   121.5   64.5    26.5    125.5   67.5    27      127     69      27
   109.5   54      18      112     55.5    18.5    114     57      19
   133     72      34.5    136     73.5    34.5    137.5   75.5    34.5
   120     62.5    26      124.5   65      26      126     66      26
   129.5   65      18.5    133.5   68      18.5    134.5   69      18.5
   122     64.5    18.5    124     65.5    18.5    125.5   66      18.5
   125     65.5    21.5    127     66.5    21.5    128     67      21.6
;
/* This data set is from Timm, N.H. (1980).  Courtesy of Dr. Thomas Zullo,
School of Dental Medicine, University of Pittsburgh. */
data b;  set a;
if _n_<10 then group='1';
else group='2';
run;
data b; set b;
subj=_n_;
y=y1; m_var='var1'; time=1; output;
y=y2; m_var='var2'; time=1; output;
y=y3; m_var='var3'; time=1; output;
y=y4; m_var='var1'; time=2; output;
y=y5; m_var='var2'; time=2; output;
y=y6; m_var='var3'; time=2; output;
y=y7; m_var='var1'; time=3; output;
y=y8; m_var='var2'; time=3; output;
y=y9; m_var='var3'; time=3; output;
drop y1-y9;
title2 'Analysis of Multivariate Repeated Measures';
title3 'Kronecker Product Covariance Structure';
proc mixed data=b method=reml covtest;
classes group subj m_var time;
model y = m_var group time group*time;
repeated m_var time/type=un@ar(1) subject=subj;
run;
```

The SAS code requires some explanation. First the data are converted to the univariate form using the standard technique discussed in several earlier examples. In addition to the variable TIME representing the repeated measures, the variable M_VAR representing three characteristics is also declared as a CLASS variable and is included in the MODEL statement of the MIXED procedure. This ensures that $E(y_{ijul}) = \mu_l$, $i = 1, 2, 3$. That is, the three multivariate responses are ensured to have a mean vector without any structure. We have used the option TYPE=UN @ AR(1) in the REPEATED statement to specify the desired covariance structure discussed earlier. Currently, UN is the only option available for the specification of matrix Σ in the MIXED procedure. However, $\mathbf{V}$ can be chosen to have any of the several covariance structures discussed earlier. The output is presented in Output 6.11.

Output 6.11

```
                             Output 6.11
                 Analysis of Multivariate Repeated Measures
                   Kronecker Product Covariance Structure

                 Covariance Parameter Estimates (REML)

        Cov Parm        Subject      Estimate      Std Error        Z

        M_VAR UN(1,1)    SUBJ       62.69183882    18.28231166     3.43
              UN(2,1)    SUBJ       19.11001566     7.01570034     2.72
              UN(2,2)    SUBJ       22.53327182     6.38954562     3.53
              UN(3,1)    SUBJ       -8.82049720     6.47187742    -1.36
              UN(3,2)    SUBJ        1.58037505     3.38585908     0.47
              UN(3,3)    SUBJ       22.50606019     5.58715855     4.03
        TIME AR(1)       SUBJ        0.94906998     0.01450582    65.43

               Pr > |Z|

                0.0006
                0.0065
                0.0004
                0.1729
                0.6407
                0.0001
                0.0001


                       Tests of Fixed Effects

             Source       NDF   DDF   Type III F   Pr > F

             M_VAR         2    137     909.21     0.0001
             GROUP         1     17       0.22     0.6439
             TIME          2    137      13.19     0.0001
             GROUP*TIME    2    137       0.01     0.9870
```

Testing for the equality of the means of the three responses may not be meaningful in this case. Thus, we note that only the TIME effect is statistically significant with a p value of 0.0001. Since, the interaction GROUP*TIME is not significant (p value = 0.9870) and since no overall differences between the groups have been observed (p value = 0.6439), the TIME trend is deemed the same for the two treatment groups.

6.9   Concluding Remarks

The models and methods considered in this chapter are limited in that (i) only models which are linear in fixed or random coefficients are considered (ii) the error component was assumed to be additive (iii) normality of error is assumed and (iv) usually some appropriate covariance structures on errors and random effects are assumed. Obviously situations where there is substantial departures from one or more of these assumptions will require some other alternative approaches. Additionally, most of the statistical tests are asymptotic and/or approximate and may not necessarily possess certain desirable properties such as a reasonably high power or the ability to maintain the type I error at the specified level. Thus caution should be exercised in selecting and evaluating the appropriate models and in making statistical inferences from these models.

References

Anderson, T. W. (1984), *An Introduction to Multivariate Statistical Analysis,* New York: John Wiley & Sons, Inc.

Andrews, D. F. (1972), "Plots of High-Dimensional Data," *Biometrics,* 28, 125–136.

Apprey, V. and Naik, D. N. (1998), "A SAS Program for Testing Multivariate Normality," Preprint, Presented at SESUG'98, Norfolk, Virginia.

Barnett, V. and Lewis, T. (1994), *Outliers in Statistical Data, Third Edition,* New York: John Wiley & Sons, Inc.

Belsley, D. A., Kuh, E. and Welsch, R. E. (1980), *Regression Diagnostics: Identifying Influential Data and Source of Collinearity,* New York: John Wiley & Sons, Inc.

Bose, R. C. (1951), *Least Square Aspects of the Analysis of Variance,* Mimeograph Series No. 9, University of North Carolina, Chapel Hill.

Box, G. E. P. (1950), "Problems in the Analysis of Growth and Wear Curves," *Biometrics,* 6, 362–389.

Box, G. E. P. (1954), "Some Theorems on Quadratic Forms Applied in the Study of Analysis of Variance Problems II. Effect of Inequality of Variance and Correlations Between Errors in the Two Way Classification," *Ann. Math. Statist.,* 25, 484–498.

Browne, M. W. (1982), "Covariance Structures," in *Topics in Multivariate Analyses,* edited by D. M. Hawkins, Cambridge: Cambridge University Press.

Chambers, J. M., Cleveland, W. S., Kleiner, B. and Tukey, P. A. (1983), *Graphical Methods for Data Analysis,* Belmont, CA: Wadsworth, Inc.

Christensen, R. and Blackwood, L. G. (1993), "Tests for Precision and Accuracy of Multiple Measuring Devices," *Technometrics,* 35, 411–420.

Cody, R. P. and Smith, J. K. (1991), *Applied Statistics and the SAS Programming Language, Third Edition,* New York: Prentice-Hall, Inc.

Cook, R. D. (1977), "Detection of Influential Observations in Regression," *Technometrics,* 19, 15–18.

Cook, R. D. and Weisberg, S. (1982), *Residuals and Influence in Regression,* London: Chapman & Hall.

Crowder, M. J. and Hand, D. J. (1990), *Analysis of Repeated Measures,* New York: Chapman & Hall/CRC Press.

Dallal, G. E. (1992), "The Computer Analysis of Factorial Experiments with Nested Factors," *Amer. Statist.,* 46, 240.

Daniel, C. and Riblett, E. W. (1954), "A Multifactor Experiment," *Industrial and Engineering Chemistry,* 46, 1465–1468.

Das, R. and Sinha, B. K. (1986), "Detection of Multivariate Outliers with Dispersion Slippage in Elliptically Symmetric Distributions," *Ann. Statist.,* 14, 1619–1624.

De Long, D. (1994), "Invited Response to Dallal, G. E. (1992)," *Amer. Statist.,* 48, 141.

Diggle, P. J., Liang, K. Y. and Zeger, S. L. (1995), *Analysis of Longitudinal Data,* London: Oxford University Press.

Dunsmore, I. R. (1981), "Growth Curves in Two Period Changeover Models," *Jour. of Royal Statist. Soc., Ser. C,* 30, 223–229.

Elston, R. C. and Grizzle, J. E. (1962), "Estimation of Time Response Curves and Their Confidence Bands," *Biometrics,* 18, 148–159.

Friendly, M. (1991), *SAS System for Statistical Graphics, First Edition,* Cary, NC: SAS Institute Inc.

Gabriel, K. R. (1971), "The Biplot Graphic Display of Matrices with Application to Principal Component Analysis," *Biometrika,* 58, 453–467.

Giri, N. C. (1977), *Multivariate Statistical Inference,* New York: Academic Press.

Gnanadesikan, R. (1980), "Graphical Methods for Internal Comparisons in ANOVA and MANOVA," in *Handbook of Statistics, Vol. 1: Analysis of Variance,* edited by P. R. Krishnaiah, 133–177, Amsterdam: North Holland.

Gnanadesikan, R. (1997), *Methods for Statistical Data Analysis of Multivariate Observations, Second Edition,* New York: John Wiley & Sons.

Goldstein, R. (1994), "Editor's Notes," *Amer. Statist.,* 48, 138–139.

Goodnight, J. H. (1976), "General Linear Model Procedure," in *SAS.ONE, Proceedings of the First International Users Conference,* 1–39.

Gower, J. C. and Hand, D. J. (1996), *Biplots,* London: Chapman & Hall.

Greenhouse, S. W. and Geisser, S. (1959), "On Methods in the Analysis of Profile Data," *Psychometrika,* 24, 95–112.

Grender, J. M. and Johnson, W. D. (1992), "Using the GLM Procedure to Analyze Multivariate Response in Crossover Designs," in *Proceedings of the Seventeenth Annual SAS Users Group International Conference,* Cary, NC: SAS Institute Inc., 17, 1355–1360.

Grizzle, J. E. and Allen, D. M. (1969), "Analysis of Growth and Dose Response Curves," *Biometrics,* 25, 357–381.

Guo, R. S. and Sachs, E. (1993), "Modelling Optimization and Control of Spatial Uniformity in Manufacturing Processes," *IEEE Transactions on Semiconductor Manufacturing,* 6, 41–57.

Gupta, R. D. and Richards, D. S. P. (1983), "Application of Results of Kotz, Johnson and Boyd to the Null Distribution of Wilks' Criterion," in *Contributions to Statistics: Essays in Honor of Norman Lloyd Johnson,* edited by P. K. Sen, 205–210, Amsterdam: North Holland.

Hargrave, B. Y. and Castle, C. C. (1995), "Effects of phenylephrine induced increase in arterial pressure and closure of the ductus arteriosclerosis on the secretion of atrial natriuretic peptide and renin in the ovine fetus," *Life Sciences,* 57, 31–43.

Hartigan J. A. (1975), *Clustering Algorithms,* New York: John Wiley & Sons, Inc.

Henderson, C. R. (1984), *Applications of Linear Models in Animal Breeding,* Guelph: University of Guelph Press.

Hossain, A. and Naik, D. N. (1989), "Detection of Influential Observations in Multivariate Regression," *Jour. Appl. Statist.,* 16, 25–37.

Huynh, H. and Feldt, L. S. (1970), "Conditions Under Which Mean Square Ratios in Repeated Measure Designs Have Exact F-Distribution," *Jour. Amer. Statist. Assoc.,* 65, 1582–1589.

Huynh, H. and Feldt, L. S. (1976), "Estimation of Box Correction for Degrees of Freedom from the Sample Data in the Randomized Block and Split-Plot Designs," *Jour. Educ. Statist.,* 1, 69–82.

Izenman, A. J. (1987), "Comments on C. R. Rao (1987)," *Statist. Sci.,* 2, 416–463.

Izenman, A. J. and Williams, J. S. (1989), "A Class of Linear Spectral Models and Analyses for the Study of Longitudinal Data," *Biometrics,* 45, 831–849.

Jackson, J. E. (1991), *A User's Guide to Principal Component Analysis,* New York: John Wiley & Sons, Inc.

Jobson, J. D. (1992), *Applied Multivariate Data Analysis, Vol. 2,* New York: Springer-Verlag, Inc.

John, P. W. M. (1971), *Statistical Designs and Analysis of Experiments,* New York: Macmillan Publishing Company, Inc.

Johnson, R. A. and Wichern, D. W. (1998), *Applied Multivariate Statistical Analysis, Fourth Edition,* Englewood Cliffs, NJ: Prentice Hall, Inc.

Jones, B. and Kenward, M. G. (1989), *Design and Analysis of Cross-over Trials,* London: Chapman and Hall.

Kennedy, W. J. and Gentle, J. E. (1980), *Statistical Computing,* New York: Marcel Dekker, Inc.

Keselman, H. J., Algina, J., Kowalchuk, R. K. and Wolfinger, R. D. (1998), "A Comparison of Two Approaches for Selecting Covariance Structures in the Analysis of Repeated Measurements," *Commun. Statist.,* Ser. B, 27, 591–604.

Khattree, R. and Naik, D. N. (1998), "Andrews Plots for Multivariate Data: Some New Suggestions and Applications," Preprint.

Kimura, D. K. (1980), "Likelihood Methods for the Von Bertalanffy Growth Curve," *U. S. Fishery Bulletin,* 77, 765–776.

Kshirsagar, A. M. (1972), *Multivariate Analysis,* New York: Marcel Dekker, Inc.

Kshirsagar, A. M. and Smith, W. B. (1995), *Growth Curves,* New York: Marcel Dekker, Inc.

Kuehl, R. O. (1994), *Statistical Principles of Research Design and Analysis,* Belmont, CA: Duxbury Press.

Lakkis, H. D. and Jones, C. M. (1992), "Comparing Von Bertalanffy Growth Curves with SAS Using the Likelihood Methods Developed by Kimura," Preprint.

Lin, C. C. and Mudholkar, G. S. (1980), "A Simple Test for Normality Against Asymmetric Alternatives," *Biometrika,* 67, 455–461.

Lindsey, J. K. (1993), *Models for Repeated Measurements,* New York: Oxford University Press, Inc.

Littell, R. C., Freund, R. J. and Spector, P. C. (1991), *SAS System for Linear Models, Third Edition,* Cary, NC: SAS Institute Inc.

Littell, R. C., Milliken, G. A., Stroup, W. W. and Wolfinger, R. D. (1996), *SAS System for Mixed Models,* Cary, NC: SAS Institute Inc.

Little, R. J. A. and Rubin, D. B. (1987), *Statistical Analysis with Missing Data,* New York: John Wiley & Sons, Inc.

Mahalanobis, P. S. (1936), "On the Generalized Distance in Statistics," *Proc. Nat. Inst. Sci. India,* 12, 49–55.

Mardia, K. V. (1970), "Measures of Multivariate Skewness and Kurtosis with Applications," *Biometrika,* 519–530.

Mardia, K. V. (1974), "Applications of Some Measures of Multivariate Skewness and Kurtosis for Testing Normality and Robustness Studies," *Sankhyā,* Ser. A 36, 115–128.

Mardia, K. V. (1980), "Tests for Univariate and Multivariate Normality," in *Handbook of Statistics, Vol. 1: Analysis of Variance,* edited by P. R. Krishnaiah, 279–320, Amsterdam: North Holland.

Mardia, K. V., Kent, J. J. and Bibby, J. M. (1979), *Multivariate Analysis,* New York: Academic Press, Inc.

Mauchly, J. W. (1940), "Significance Test for Sphericity of a Normal n-Variate Distribution," *Ann. Math. Statist.,* 29, 204–209.

McLachlan, G. J. and Krishnan, T. (1997), *The EM Algorithm and Extensions,* New York: John Wiley & Sons, Inc.

Milliken, G. A. (1989), "Analysis of Covariance: Multiple Covariates," in *Proceedings of the Fourteenth Annual SAS Users Group International Conference,* 51–60.

Milliken, G. A. (1990), "Analysis of Covariance: Repeated Measures and Split Plot Designs," in *Proceedings of the Fifteenth Annual SAS Users Group International Conference,* 1268–1277.

Milliken, G. A. and Johnson, D. E. (1989), *Analysis of Messy Data: Analysis of Covariance,* Arlington, VA: The Institute of Professional Education.

Milliken, G. A. and Johnson, D. E. (1991), *Analysis of Messy Data, Vol. 1,* New York: Van Nostrand Reinhold.

Montgomery, D. C. (1991), *Design and Analysis of Experiments,* New York: John Wiley & Sons, Inc.

Morrison, D. F. (1976), *Multivariate Statistical Methods,* New York: McGraw Hill, Inc.

Mudholkar, G. S., McDermott, M. and Srivastava, D. K. (1992), "A Test of p Variate Normality," *Biometrika,* 79, 850–854.

Muirhead, R. J. (1982), *Aspects of Multivariate Statistical Theory,* New York: John Wiley & Sons, Inc.

Naik, D. N. (1989), "Detection of Outliers in the Multivariate Linear Regression Model," *Commun. Statist.,* Ser. A, 18, 2225–2232.

Naik, D. N. and Rao, S. (1997), "Analysis of Multivariate Repeated Measures Data with a Kronecker Product Structured Covariance Matrix," Preprint.

Nunez-Anton, V. and Woodworth, G. G. (1994), "Analysis of Longitudinal Data with Unequally Spaced Observations and Time-dependent Correlated Errors," *Biometrics,* 50, 445–456.

Obenchain, R. L. (1990), *STATBLSIM.EXE Version 9010,* unpublished C Code, Eli Lily & Company, Indianapolis, IN.

Olkin, I. and Press, S. J. (1969), "Testing and Estimation for a Circular Stationary Model," *Ann. Math. Statist.,* 40, 1358–1373.

Patel, H. I. (1986), "Analysis of Repeated Measures Designs with Changing Covariates in Clinical Trials," *Biometrika,* 73, 707–715.

Pillai, K. C. S. (1960), *Statistical Tables for Tests of Multivariate Hypotheses,* Manila: Statistical Center, University of Phillipines.

Pillai K. C. S. and Jayachandran, K. (1967), "Power Comparison of Tests of Two Multivariate Hypotheses Based on Four Criteria," *Biometrika,* 54, 195–210.

Pillai, K. C. S. and Jayachandran, K. (1968), "Power Comparison of Tests of Equality of Two Covariance Matrices Based on Four Criteria," *Biometrika,* 55, 335–342.

Potthoff, R. F. and Roy, S. N. (1964), "A Generalized Multivariate Analysis of Variance Model Useful Especially for Growth Curve Problems," *Biometrika,* 51, 313–326.

Rao, C. R. (1948), "Tests of Significance in Multivariate Analysis," *Biometrika,* 35, 58–79.

Rao, C. R. (1951), "An Asymptotic Expansion of the Distribution of Wilks' Criterion," *Bulletin of the International Statistical Institute,* 33, 177–180.

Rao, C. R. (1964), "The Use and Interpretation of Principal Components in Applied Research," *Sankhyā,* Ser. A, 26, 329–358.

Rao, C. R. (1972), "Estimation of Variance and Covariance Components in Linear Models," *Jour. Amer. Statist. Assoc.,* 67, 112–115.

Rao, C. R. (1973), *Linear Statistical Inference and Its Applications,* New York: John Wiley & Sons, Inc.

Rao, C. R. (1987), "Prediction of Future Observations in Growth Curve Models," *Statist. Sci.,* 2, 434–471.

Rao, S. (1995), *Linear Models for Multivariate Repeated Measures Data,* Ph.D. diss., Old Dominion Univ., Norfolk, Virginia.

Ratkowsky, D., Evans, M. and Alldredge, J. (1993), *Cross-over Experiments,* New York: Marcel Dekker, Inc.

Rawlings, J. O. (1988), *Applied Regression Analysis: A Research Tool,* Pacific Grove CA: Wadsworth & Brooks.

Rencher, A. C. (1995), *Methods of Multivariate Analysis,* New York: John Wiley & Sons, Inc.

Rouanet, H. and Lépine, D. (1970), "Comparison Between Treatments in a Repeated Measures Design: ANOVA and Multivariate Methods," *British Jour. of Math. and Stat. Psych.,* 23, 147–163.

Roy, S. N., Gnanadesikan, R. and Srivastava, J. N. (1971), *Analysis and Design of Certain Quantitative Multiresponse Experiments,* New York: Oxford University Press.

SAS Institute Inc. (1989), *SAS/IML Software: Usage and Reference, Version 6, First Edition,* Cary, NC: SAS Institute Inc.

SAS Institute Inc. (1989), *SAS/QC Software: Usage and Reference, Version 6, First Edition,* Cary, NC: SAS Institute Inc.

SAS Institute Inc. (1989), *SAS/STAT User's Guide, Version 6, Fourth Edition, Volume 1,* Cary, NC: SAS Institute Inc.

SAS Institute Inc. (1989), *SAS/STAT User's Guide, Version 6, Fourth Edition, Volume 2,* Cary, NC: SAS Institute Inc.

SAS Institute Inc. (1990), *SAS Language: Reference, Version 6, First Edition,* Cary, NC: SAS Institute Inc.

SAS Institute Inc. (1990), *SAS Procedures Guide, Version 6, Third Edition,* Cary, NC: SAS Institute Inc.

SAS Institute Inc. (1993), *SAS/INSIGHT User's Guide, Version 6, Second Edition,* Cary, NC: SAS Institute Inc.

SAS Institute Inc. (1996), *SAS/STAT Software: Changes and Enhancements through Release 6.12* Cary, NC: SAS Institute Inc.

Schaefer, R. L. (1994), "Using Default Tests in Repeated Measures: How Bad Can It Get?," *Commun. Statist.,* Ser. B, 23, 109–127.

Schwager, J. J. and Margolin, B. H. (1982), "Detection of Multivariate Normal Outliers," *Ann. Statist.,* 10, 943–954.

Searle, S. R. (1971), *Linear Models,* New York: John Wiley & Sons, Inc.

Searle, S. R. (1987), *Linear Models for Unbalanced Data,* New York: John Wiley & Sons, Inc.

Searle, S. R. (1994), "Analysis of Variance Computing Package Output for Unbalanced Data from Fixed-Effects Models with Nested Factors," *Amer. Statist.,* 48, 148–153.

Seber, G. A. F. (1984), *Multivariate Observations,* New York: John Wiley & Sons, Inc.

Sen, P. K. and Singer, J. M. (1993), *Large Sample Methods in Statistics: An Introduction with Applications,* New York: Chapman & Hall.

Smith, W. B. and Hocking, R. R. (1972), "Wishart Variate Generator," *Appl. Statist.,* 21, 341–345.

Spector, P. (1987), "Strategies for Repeated Measures Analysis of Variance," in *Proceedings of the Twelfth Annual SAS Users Group International Conference,* 1174–1177.

Srivastava, M. S. and Carter, E. M. (1983), *An Introduction to Applied Multivariate Statistics,* New York: North Holland.

Tatsuoka, M. M. (1988), *Multivariate Analysis: Techniques for Education and Psychological Research,* New York: John Wiley & Sons, Inc.

Theil, H. (1971), *Principles of Econometrics,* New York: John Wiley & Sons, Inc.

Timm, N. (1975), *Multivariate Analysis with Applications in Education and Psychology*, California: Brooks/Cole.

Timm, N.H. (1980), "Multivariate analysis of variance of repeated measurements," in *Handbook of Statistics, Vol. 1*, edited by P. R. Krishnaiah, 41–87, New York: North-Holland.

Timm, N. H. and Mieczkowski, T. A. (1997), *Univariate and Multivariate General Linear Models: Theory and Applications Using SAS Software*, Cary, NC: SAS Institute Inc.

Tong, Y. L. (1990), *Multivariate Normal Distribution,* New York: Springer-Verlag, Inc.

Tseo, C. L., Deng, J. C., Cornell, J. A., Khuri, A. I. and Schmidt, R. H. (1983), "Effect of Washing Treatment on Quality of Minced Mullet Flesh," *Jour. Food Sci.,* 48, 163–167.

Verbyla, A. P. (1988), "Analysis of Repeated Measures Designs with Changing Covariates," *Biometrika*, 75, 172–174.

Welsch, R. E. and Kuh, E. (1977), "Linear Regression Diagnostics," in *MIT Technical Report*, 923–77.

A Brief Introduction to the IML Procedure

Sometimes we may need to do certain data analyses which are not readily provided by any of the SAS procedures. The IML procedure is a very helpful matrix language within SAS for such situations. One can conveniently perform many of the matrix calculations by using simple statements and by calling various subroutines and functions within PROC IML.

We will briefly describe here a few of these choices which are likely to be encountered in various multivariate analyses and other related matrix manipulations. For further details on other relatively more complex manipulations, the user is referred to *SAS/IML Software: Usage and Reference, Version 6, First Edition*.

A.1 The First SAS Statement

The first SAS statement of the code is

```
proc iml;
```

A.2 Scalars

Scalars are specified in the usual way. For example,

```
c = 7.856;
```

A.3 Matrices

Matrices can be specified within braces ({}) row by row. Rows are separated by commas (,). For example the matrix,

$$\mathbf{A} = \begin{bmatrix} 1 & 3 \\ 9 & 4 \end{bmatrix}$$

is specified as

```
a = { 1   3, 9   4};
```

or

```
a = { 1 3,
      9 4};
```

A.4   Printing of Matrices

One or more matrices, say **A** and **B**, can be printed using one of the following statements

```
print a b ;
```

or

```
print a, b ;
```

Note that the use of the comma (,) results in the two matrices printing one below the other, while when no comma is used, they are printed side by side.

A.5   Algebra of Matrices

The algebraic operators are defined in the usual way, that is,

Addition: $+$

Substraction: $-$

Multiplication: $*$

For example, the matrix **E** defined as

$$\mathbf{E} = \mathbf{A} - \mathbf{B} * (\mathbf{C} + \mathbf{D})$$

is specified as

```
e = a-(b*(c+d));
```

Of course, the order of the matrices involved should be conformable.

A.6   Transpose

PROC IML uses the leading quote (') for transpose. For example,

```
a_trans =a';
```

Since certain keyboards do not support this key, an alternative way is to use the function T. For example,

```
trans_a =t(a);
```

A.7   Inverse

The inverse of a square matrix, if defined, can be obtained by using the function INV. For example, for

$$\mathbf{A} = \begin{bmatrix} 1 & 3 \\ 9 & 4 \end{bmatrix}$$

the inverse denoted as **A_INV** is obtained by

```
inv_a = inv(a);
```

and results in

$$\mathbf{A\_INV} = \begin{bmatrix} -0.173913 & 0.1304348 \\ 0.3913043 & -0.043478 \end{bmatrix}.$$

A.8   Finding the Number of Rows and Columns

Often, it is convenient to alllow the program to determine the number of rows and columns in a matrix rather than explicitly specifying it. Functions NROW and NCOL respectively perform the desired task. For example, for a 2 by 3 matrix

$$\mathbf{B} = \begin{bmatrix} 2 & 4 & 8 \\ 0 & 5 & 9 \end{bmatrix},$$

the statements

```
row_in_b = nrow(b);
col_in_b = ncol(b);
print row_in_b, col_in_b;
```

respectively result in $ROW\_IN\_B = 2$ and $COL\_IN\_B = 3$.

A.9   Trace and Determinant

For a square matrix, functions TRACE and DET will perform the respective tasks of finding the trace and the determinant. For **A** defined earlier, the statements

```
trace_a = trace(a);
det_a = det(a);
print trace_a, det_a;
```

result in $tr(\mathbf{A}) = 5$ and $det(\mathbf{A}) = -23$.

A.10   Eigenvalues and Eigenvectors

There are two SAS functions—one for each—to compute the eigenvalues and to compute the eigenvectors of a *symmetric* matrix. These are EIGVAL and EIGVEC respectively. For the symmetric matrix denoted by **SYM** and given as,

$$\mathbf{SYM} = \begin{bmatrix} 4 & 2 \\ 2 & 4 \end{bmatrix}$$

the statements

```
sym ={4 2, 2 4};
eval_sym = eigval(sym)
evec_sym = eigvec(sym);
print eval_sym,evec_sym;
```

produce the desired eigenvalues and eigenvectors.

The subroutine EIGEN will achieve both tasks in a single call. The corresponding statement is

```
call eigen(lambda,p,sym);
print lambda,p;
```

The eigenvalues and *respective* eigenvectors are stored in $\mathbf{\Lambda}$ and $\mathbf{P}$. Columns of $\mathbf{P}$ are the eigenvectors.

A.11   Square Root of a Symmetric Nonnegative Definite Matrix

For a symmetric nonnegative definite matrix (that is, all eigenvalues of the matrix are nonnegative), $\mathbf{A}$, one can find an upper triangular matrix $\mathbf{U}$ such that $\mathbf{A} = \mathbf{U}'\mathbf{U}$. This is called the Cholesky decomposition. The function ROOT can find this matrix $\mathbf{U}$. The corresponding SAS statement is

```
u = root(a);
```

For the symmetric nonnegative definite (since its eigenvalues were found to be nonnegative) matrix, **SYM** defined earlier, we can compute $\mathbf{U}$ as

$$\mathbf{U} = \begin{bmatrix} 2 & 1 \\ 0 & 1.7320508 \end{bmatrix}.$$

A.12   Generalized Inverse of a Matrix

For any matrix say $\mathbf{B}$, one can find a matrix say $\mathbf{G}$ such that $\mathbf{BGB} = \mathbf{B}$. The matrix $\mathbf{G}$ is called a generalized inverse (or g-inverse for short) of $\mathbf{B}$. A g-inverse always exists but is not unique unless the matrix $\mathbf{B}$ is square and nonsingular (in which case, the regular inverse is the unique g-inverse).

SAS produces one particular g-inverse called the Moore Penrose inverse which in addition to above requirement satisfies some more conditions (Rao, 1973). It is obtained by the statement,

```
g = ginv(b);
```

For the 2 by 3 matrix $\mathbf{B}$ defined as,

$$\mathbf{B} = \begin{bmatrix} 2 & 4 & 8 \\ 0 & 5 & 9 \end{bmatrix},$$

the Moore Penrose g-inverse is obtained as a 3 by 2 matrix,

$$\mathbf{G} = \begin{bmatrix} 0.4818182 & -0.418182 \\ -0.081818 & 0.1181818 \\ 0.0454545 & 0.0454545 \end{bmatrix}.$$

A.13   Singular Value Decomposition

Any matrix $\mathbf{B}$ of order m by n can be partitioned as $\mathbf{B} = \mathbf{UQV'}$, where $\mathbf{U}$ and $\mathbf{V}$ are orthogonal or suborthogonal. If m is larger than n then $\mathbf{U}$ is suborthogonal and $\mathbf{V}$ is orthogonal. If m is smaller than n, then it is the other way around. If $\mathbf{B}$ is square then both $\mathbf{U}$ and $\mathbf{V}$ are orthogonal. The matrix $\mathbf{Q}$ contains the singular values of $\mathbf{B}$. Denoting $\mathbf{U}$, $\mathbf{Q}$ and $\mathbf{V}$ by LEFT, MID and RIGHT, the following subroutine call will result in their computation,

```
call svd(left,mid,right,b);
print left,mid,right;
```

A.14   Symmetric Square Root of a Symmetric Nonnegative Definite Matrix

For a symmetric nonnegative definite matrix $\mathbf{A}$, a symmetric square root denoted by $\mathbf{A}^{1/2}$ can be obtained by using the ROOT function and the subroutine EIGEN. Specifically, since $\mathbf{A}$ is symmetric, we must have $\mathbf{A} = \mathbf{P\Lambda P'} = (\mathbf{P\Lambda}^{1/2}\mathbf{P'})(\mathbf{P\Lambda}^{1/2}\mathbf{P'}) = \mathbf{A}^{1/2}\mathbf{A}^{1/2}$ where $\mathbf{P}$ is orthogonal. The diagonal matrix $\mathbf{\Lambda}$ contains the eigenvalues of $\mathbf{A}$ in the diagonal places, which are nonnegative since the matrix $\mathbf{A}$ is nonnegative definite. Thus, $\mathbf{\Lambda}^{1/2}$ is just a diagonal matrix with diagonal elements as the nonnegative square roots of the corresponding elements of $\mathbf{\Lambda}$. Accordingly, we define $\mathbf{A}^{1/2}$ as $\mathbf{A}^{1/2} = \mathbf{P\Lambda}^{1/2}\mathbf{P'}$. Thus $\mathbf{A}^{1/2}$ is also symmetric. However, it may not be unique.

The needed SAS statements to accomplish this task are

```
proc iml;
a = {
10 3 9,
3 40 8,
9 8 15};
call eigen(d,p,a);
lam_half = root(diag(d));
a_half = p*lam_half*p`;
print a, p, lam_half;
print a_half ;
```

The symmetric square root matrix $\mathbf{A}^{1/2}$ in the above code is denoted by A_HALF. It may be pointed out that $\mathbf{A}^{-1/2}$ may be computed by taking the inverse of $\mathbf{A}^{1/2}$ or by directly computing the symmetric square root of $\mathbf{A}^{-1}$ instead of $\mathbf{A}$ using the above code.

A.15   Kronecker Product

One defines the Kronecker product of $\mathbf{C}$ *with* $\mathbf{D}$ (denoted by $\mathbf{C} \otimes \mathbf{D}$) by multiplying every entry of $\mathbf{C}$ by matrix $\mathbf{D}$ and then creating a matrix out of these block matrices. In notations, the Kronecker product is defined as $\mathbf{C} \otimes \mathbf{D} = (c_{ij}\mathbf{D})$. In SAS, the operator @ does this job. For example, the Kronecker product matrix $KRON\_CD$ is obtained by writing

```
kron_cd = c @ d;
```

With

$$\mathbf{C} = \begin{bmatrix} 1 & 0 & 3 & 4 \\ 0 & 4 & 1 & -1 \\ 1 & 1 & -3 & 2 \end{bmatrix},$$

and

$$\mathbf{D} = \begin{bmatrix} 1 & 3 & 7 \end{bmatrix},$$

the Kronecker product $\mathbf{C} \otimes \mathbf{D}$ (in SAS notation $\mathbf{C}@\mathbf{D}$) is equal to

$$\mathbf{C} \otimes \mathbf{D} = \begin{bmatrix} 1 & 0 & 3 & 4 \\ 3 & 0 & 9 & 12 \\ 7 & 0 & 21 & 28 \\ 0 & 4 & 1 & -1 \\ 0 & 12 & 3 & -3 \\ 0 & 28 & 7 & -7 \\ 1 & 1 & -3 & 2 \\ 3 & 3 & -9 & 6 \\ 7 & 7 & -21 & 14 \end{bmatrix}.$$

A.16 Augmenting Two or More Matrices

Several matrices can be lined up one after the other or one atop the other or in both directions to obtain a single matrix of larger size. Of course in such augmentations, one needs to be careful about the respective orders of various matrices.

In SAS the operator || (two vertical lines) will arrange matrices side by side and the operator // (two division signs) will do so in the order of one below the other. For example, for the matrices $\mathbf{C}$ and $\mathbf{D}$ defined above, $\mathbf{SIDE\_CD} = \mathbf{C}||\mathbf{D}$ results in the matrix $\mathbf{SIDE\_CD}$ equal to

$$\mathbf{SIDE\_CD} = \begin{bmatrix} 1 & 0 & 3 & 4 & 1 \\ 0 & 4 & 1 & -1 & 3 \\ 1 & 1 & -3 & 2 & 7 \end{bmatrix},$$

and $\mathbf{BLO\_CTDT} = \mathbf{C}'//\mathbf{D}'$ results in the matrix $\mathbf{BLO\_CTDT}$ as

$$\mathbf{BLO\_CTDT} = \begin{bmatrix} 1 & 0 & 1 \\ 0 & 4 & 1 \\ 3 & 1 & -3 \\ 4 & -1 & 2 \\ 1 & 3 & 7 \end{bmatrix}.$$

A.17 Construction of a Design Matrix

Often, we may need to create a design matrix (usually denoted by $\mathbf{X}$ in linear models contexts) corresponding to a given experimental design. The function DESIGN will provide such a matrix. This is best illustrated by an example.

Suppose we want the design matrix corresponding to a one-way classification model with three groups. The sample sizes are $n_1 = 3$, $n_2 = 2$ and $n_3 = 5$. As a result, $n = n_1 + n_2 + n_3 = 10$. What one needs to do is to define a column vector titled ADDRESS which provides the addresses for all nonzero entries (all of which would be 1) in the individual rows. For instance, if the seventh entry in the column ADDRESS is 3 then in the seventh row of resulting matrix $\mathbf{X}$ a '1' will be placed in the third column and zero elsewhere.

Suppose we want the design matrix without a column corresponding to the intercept. Then, for our example, we define the vector ADDRESS as a column vector by using the SAS statement

```
address = {1,1,1,  2,2,  3,3,3,3,3};
```

and then define

```
x_w_out = design(address);
```

An alternative way to obtain the column address would be to use the J function, which can be very handy if matrix $\mathbf{X}$ were too big. Specifically, we could use

```
address = j(n1,1,1)  //  j(n2,1,2)  //  j(n3,1,3);
```

To obtain the design matrix with a column for intercept, we could pre-augment the matrix X_W_OUT with the corresponding column, namely a column of 1 everywhere or in SAS notation, with the matrix J(N,1,1). Specifically, the new design matrix is X_WITH given by the SAS statement

```
x_with = j(n,1,1)||x_w_out;
```

A.18   Checking the Estimability of a Linear Function p'β

Often, one needs to check if a given linear function of $\boldsymbol{\beta}$ say, $\mathbf{p}'\boldsymbol{\beta}$, in the linear model set up is estimable. There is a theorem (Searle, 1972) which says that it is estimable if and only if $\mathbf{X}'\mathbf{X}(\mathbf{X}'\mathbf{X})^{-}\mathbf{p} = \mathbf{p}$ where $(\mathbf{X}'\mathbf{X})^{-}$ is any g-inverse of $(\mathbf{X}'\mathbf{X})$. Also, it so happens that checking the above requirement for a given g-inverse is as good as checking it for all g-inverses and hence with just one choice of a g-inverse, we can determine, using above, the estimability of a linear function.

Consider the one way classification model, described above. Suppose we want to know (a) if $\mu + \tau_2$ is estimable and (b) if $\tau_1 + \tau_2$ is estimable. The corresponding choices for $\mathbf{p}'$ are $\mathbf{p}_1' = (1\ 0\ 1\ 0)$ and $\mathbf{p}_2' = (0\ 1\ 1\ 0)$. With $\mathbf{X}$ as obtained earlier, for $\mathbf{p}_1$ and $\mathbf{p}_2$, the left sides of the estimability condition can be calculated using the SAS statements

```
left_p1 = (x_with)'*x_with*(ginv( (x_with)'*x_with ) )*p1;
left_p2 = (x_with)'*x_with*(ginv( (x_with)'*x_with ) )*p2;
```

resulting in

$$\mathbf{LEFT\_P1} = \begin{bmatrix} 1 \\ 0 \\ 1 \\ 0 \end{bmatrix},$$

and

$$\mathbf{LEFT\_P2} = \begin{bmatrix} .5 \\ .5 \\ .5 \\ -.5 \end{bmatrix}.$$

Since **LEFT_P1** is identically equal to $\mathbf{p}_1$, estimability of $\mu + \tau_2$ is established. However, since **LEFT_P2** is not equal to $\mathbf{p}_2$, $\tau_1 + \tau_2$ is not estimable.

A.19   Creating a Matrix from a SAS Data Set

We can create a matrix out of a SAS data set. It is very helpful because many times after runnuing a SAS program, one may, for further calculations, need to use PROC IML. An example for the intended task is presented here. Suppose we have a data set called MY-DATA with three variables X1, X2 and X3 and five data points, from which we want to create a matrix called MYMATRIX. To do so, we use the following SAS statements,

```
data mydata;
input x1 x2 x3;
lines;
2 4 8
3 9 1
9 4 8
1 1 1
2 7 8
;
proc iml;
use mydata;
read all into mymatrix;
print mymatrix;
```

If we want a matrix consisting of only a few variables, say in this case X3 and X1 (in that specific order) from the data set, then the appropriate READ statement needs to be slightly more specific as shown below.

```
read all var {x3 x1} into mymatrix;
```

A.20   Creating a SAS Data Set from a Matrix

Conversely, we can create a SAS data set out of a matrix. An example is presented here. Suppose we have a 5 by 3 matrix titled MYMATRIX containing five observations on three variables for which we will use the default names COL1, COL2 and COL3. From this, we wish to create a data set named NEWDATA. It is done as follows.

```
proc iml;
mymatrix  = {
2 4 8,
3 9 1,
9 4 8,
1 1 1,
2 7 8};
create newdata from  mymatrix;
append from mymatrix;
close newdata;
proc print data = newdata;
```

A.21   Generation of Normal Random Numbers

The standard normal random numbers can be generated by specifying SEEDs as a matrix and then using the function NORMAL. For example,

```
seed = {12490,129479,69737};
nor_rand = normal(seed);
```

Through these, one can generate the multivariate normal random vectors with the desired mean and variance covariance matrix, as shown in Chapter 1.

A.22   Computation of Cumulative Probabilities

Cumulative probabilities or distribution functions for various distributions at the given CUTOFF values can be obtained as shown below for standard normal, chi-square, t and F distributions. The notations are self explanatory.

```
cutoff = {.3 .5 1, 2 3 4};
df = 3;
nc = 0; /*This choice corresponds to a CentraL chi-sq/f/t */
f_ndf = 1;
f_ddf = 3;
dist_nor = probnorm(cutoff);
dist_chi = probchi(cutoff,df,nc);
dist_f = probf(cutoff,f_ndf,f_ddf,nc);
dist_t = probt(cutoff,df,nc);
print dist_nor,dist_chi,dist_f,dist_t;
```

These probabilities are frequently needed in various p value calculations.

A.23   Computation of Percentiles and Cut Off Points

Various percentiles and the cut off points are also needed as intermediate or final calculations in many statistical analyses. The percentiles corresponding to given probabilities (PROB) can be obtained for various distributions as shown below for standard normal, chi-square, t and F distributions. The notations are self explanatory.

```
prob = {.3 .5 .7, .95 .975 .99 };
df = 3;
nc = 0; /*This choice corresponds to a CentraL chi-sq/f/t */
f_ndf = 1;
f_ddf = 3;
nor_inv = probit(prob);
c_inv = cinv(prob,df,nc);
f_inv = finv(prob,f_ndf,f_ddf,nc);
t_inv = tinv(prob,df,nc);
print nor_inv, c_inv, f_inv, t_inv;
```

Any 100α percent upper cut off point can be obtained as the corresponding $100(1-\alpha)^{th}$ percentile.

Data Sets

```
/*CORK DATA SET: cork.dat*/

72 66 76 77
60 53 66 63
56 57 64 58
41 29 36 38
32 32 35 36
30 35 34 26
39 39 31 27
42 43 31 25
37 40 31 25
33 29 27 36
32 30 34 28
63 45 74 63
54 46 60 52
47 51 52 43
91 79 100 75
56 68 47 50
79 65 70 61
81 80 68 58
78 55 67 60
46 38 37 38
39 35 34 37
32 30 30 32
60 50 67 54
35 37 48 39
39 36 39 31
50 34 37 40
43 37 39 50
48 54 57 43
```

```
/* Cork Boring Data: Source: C. R. Rao (1948). Reproduced
   with permission of the Biometrika Trustees. */
```

```
/* NEWCORK DATA SET: newcork.dat */

T1 72 66 76 77
T2 60 53 66 63
T3 56 57 64 58
T4 41 29 36 38
T5 32 32 35 36
T6 30 35 34 26
T7 39 39 31 27
T8 42 43 31 25
T9 37 40 31 25
T10 33 29 27 36
T11 32 30 34 28
T12 63 45 74 63
```

```
T13 54 46 60 52
T14 47 51 52 43
T15 91 79 100 75
T16 56 68 47 50
T17 79 65 70 61
T18 81 80 68 58
T19 78 55 67 60
T20 46 38 37 38
T21 39 35 34 37
T22 32 30 30 32
T23 60 50 67 54
T24 35 37 48 39
T25 39 36 39 31
T26 50 34 37 40
T27 43 37 39 50
T28 48 54 57 43
```

/* In the cork boring data of C. R. Rao (1948) the trees have
   been numbered as T1-T28 */

/* FISH DATA SET: fish.dat */

```
0 0. .25 .25 .25 270 .6695
0. .10 .30 .30 .30 410 .6405
0. .5 .75 .9 .9 610 .729
.15 .65 1.0 1.0 1.0 940 .77
.45 1. 1. 1. 1. 1450 .5655
0. .05 .20 .20 .2 270 .782
.05 .1 .3 .3 .3 410 .812
.05 .45 .95 1. 1. 610 .8215
.1 .7 1. 1. 1. 940 .869
.2 .85 1. 1. 1. 1450 .8395
0. 0. 0. 0. .05 270 .8615
0. .05 .15 .25 .30 410 .9045
0. .15 .95 .95 .95 610 1.028
0. .55 .95 1. 1. 940 1.0445
.1 .85 1. 1. 1. 1450 1.0455
0. 0. 0. .05 .10 270 .6195
0. .05 .15 .20 .25 410 .5305
.1 .45 .95 .95 .95 610 .597
.1 .7 1 1 1 940 .6385
.35 .95 1. 1. 1. 1450 .6645
0 .05 .20 .20 .20 270 .5685
0 0 .15 .25 .25 410 .604
0 .4 .9 1. 1. 610 .6325
.05 .65 1 1. 1. 940 .6845
.3 .85 1. 1. 1. 1450 .723
```

/* Source: Srivastava and Carter (1983, p. 143). */

/* THERMOCOUPLES DATA SET: thermoco.dat */

```
326.06   321.92 326.03 323.59    322.84
326.09 322.00 326.06 323.63    322.92
326.07   321.98 326.03 323.62    322.88
326.08 321.99 326.06 323.64   322.91
326.05   321.96 326.02 323.64   322.89
326.05 321.96 326.02 323.60    322.89
326.03   321.94 326.01 323.62    322.87
```

```
326.08 321.86 326.01 323.64   322.89
326.00   321.85 325.99 323.58   322.85
326.16 322.05 326.13 323.70   322.98
326.00   321.90 325.97 323.55   322.84
326.20 322.12 326.20 323.76   323.03
325.97   321.89 325.95 323.55   322.82
326.20 322.10 326.18 323.74   323.02
326.07   321.99 326.04 323.66   322.93
326.11 322.02 326.08 323.66   322.93
326.00   321.91 325.98 323.57   322.82
326.20 322.11 326.16 323.75   323.03
326.13   322.04 326.12 323.70   322.95
326.12 322.03 326.08 323.68   322.95
326.14   322.04 326.11 323.69   322.94
326.15 322.05 326.12 323.70   322.97
326.07   321.96 326.03 323.62   322.89
326.11 322.00 326.08 323.66   322.93
326.07   321.97 326.03 323.62   322.89
326.13 322.02 326.07 323.67   323.02
326.01   321.92 325.99 323.58   322.85
326.22 322.25 326.19 323.77   323.03
326.08   321.97 326.05 323.63   322.90
326.16 322.06 326.12 323.70   322.97
325.99   321.91 325.96 323.55   322.81
326.14 322.03 326.11 323.68   322.93
325.97   321.86 325.94 323.51   322.80
326.17 322.06 326.14 323.70   322.99
326.02   321.93 325.99 323.58   322.84
326.14 322.04 326.11 323.71   322.96
326.03   321.93 325.98 323.59   322.86
326.10 321.99 326.08 323.66   322.91
326.02   321.91 325.98 323.56   322.93
326.15 322.04 326.13 323.69   322.96
326.10   322.00 326.03 323.64   322.91
326.05 321.96 326.02 323.63   322.88
326.07   321.96 326.04 323.61   322.88
326.00 321.90 325.97 323.57   322.84
326.07   321.98 326.04 323.62   322.89
325.96 321.86 325.90 323.51   322.80
326.08   321.98 326.05 323.63   322.88
326.09 321.99 326.06 323.65   322.92
326.06   321.96 326.01 323.61   322.86
326.11 322.01 326.00 323.65   322.93
326.10   321.99 326.06 323.65   322.90
326.12 322.02 326.07 323.67   322.94
326.05   322.11 326.02 323.62   322.87
326.03 321.95 326.00 323.60   322.87
326.07   321.98 326.05 323.62   322.89
326.14 321.95 326.01 323.62   322.89
326.09   322.00 326.06 323.65   322.91
326.04 321.93 326.01 323.60   322.87
326.17   322.08 326.14 323.72   322.99
326.10 321.88 326.07 323.65   322.92
326.02   321.95 326.00 323.59   322.86
326.11 322.02 326.08 323.67   322.94
326.07   321.96 326.04 323.65   322.90
326.07 321.98 326.03 323.62   322.89
```

```
/* Source: Christensen and Blackwood (1993).  Reprinted with
   permission from Technometrics.  Copyright 1993 by the American
   Statistical Association and the American Society for Quality
   Control.  All rights reserved. */


/* ROHWER's DATA SET: rohwer.dat */
```

OBS	PPVT	RPMT	SAT	N	S	NS	NA	SS
1	68	15	24	0	10	8	21	22
2	82	11	8	7	3	21	28	21
3	82	13	88	7	9	17	31	30
4	91	18	82	6	11	16	27	25
5	82	13	90	20	7	21	28	16
6	100	15	77	4	11	18	32	29
7	100	13	58	6	7	17	26	23
8	96	12	14	5	2	11	22	23
9	63	10	1	3	5	14	24	20
10	91	18	98	16	12	16	27	30
11	87	10	8	5	3	17	25	24
12	105	21	88	2	11	10	26	22
13	87	14	4	1	4	14	25	19
14	76	16	14	11	5	18	27	22
15	66	14	38	0	0	3	16	11
16	74	15	4	5	8	11	12	15
17	68	13	64	1	6	10	28	23
18	98	16	88	1	9	12	30	18
19	63	15	14	0	13	13	19	16
20	94	16	99	4	6	14	27	19
21	82	18	50	4	5	16	21	24
22	89	15	36	1	6	15	23	28
23	80	19	88	5	8	14	25	24
24	61	11	14	4	5	11	16	22
25	102	20	24	5	7	17	26	15
26	71	12	24	0	4	8	16	14
27	102	16	24	4	17	21	27	31
28	96	13	50	5	8	20	28	26
29	55	16	8	4	7	19	20	13
30	96	18	98	4	7	10	23	19
31	74	15	98	2	6	14	25	17
32	78	19	50	5	10	18	27	26

```
/* Source: Timm (1975). Data courtesy of Dr. William D.
   Rohwer, University of California, Berkeley.*/


/* AIR-POLLUTION DATA SET: airpol.dat */

8 98 7 2 12 8 2
7 107 4 3 9 5 3
7 103 4 3 5 6 3
10 88 5 2 8 15 4
6 91 4 2 8 10 3
8 90 5 2 12 12 4
9 84 7 4 12 15 5
5 72 6 4 21 14 4
7 82 5 1 11 11 3
8 64 5 2 13 9 4
6 71 5 4 10 3 3
6 91 4 2 12 7 3
7 72 7 4 18 10 3
```

```
10 70 4 2 11 7 3
10 72 4 1 8 10 3
9 77 4 1 9 10 3
8 76 4 1 7 7 3
8 71 5 3 16 4 4
9 67 4 2 13 2 3
9 69 3 3 9 5 3
10 62 5 3 14 4 4
9 88 4 2 7 6 3
8 80 4 2 13 11 4
5 30 3 3 5 2 3
6 83 5 1 10 23 4
8 84 3 2 7 6 3
6 78 4 2 11 11 3
8 79 2 1 7 10 3
6 62 4 3 9 8 3
10 37 3 1 7 2 3
8 71 4 1 10 7 3
7 52 4 1 12 8 4
5 48 6 5 8 4 3
6 75 4 1 10 24 3
10 35 4 1 6 9 2
8 85 4 1 9 10 2
5 86 3 1 6 12 2
5 86 7 2 13 18 2
7 79 7 4 9 25 3
7 79 5 2 8 6 2
6 68 6 2 11 14 3
8 40 4 3 6 5 2
```

```
/* These are 42 measurements on air-pollution variables recorded
   at 12:00 noon in the Los Angeles area on different days.
   The variables respectively are Wind, Solar rad., CO, NO,
   NO_2, O_3, and HC. (Data Courtesy Prof. G. C. Tiao,
   Ref. Johnson and Wichern (1998, p. 30)).*/
```

```
/* CHEMISTRY DATA: chemist.dat */
```

```
1  1  1  1  1  1  1 -1 4.99 92.2
1  1  1  1  1  1 -1  1 5.00 93.9
1  1  1  1 -1 -1  1  1 5.61 94.6
1  1  1  1 -1 -1 -1 -1 4.76 95.1
1  1 -1 -1  1  1  1  1 5.23 91.8
1  1 -1 -1  1  1 -1 -1 4.77 94.1
1  1 -1 -1 -1 -1  1 -1 4.99 95.4
1  1 -1 -1 -1 -1 -1  1 5.17 93.4
1 -1  1 -1  1 -1  1  1 4.90 94.1
1 -1  1 -1  1 -1 -1 -1 4.90 93.2
1 -1  1 -1 -1  1  1 -1 5.24 92.8
1 -1  1 -1 -1  1 -1  1 4.95 93.8
1 -1 -1  1  1 -1  1 -1 4.96 91.6
1 -1 -1  1  1 -1 -1  1 5.03 92.3
1 -1 -1  1 -1  1  1  1 5.14 90.6
1 -1 -1  1 -1  1 -1 -1 5.05 93.4
-1  1  1 -1  1 -1  1 -1 4.97 93.1
-1  1  1 -1  1 -1 -1  1 4.83 93.3
-1  1  1 -1 -1  1  1  1 5.27 92.0
-1  1  1 -1 -1  1 -1 -1 5.20 92.5
-1  1 -1  1  1 -1  1  1 5.34 91.9
```

```
-1  1 -1  1  1 -1 -1 -1 5.00 92.1
-1  1 -1  1 -1  1  1 -1 5.28 91.9
-1  1 -1  1 -1  1 -1  1 4.93 93.7
-1 -1  1  1  1  1  1  1 4.91 91.0
-1 -1  1  1  1  1 -1 -1 4.71 92.9
-1 -1  1  1 -1 -1  1 -1 4.99 94.8
-1 -1  1  1 -1 -1 -1  1 4.91 94.1
-1 -1 -1 -1  1  1  1 -1 4.86 91.7
-1 -1 -1 -1  1  1 -1  1 4.65 89.4
-1 -1 -1 -1 -1 -1  1  1 5.24 92.8
-1 -1 -1 -1 -1 -1 -1 -1 5.05 93.7
```

/* Source: Daniel and Riblett (1954). Reprinted with
   permission from American Chemical Society. Copyright 1954,
   American Chemical Society. */

/* HEART RATE DATA: heart.dat */

```
ax23     72 86 81 77
ax23     78 83 88 82
ax23     71 82 81 75
ax23     72 83 83 69
ax23     66 79 77 66
ax23     74 83 84 77
ax23     62 73 78 70
ax23     69 75 76 70
bww9     85 86 83 80
bww9     82 86 80 84
bww9     71 78 70 75
bww9     83 88 79 81
bww9     86 85 76 76
bww9     85 82 83 80
bww9     79 83 80 81
bww9     83 84 78 81
control 69 73 72 74
control 66 62 67 73
control 84 90 88 87
control 80 81 77 72
control 72 72 69 70
control 65 62 65 61
control 75 69 69 68
control 71 70 65 65
```

/* Source: Spector (1987, pp. 1174-1177). "Strategies for
   Repeated Measures Analysis of Variance," SUGI 1987. */

/* DATA SET FROM BOX (1950): box.dat */

```
yes a 25 194 192 141
yes a 25 208 188 165
yes a 50 233 217 171
yes a 50 241 222 201
yes a 75 265 252 207
yes a 75 269 283 191

yes b 25 239 127 90
yes b 25  187 105 85
yes b 50 224 123 79
yes b 50 243 123 110
```

```
yes  b  75  243  117  100
yes  b  75  226  125  75


no  a  25  155  169  151
no  a  25   173  152  141
no  a  50  198  187  176
no  a  50  177  196  167
no  a  75  235  225  166
no  a  75  229  270  183


no  b  25  137  82  77
no  b  25   160  82  83
no  b  50  129  94  78
no  b  50  98  89  48
no  b  75  155  76  91
no  b  75  132  105  67
```

/* Source: Box (1950). Reproduced by permission of the
   International Biometric Society. */

/* TASK DATA SET: task.dat */

```
control  4.1   6.1   7.6   7.5   8.9   9.5   8.7   8.8    .    7.0    .    6.5
control  5.8   7.5  10.1  10.4  10.4   8.9   8.9   8.4   9.9   8.6    .    6.9
control  7.0   8.4  11.2  12.8  10.0  10.3   9.5   9.2   9.0   9.4    .    8.4
control  9.0   7.8  10.8  10.3   9.3  10.3  11.5  12.3  10.0  11.4    .    5.9
control  3.6   4.3   3.9   3.9   4.5   3.2   4.1   4.0   3.5   3.7   3.0   2.8
control  7.7   7.0   6.7   7.0   7.9   7.4   7.3   7.2   6.6   6.6   8.3   7.9
control  3.4   2.1   2.2   2.0   2.2   2.2   2.5   2.3   2.5   2.4   2.0   2.2
control  1.8   1.4   2.1   2.4   2.5   2.3   2.0   2.0   1.9   2.0   2.0   1.4
dinocom  7.6   8.9   8.5   8.4   8.5   8.2   5.6   8.8   8.8   8.4   8.0   8.2
dinocom  4.2   6.5   7.5   7.1   7.2   7.0   5.0   4.2   6.9   9.5    .     .
dinocom  6.9  13.3  12.9  13.5  13.4  13.1  13.6  13.1  14.8  15.3  16.1  16.9
dinocom  8.1   7.4   8.8   9.2   8.4   9.2   7.9   7.9   7.9   7.3    .    7.2
dinocom  4.5   4.9   5.5   5.6   5.2   5.3   6.4   6.0   6.4   6.4    .    6.9
dinocom  4.2   3.2   3.2   4.0   3.2   3.4   3.4   3.2   3.2   3.2   2.8   2.8
dihypot  5.9   5.5   5.5   5.5   5.3   5.0   4.5   4.1   4.3   3.9   3.7   3.5
dihypot   .    0.8   0.4   0.6   0.4   0.4   0.5   0.6   0.5   0.5   0.8   0.7
dihypot 10.1   6.5   6.2   6.3   6.6   5.9   6.5   5.5   5.7   5.1   4.4   4.9
dihypot  5.7   4.3   4.6   3.8   3.9   3.6   3.0   3.7   3.2   3.1   2.7   2.4
dihypot  2.1   2.9   3.2   3.2   2.7   2.7   2.4   2.2   1.8   1.7   1.7   1.5
dihypot  5.5  11.1  10.8   8.7   9.3  10.5  12.7  11.3  19.1  18.9  37.0  39.0
dihypot  0.9   4.9   5.7   7.0   7.0   5.8   6.9   7.7   7.5   8.8   8.1   9.9
dihyper  5.0   5.2   3.4   3.0   3.1   3.6   3.2   2.6   4.6   3.8   4.9   2.7
dihyper  4.2   4.3   4.1   3.5   2.8   2.8   4.7   3.7   3.7   4.2    .    4.4
dihyper  3.2   3.0   3.3   3.5   3.4   3.3   3.3   3.3   3.4   3.2   3.1   3.2
dihyper  5.0   6.9   7.5   5.9    .    7.7   7.3   7.6   7.5   7.5   7.0   7.5
dihyper  2.5  12.0  12.2  11.4  11.6  11.7  12.6  10.1  11.4  12.8  11.5  10.7
dihyper  1.6   1.6   2.1   1.9   1.7   2.5   1.6   1.3   3.5   0.6    .     .
```

/* Source: Crowder and Hand (1990, p.8).   */

/* DOG DATA SET: dog.dat */

```
4.  4.  4.1 3.6 3.6 3.8 3.1
4.2 4.3 3.7 3.7 4.8 5.0 5.2
4.3 4.2 4.3 4.3 4.5 5.8 5.4
4.2 4.4 4.6 4.9 5.3 5.6 4.9
4.6 4.4 5.3 5.6 5.9 5.9 5.3
```

```
3.1 3.6 4.9 5.2 5.3 4.2 4.1
3.7 3.9 3.9 4.8 5.2 5.4 4.2
4.3 4.2 4.4 5.2 5.6 5.4 4.7
4.6 4.6 4.4 4.6 5.4 5.9 5.6
3.4 3.4 3.5 3.1 3.1 3.7 3.3
3.0 3.2 3.0 3.0 3.1 3.2 3.1
3.0 3.1 3.2 3.0 3.3 3.0 3.0
3.1 3.2 3.2 3.2 3.3 3.1 3.1
3.8 3.9 4.0 2.9 3.5 3.5 3.4
3.0 3.6 3.2 3.1 3.0 3.0 3.0
3.3 3.3 3.3 3.4 3.6 3.1 3.1
4.2 4.0 4.2 4.1 4.2 4.0 4.0
4.1 4.2 4.3 4.3 4.2 4.0 4.2
4.5 4.4 4.3 4.5 5.3 4.4 4.4
3.2 3.3 3.8 3.8 4.4 4.2 3.7
3.3 3.4 3.4 3.7 3.7 3.6 3.7
3.1 3.3 3.2 3.1 3.2 3.1 3.1
3.6 3.4 3.5 4.6 4.9 5.2 4.4
4.5 4.5 5.4 5.7 4.9 4.0 4.0
3.7 4.0 4.4 4.2 4.6 4.8 5.4
3.5 3.9 5.8 5.4 4.9 5.3 5.6
3.9 4.0 4.1 5.0 5.4 4.4 3.9
3.1 3.5 3.5 3.2 3.0 3.0 3.2
3.3 3.2 3.6 3.7 3.7 4.2 4.4
3.5 3.9 4.7 4.3 3.9 3.4 3.5
3.4 3.4 3.5 3.3 3.4 3.2 3.4
3.7 3.8 4.2 4.3 3.6 3.8 3.7
4.0 4.6 4.8 4.9 5.4 5.6 4.8
4.2 3.9 4.5 4.7 3.9 3.8 3.7
4.1 4.1 3.7 4.0 4.1 4.6 4.7
3.5 3.6 3.6 4.2 4.8 4.9 5.0
```

/* Source: Grizzle and Allen (1969). Reproduced by permission
of the International Biometric Society. */

/* BLOOD PRESSURE DATA: bpress.dat */

```
1 206 c 0 1 1 1
1 220 a c 2 1 1
1 210 b a 3 1 1
2 174 a 0 1 1 2
2 146 b a 2 1 2
2 164 c b 3 1 2
3 192 a 0 1 1 3
3 150 c a 2 1 3
3 160 b c 3 1 3
4 184 b 0 1 1 4
4 192 a b 2 1 4
4 176 c a 3 1 4
5 136 b 0 1 1 5
5 132 c b 2 1 5
5 138 a c 3 1 5
1 190 c 0 1 4 6
1 145 b c 2 4 6
1 160 a b 3 4 6
1 145 a 0 1 2 7
1 125 b a 2 2 7
1 130 c b 3 2 7
2 160 c 0 1 2 8
```

```
2 180 a c 2 2 8
2 145 b a 3 2 8
3 145 b 0 1 2 9
3 154 c b 2 2 9
3 166 a c 3 2 9
1 230 a 0 1 3 10
1 174 b a 2 3 10
1 200 c b 3 3 10
2 194 b 0 1 3 11
2 210 c b 2 3 11
2 190 a c 3 3 11
3 180 c 0 1 3 12
3 180 b c 2 3 12
3 208 a a 3 3 12
4 140 b 0 1 3 13
4 150 a b 2 3 13
4 150 c a 3 3 13
5 194 a 0 1 3 14
5 208 c a 2 3 14
5 160 b c 3 3 14
6 188 c 0 1 3 15
6 200 a c 2 3 15
6 190 b a 3 3 15
7 240 a 0 1 3 16
7 130 b a 2 3 16
7 195 c b 3 3 16
8 180 b 0 1 3 17
8 180 c b 2 3 17
8 190 a c 3 3 17
9 210 c 0 1 3 18
9 160 b c 2 3 18
9 226 a b 3 3 18
10 175 a 0 1 3 19
10 152 c a 2 3 19
10 175 b c 3 3 19
11 155 b 0 1 3 20
11 230 a b 2 3 20
11 226 c a 3 3 20
12 202 a 0 1 3 21
12 160 c a 2 3 21
12 180 b c 3 3 21
13 180 b 0 1 3 22
13 185 a b 2 3 22
13 190 c a 3 3 22
14 185 c 0 1 3 23
14 180 b c 2 3 23
14 200 a b 3 3 23
```

/* Source: Jones and Kenward (1989, p. 230). Reprinted
   by permission of Chapman and Hall, Andover, England. */

/* ONION DATA SET: onion.dat */

```
1 a 0146 3756 2172 3956 3806 4990 1
1 b 9666 8496 12056 8216 5076 -1568 1
2 a 5767 12321 16440 15736 12760 8504 1
2 b 2784 9668 9138 8096 9286 7413  1
3 a 2723 3177 3427 3549 4084 4754 1
3 b 2595 4959 6993 5614 5108 2690 1
```

```
4 a 3365 3567 3646 5120 5390 5051 1
4 b -2191 1281 1184 1750 0532 1328 1
5 a 3238 8804 6480 7221 5680 4439 1
5 b 1872 3655 4810 4737 3776 0682 1
6 a 3567 2820 5878 4410 6988 7962 1
6 b -1640 1063 0430 0476 0145 -3326 1
7 a 1878 3832 3459 4300 3010 2541 1
7 b 2126 2041 0771 1985 -1914 -0537 1
8 a -0408 0392 1310 -0619 0770 1310 1
8 b 2412 0000 3102 3953 3536 1671 1
9 a 3933 5674 5878 9235 8089 7676 2
9 b -0119 -3483 0000 2208 0791 5132 2
10 a 6373 5902 7357 7357 5021 4888 2
10 b -1424 0601 0546 3289 2284 0000 2
11 a  0426 0834 8285 7357 6022 6101 2
11 b 1773 8579 6931 6473 4774 2839 2
12 a 8249 12011 13649 15364 14803 9191 2
12 b 2800 6698 8755 8228 6737 -1947 2
13 a 5623 7302 7482 0370 -1861 0370 2
13 b 3716 4055 9555 15041 11151 2231 2
14 a -0089 4361 4893 5001 7493 6025 2
14 b 0517 0785 4068 3931 4776 2918 2
;

/* Source: Dunsmore (1981).  Reprinted by permission of the
   Royal Statistical Society. */


/* GLUCOSE DATA SET: glucose.dat */

 1    g1   4.90   4.50   7.84    5.46    5.08   4.32   3.91   3.99   4.15   4.41
 2    g1   4.61   4.65   7.90    6.13    4.45   4.17   4.96   4.36   4.26   4.13
 3    g1   5.37   5.35   7.94    5.64    5.06   5.49   4.77   4.48   4.39   4.45
 4    g1   5.10   5.22   7.20    4.95    4.45   3.88   3.65   4.21   4.38   4.44
 5    g1   5.34   4.91   5.69    8.21    2.97   4.30   4.18   4.93   5.16   5.54
 6    g1   5.24   5.04   8.72    4.85    5.57   6.33   4.81   4.55   4.48   5.15
 7    g2   4.91   4.18   9.00    9.74    6.95   6.92   4.66   3.45   4.20   4.63
 8    g2   4.16   3.42   7.09    6.98    6.13   5.36   6.13   3.67   4.37   4.31
 9    g2   4.95   4.40   7.00    7.80    7.78   7.30   5.82   5.14   3.59   4.00
10    g2   3.82   4.00   6.56    6.48    5.66   7.74   4.45   4.07   3.73   3.58
11    g2   3.76   4.70   6.76    4.98    5.02   5.95   4.90   4.79   5.25   5.42
12    g2   4.13   3.95   5.53    8.55    7.09   5.34   5.56   4.23   3.95   4.29
13    g3   4.22   4.92   8.09    6.74    4.30   4.28   4.59   4.49   5.29   4.95
14    g3   4.52   4.22   8.46    9.12    7.50   6.02   4.66   4.69   4.26   4.29
15    g3   4.47   4.47   7.95    7.21    6.35   5.58   4.57   3.90   3.44   4.18
16    g3   4.27   4.33   6.61    6.89    5.64   4.85   4.82   3.82   4.31   3.81
17    g3   4.81   4.85   6.08    8.28    5.73   5.68   4.66   4.62   4.85   4.69
18    g3   4.61   4.68   6.01    7.35    6.38   6.16   4.41   4.96   4.33   4.54
19    g4   4.05   3.78   8.71    7.12    6.17   4.22   4.31   3.15   3.64   3.88
20    g4   3.94   4.14   7.82    8.68    6.22   5.10   5.16   4.38   4.22   4.27
21    g4   4.19   4.22   7.45    8.07    6.84   6.86   4.79   3.87   3.60   4.92
22    g4   4.31   4.45   7.34    6.75    7.55   6.42   5.75   4.56   4.30   3.92
23    g4   4.30   4.71   7.44    7.08    6.30   6.50   4.50   4.36   4.83   4.50
24    g4   4.45   4.12   7.14    5.68    6.07   5.96   5.20   4.83   4.50   4.71
25    g5   5.03   4.99   9.10   10.03    9.20   8.31   7.92   4.86   4.63   3.52
26    g5   4.51   4.50   8.74    8.80    7.10   8.20   7.42   5.79   4.85   4.94
27    g5   4.87   5.12   6.32    9.48    9.88   6.28   5.58   5.26   4.10   4.25
28    g5   4.55   4.44   5.56    8.39    7.85   7.40   6.23   4.59   4.31   3.96
29    g5   4.79   4.82   9.29    8.99    8.15   5.71   5.24   4.95   5.06   5.24
30    g5   4.33   4.48   8.06    8.49    4.50   7.15   5.91   4.27   4.78   5.72
```

```
31    g6    4.60  4.72  9.53  10.02  10.25  9.29  5.45  4.82  4.09  3.52
32    g6    4.33  4.10  4.36   6.92   9.06  8.11  5.69  5.91  5.65  4.58
33    g6    4.42  4.07  5.48   9.05   8.04  7.19  4.87  5.40  4.35  4.51
34    g6    4.38  4.54  8.86  10.01  10.47  9.91  6.11  4.37  3.38  4.02
35    g6    5.06  5.04  8.86   9.97   8.45  6.58  4.74  4.28  4.04  4.34
36    g6    4.43  4.75  6.95   6.64   7.72  7.03  6.38  5.17  4.71  5.14
```

```
/*
The data are ID, group, y1-y10 (the observations taken at the time points
-15 0 30 60 90 120 180 240 300 and 360  minutes.
From Crowder and Hand (1990) p. 14.
*/
```

```
/* AUDIOLOGY DATA SET: audiology.dat */
```

```
a 28.57 53.00 57.83 59.22
a  .     13.00 21.00 26.50
a 60.37 86.41  .     .
a 33.87 55.60 61.06  .
a 26.04 61.98 67.28  .
a  .     59.00 66.80 83.20
a 11.29 38.02  .     .
a  .     35.10 37.79 54.80
a 16.00 33.00 45.39 40.09
a 40.55 50.69 41.70 52.07
a  3.90 11.06  4.15 14.90
a   .00 17.74 44.70 48.85
a 64.75 84.50 92.40 95.39
a 38.25 81.57 89.63  .
a 67.50 91.47 92.86  .
a 45.62 58.00  .     .
a  .00   .00 37.00  .
a 51.15 66.13  .     .
a  .00  48.16  .     .
b  8.76 24.42  .     .
b   .00 20.79 27.42 31.80
b  2.30 12.67 28.80 24.42
b 12.90 28.34  .     .
b  .     45.50 43.32 36.80
b 68.00 96.08 97.47 99.00
b 20.28 41.01 51.15 61.98
b 65.90 81.30 71.20 70.00
b   .00  8.76 16.59 14.75
b  9.22 14.98  9.68  .
b 11.29 44.47 62.90 68.20
b 30.88 29.72  .     .
b 29.72 41.40 64.00  .
b   .00 43.55 48.16  .
b  8.76 60.00  .     .
b  8.00 25.00 30.88 55.53
```

```
/* Nunez-Anton and Woodworth (1994).
   Biometrics, 50, 445-456).  */
```

```
/* SHEEP DATA: sheep.dat */
```

```
young 528  5   66      135    -20
young 528 15  -12      545    -28
young 528 30   60      840    -24
```

```
young 502  5    59       21      9.8
young 502 15    59       97     20
young 502 30    46      224     45
young 505  5    47       55     20
young 505 15     6.7     78     41
young 505 30     6.7    210     39
young 24   5   -16       22     13
young 24  15     2.6    -28     36
young 24  30   -18.4    -33     49
young 599  5   -29      -22     19
young 599 15   -53      155     45
young 599 30   -61      629     50
young 49   5     6.9      7.6   16
young 49  15   -29      169     44
young 49  30   -35      754     60
old   717  5   -16       55     18
old   717 15   -40      158     18
old   717 30   -38      133     24
old   722  5    29      875     35
old   722 15   -40      158     18
old   722 30   -38      133     24
old   617  5     5.8     23      4.3
old   617 15   -29       56     -2.1
old   617 30     5.8    177     -2.1
old   618  5   -40       86      7.4
old   618 15   -60      121     22
old   618 30   -58       82     28
```

```
/* Data courtesy of Dr.Barbara Hargrave, Department of
Biological Sciences, Old Dominion University. */
```

Index

A

abrasion data example 196-200
ADJUST option, LSMEANS statement 254
AIC (Akaike's information criterion) 253, 264
air pollution data example, D-D plot 107-110
Akaike's information criterion (AIC) 253, 264
algebra of matrices 306
ALPHA= option, MIXED procedure 254
analysis of covariance 145-149
analysis of repeated measures data 151-246
    analysis in presence of covariates 207-219
    crossover designs 236-246
    factorial designs 195-207
    growth curve models 219-236
    k populations 176-195
    single population 152-176
    treatment combinations/conditions 170-176
Andrews function plots 33-38
ANOVA F test
    See F statistic
ANOVA partitioning 64-65
    sphericity of orthogonal contrasts, guaranteeing 161
    unbalanced data and 120-123
ANOVA partitioning, multivariate
    See MANOVA partitioning
AR(1) covariance structure 256, 265-270
    repeated measure with time-fixed covariates,
        example 277-278
arranging matrices 310
ARRAY statement 31
assumption of compound symmetry 162
audiology data example 270-274
augmenting matrices 310
autoregressive covariance structure 256, 265-270
    repeated measure with time-fixed covariates,
        example 277-278

B

balanced data 120-123
    one-way laboratories comparison data example 123-126
    two-way classification, mice weight loss example 130-132
banded main diagonal covariance structure 256
Bartlett-Hotelling-Lawley trace 67
Bartlett-Nanda-Pillai's trace
    See Pillai's trace statistic
Bartlett's correction 255
Best Linear Unbiased Estimator (BLUE) 63, 249
Best Linear Unbiased Predictor (BLUP) 249

best linear unbiased scalar (BLUS) residuals 111
BETAINV function 115
Beta Type 1 matrix variate distribution 8
    generating Beta matrices 17-19
Beta Type 2 matrix variate distribution 8
    generating Beta matrices 17-19
between-subject hypotheses 178-188
BIC (Schwartz's Bayesian criterion) 253, 264
BIPLOT macro 40-44
biplots 38-45
bivariate normal distribution 53-58
    See also multivariate normal distribution
    contour plot of density 56-58
    pdf, plotting 55
block design structure 69, 137-139
    profile analysis of single population 154
    testing for covariance structures 156
BLUE (Best Linear Unbiased Estimator) 63, 249
BLUP (Best Linear Unbiased Predictor) 249
BLUS (best linear unbiased scalar) residuals 111
Bonferroni's inequality 86, 195

C

cabbage data example 229-231
calibration problems 98-101
CALIS procedure, computing multivariate kurtosis 11-13
canonical correlation coefficients 5
CAPABILITY procedure 47-48, 52
    GRAPHICS option 48
car door panels warpage data example 262-264
chemical process model, example 140-145
child test performance data, example
    detection of outliers 110-111
    influential observations, detecting 114-116
    multivariate normality test 102-107
CHISQ (chi-square quantiles) 52
CHISQ option, MODEL statement 254, 270
chi-square distributions
    approximation of Wilks' ratio 68
    cumulative probabilities, percentiles, cut off points 313
    Q-Q plots 45
chi-square quantiles (CHISQ) 52
chi-square statistic, MIXED procedure for 254
Cholesky decomposition 17-18, 308
CINV function 46
circular covariance structures, example 260-262
circular covariance testing 158
coefficients
    canonical correlation coefficients 5

coefficient of determination 65
compound symmetry correlation coefficient 287
correlation coefficients 4-5
homogeneity tests for regression coefficients 228-231
Pearson's correlation coefficient 4-5
population coefficient of determination 5
random coefficient model 288-294
regression coefficients homogeneity testing 228
square of multiple correlation coefficient 5
coincidental profiles, *k* populations 186
columns in matrix, determining 307
compound symmetry testing 157-158
concomitant variables, comparisons in presence of 145-149
flammability study example 146-149
conditional distribution 4
confidence intervals
simultaneous 84-87
variance-covariance matrix parameters 254
confounded variables 140
contours of bivariate pdf 54
plotting 56-58
contrast matrix 163
CONTRAST option, REPEATED statement 163
contrasts
sphericity of orthogonal contrasts, guaranteeing 161
testing for, cork data example 14
CONTRAST statement, GLM procedure
crossover design analysis 239
one-way balanced MANOVA partitioning, example 126
pairwise comparisons of SS&CP matrix types 138
parallel profiles testing, example 185
spatial uniformity semiconductor processes,
example 94, 97
SS&CP matrices, analysis of covariances 147-149
two-way factorial experiment 196
univariate analysis of *k* populations, example 193
CONTRAST statement, MIXED procedure 254
Cook type distance 113
test performance data example 114-116
cork data example
Andrews function plots 35-37
circular covariance testing 159-161
Hotelling T^2 statistic 69-73
outliers, detecting with plots 52-53
profile plots 31-33
Q-Q plots 48-50
simultaneous confidence intervals 84
testing circular covariance structure 260-262
testing for multivariate normality 10-16
three-dimensional scatter plots 26-28
two-dimensional scatter plots 22-26
corn varieties comparison, example 137-139
corrected sums of squares, unbalanced data 120

correlation, multiple 5
correlation coefficient matrix 4
correlation coefficients 4-5
covariables 69
covariance ratio 113
influential observations, example 114-116
covariances 3-5
assessing dispersion homogeneity 107-110
circular, testing for 158-161
inference for covariance parameters 253
sphericity of orthogonal contrasts, guaranteeing 161
covariance structures
choosing for tests 264-265
circular, example 260-262
circular covariance tests 158-161
compound symmetry tests 157-158
glucose data example 257-259
linear 256-262
Markov structure, example 270-274
spatial power structure 257
sphericity of orthogonal contrasts 161
sphericity tests 156
statistical tests 255-265
Type H structure 161
covariate effects 283
covariates 207-209
choosing in growth curve model, example 225-228
comparisons in presence of 145-149
fixed over time 275-278
repeated measures analysis 274-287
subject-specific covariates 208-214
univariate approach of split plot design 215-219
COVTEST option, MIXED procedure 253, 267
cross-classified unbalanced data 121-122
crossover designs 236-246
constructing 242-246
multivariate analysis, example 239-242
univariate analysis, example 237-239
cumulative probabilities 313
curve fitting, polynomial 165-170
cut off points, computing 313
cyclic Latin-square crossover designs 244-246

D

DATA step, fish data example 73
DDF= option, MODEL statement 254
DDFM= option, MODEL statement 254
D-D plot, air pollution data example 107-110
degree of polynomial curve fit 165-170
density
See probability distribution
dental study data example 295-297

DESIGN function 310
design matrix, constructing 310
detecting outliers
    See outliers
determinant of matrices 307
DET function 307
DEV= specification, GPLOT procedure 22
diabetic patients study example 126-129
    subject-specific covariates 208-214, 275-278
    univariate approach to split plot design 216-219
diagonal matrix 4
dietary treatment study example 174-176
direct product
    See Kronecker product
dispersion homogeneity, assessing 107-110
dispersion matrices
    See variance-covariance matrices
distribution functions 313
distributions
    See also multivariate normal distribution
    Beta Type 1 matrix variate distribution 8
    Beta Type 2 matrix variate distribution 8
    bivariate normal distribution 53-58
    chi-square distributions 45, 68, 313
    conditional distribution 4
    F distributions 313
    joint probability distribution 4
    outlier detection in multivariate normal distribution
        110-111
    probability distribution 4
    sample statistics for 8-9
    sampling from multivariate normal distributions 6-8
    symmetry of, scatter plots to examine 23-26
    t distributions 313
    Wishart distribution 6-8, 17-19
DMM analysis 295
dog data example 170-174
door panels warpage, example 262-264
doubly multivariate model (DMM) analysis 295
DROP statement 55
drug comparison data example 237-239
drug response time data example 221-225

E

EIGEN function 308, 309
eigenvalues and eigenvectors 307
EIGVAL function 307
EIGVEC function 307
equality of variances in calibration of thermocouples,
    example 99-101

error sums of squares and crossproducts
    See SS&CP matrices
estimability, checking
    laboratories comparison data example 118-119
    linear functions 311
estimable functions 64
ESTIMATE statement, GLM procedure 144, 225
ESTIMATE statement, MIXED procedure 254
etching process data example 132-137
Euclidean distance 34

F

FACTEX procedure 140, 243
factorial designs 195-207
    fractional 139-145, 243
    fractional, example 140-145
    three-factor experiment with two repeated measures
        factors 202-207
    two-way experiment, example 196-200
    two-way factorial experiment 196
    two-way factorial experiment, example 170-174
F distribution 313
fish data example
    multivariate regression 73-80
    polynomial fitting 165-170
    stepdown analysis 81-83
Fisher information matrix 251
fish growth data example 233-236
fitting polynomial curves 165-170
fixed effects models 265-274
    repeated measures 265-270
    repeated measures, example 266-270
    unbalanced and unequally spaced data 270-274
fixed effects parameters 251, 254
fixed-time covariates 275-278
flammability study example 146-149
FOOTNOTE statement, J= option 28
fractional factorial designs 139-145
    chemical process modeling, example 140-145
    crossover designs 243
F statistic
    approximations to 67
    covariance structures 156
    fitting Markov covariance structure 272
    fixed effect parameter tests 251
    k populations 189-193
    MIXED procedure for 254
    simultaneous confidence intervals 84
    sphericity and compound symmetry tests 158
    univariate analysis adjustments 162-164

G

GCONTOUR procedure 56-58
G3D procedure
    plotting pdf of bivariate normal distribution 55
    three-dimensional scatter plots 26
generalized inverse of matrix 63, 308
generalized variance 3
general linear hypotheses 91-98
    spatial uniformity example 92-98
    time-varying covariates 284-287
generating normal random numbers 312-313
g-inverse of matrix 63, 308
GLM procedure
    See also ESTIMATE statement, GLM procedure
    cubic growth model, example 225
    fractional factorial experiment, example 144-145
    missing observations, handling 122
    multivariate approach to time-varying covariates,
        example 282
    multivariate regression, fish data example 73-80
    multivariate tests as options 68-69
    one-way classification 123-126
    profile analysis of k population, example 185, 188
    profile analysis of single population 154-156
    regression coefficients homogeneity testing 228
    repeated measures with fixed effects, example 267-270
    simultaneous confidence intervals, fish data example 84
    univariate analysis of k populations, example 189
    variance and bias analysis 98
glucose data example, testing covariance structure 257-259
GPLOT procedure
    Andrews function plots 35-37
    biplots 40-44
    two-dimensional scatter plots 22-26
graphical representation of multivariate data 21-59
    Andrews function plots 33-38
    biplots 38-45
    bivariate normal distribution 53-58
    contour plots 56-58
    D-D plot, air pollution data example 107-110
    outlier detection plots for 50-53
    pdf, plotting 55
    P-P plots 48
    profile plots 31-33
    Q-Q plots 45-51
    Q-Q plot to detect outliers 110-111
    Q-Q plot to assess normality, example 105-106
    SAS/INSIGHT software 58
    scatter plots 22-31
GRAPHICS option, CAPABILITY procedure 48
Greenhouse-Geisser procedure 163
growth curve models 219-236

growth as nonlinear regression model 231-236
polynomial growth 219
Rao-Khatri reduction 220-228
regression coefficient homogeneity tests 228-231

H

hat (projection) matrix 113
    test performance data example 114-116
heart rate data example
    comparison of treatments 178
    repeated measures with fixed effects only 266-270
    significance of orthogonal contrasts 188
    testing for parallel profiles 180-186
    time trend analysis 194-195
    univariate analysis 189-193
HELMERT option, REPEATED statement 164
homogeneity of dispersion, assessing 107-110
homogeneity tests for regression coefficients 228-231
H= option, MANOVA statement 127, 129
horizontal profiles, k populations 187
Hotelling-Lawley trace criterion 67
Hotelling's T^2 statistic
    cork data example 69-73
    simultaneous confidence intervals 86
    squared Mahalanobis distance and robust version 110
    treatments comparison 170

I

IML procedure 305-313
    circular symmetry tests 159-161
    compound symmetry testing 157-158
    generating Wishart random matrix, example 17-19
    Mardia's kurtosis measure of BLUS residuals 111
    matrices and SAS data sets 312
    multivariate approach to time-varying covariates 279-283
    ORPOL function 165-170, 224
    Q-Q plots 45-46
    simultaneous confidence intervals, fish data example 86
    skewness and kurtosis calculations 11-13
influential observations 111-116
INTERCEPT keyword, MTEST statement 75
    excluding for testing intercept differences 97
inverse of matrices 306, 308
INV function 306

J

J function 311
joint probability distribution 4
J= option 28
JUSTIFY= (J=) option 28

K

KEEP statement 55
k populations 176-195
    comparison of treatments 178
    profile analysis 178-188
    time trend studies 194-195
    univariate analysis 189-193
Kronecker product 249, 252, 309
kurtosis, multivariate
    CALIS procedure to calculate, example 11-13
    Mardia's measures for 10
    outliers, detecting 53

L

laboratories comparison data example
    checking estimability 118-119
    hypothesis testing 123-126
Latin-square crossover designs 243-246
LDATA= option, REPEATED statement, example 260
L_2-distance 34
leading quote (') for transpose 306
least squares analysis 63
    NLIN procedure for 80
    nonlinear 232
LEVELS option, PLOT statement 57
leverage points, hat matrices 113
likelihood ratio test statistic 9
    circular covariance structure 159, 161
    covariance structure tests 255
    nonlinear regression growth model 231
    testing sphericity 156-157
linear covariance structure 256-262
linear functions, checking estimability of 311
linear hypotheses, testing 66-83, 119
    multivariate tests 66-80
    stepdown analysis 80-83
linear regression model 61
LINESIZE= option 23
log-likelihood ratio test statistic 254
LRT statistic, mixed effects model 250
LSMEANS statement, MIXED procedure 147, 254
LSMEAN statement, MIXED procedure
    ADJUST option 254
LS= option 23

M

Mahalanobis distance, squared
    approximately with biplots 39
    assessing multivariate normality 45
    outlier detection 50, 110-111
sample version 10
MANOVA partitioning 121-123
    blocking, corn varieties comparison example 137-139
    one-way balanced, laboratories classification example 123-126
    one-way unbalanced, diabetic patients example 126-129
    two-way balanced, mice weight loss example 130-132
    two-way unbalanced, etching process example 132-137
MANOVA statement, GLM procedure
    See also M= specification, MANOVA statement
    See also PRINTE option, MANOVA statement
    See also PRINTH option, MANOVA statement
    H= option 127, 129
    laboratories comparison data example 124
    mice weight loss example 131
    MNAMES= option 72
    multivariate hypothesis testing, cork data example 71-72
    NOUNI option 186
    polynomial fitting 169
    PREFIX option 186
    spatial uniformity semiconductor processes, example 94, 97
    two-way factorial experiment, example 171
Mardia's multivariate skewness and kurtosis measures 10
    computing, example 11-13
    outliers, detecting 53
Markov covariance structure 256, 257
    fitting audiology data example 270-274
    testing, glucose data example 257
matrices
    See also SS&CP matrices
    See also variance-covariance matrices
    algebra of matrices 306
    arranging and augmenting 310
    Beta matrices 8
    Beta matrices, generating 17-19
    contrast matrix 163
    converting from/into SAS data sets 312
    correlation coefficient matrix 4
    creating from/into SAS data sets 312
    design matrix, constructing 310
    determinants of 307
    diagonal matrix 4
    Fisher information matrix 251
    generalized inverse 63, 308
    IML procedure, syntax for 305-313
    inverse of 306, 308
    mixed effects model, estimating 250-252
    multivariate linear model representation 62
    of partial correlation coefficients 4
    projection (hat) matrix 113
    projection (hat) matrix, example 114-116

random matrix generation, example 17-19
rows in matrix, determining 307
scatter plot matrices 28-31
square roots of 4
symmetric, eigenvalues and eigenvectors 307
symmetric nonnegative definite matrix 308
traces of 63, 307
transposes of 306
Wishart random matrix, generating 17-19
matrix variate Beta Type 1 distribution 8
    generating Beta matrices 17-19
matrix variate Beta Type 2 distribution 8
    generating Beta matrices 17-19
Mauchly's sphericity test 156
    orthogonal contrasts, example 161
    univariate analysis of k populations, example 192
maximum likelihood estimates 9
    See also likelihood ratio test statistic
    mixed effects linear model 249-251
    nonlinear regression growth model 231
MAXITER= option, NLIN procedure 233
means 2-5
MEANS procedure 46
memory data profile, example
    compound symmetry testing 157-158
    profile analysis 152-156
    Type H conditions testing 163-164
METHOD= option, MIXED procedure 253
    likelihood ratios for covariance structures 259
METHOD= option, MODEL statement, example 292
mice data example
    choosing covariates in growth curve model 225-228
    two-way balanced MANOVA partitioning 130-132
minimum variance quadratic unbiased estimator 252
missing observations 122
MIVQUE0 estimator 252
mixed effects linear model 248-252
mixed model equations 249, 251
mixed models for repeated measures analysis 247-297
    analysis in presence of covariates 274-287
    fixed effects only 265-274
    mixed effects linear model 248-252
    multivariate repeated measures 294-297
    random coefficient model 288-294
    statistical tests for covariance structures 255-265
MIXED procedure 247, 252-255
    ALPHA= option 254
    COVTEST option 253, 267
    METHOD= option 253, 259
    random coefficient model, example 289
    repeated measures with fixed effects, example
        267-270

repeated measures with time-fixed covariates, example
    275-278
repeated measures with time-varying covariates, example
    284-287
testing covariance structure 256-257
testing covariance structure of variance-covariance matrix
    259
testing prespecified covariance structures, example
    262-264
Toeplitz structure 161
ML estimator 9
    See also likelihood ratio test statistic
    mixed effects linear model 249-251
    MIXED procedure for 252
    nonlinear regression growth model 231
ML estimator, restricted
    See REML estimator
MNAMES= option, MANOVA statement 72
modeling chemical processes, example 140-145
MODEL statement, GLM procedure
    analysis of covariance 147-149
    blocking variables in 137-139
    CHISQ option 254
    crossover design analysis 237-239
    DDF= option 254
    DDFM= option 254
    fish data example 73
    fitting Markov covariance structure, example 272
    multivariate regression, fish data example 82
    NOINT option 94
    NOUNI option 70, 72
    polynomial fitting 169
    repeated measures with fixed effects, example 267
    specifying SS&CP matrix type 125
    two-way unbalanced, etching process example 133
    Type I through Type IV sums of squares 120
    univariate analysis of covariates, example 216-219
    univariate analysis of k populations, example 189
MODEL statement, MIXED procedure
    CHISQ option 270
    METHOD= option, example 292
    multivariate repeated measures data, example 296
    repeated measures with fixed effects, example 267
    S option 292
    testing covariance structure, example 259
MODEL statement, NLIN procedure 233
MODEL statement, REG procedure, example 92-98
Moore Penrose inverse 308
M= specification, MANOVA statement 71-72, 127, 129
    parallel profiles testing, example 185
    two-way factorial experiment 196
    two-way factorial experiment, example 174

MTEST statement, REG procedure 75
    fractional factorial experiment, example 141-142
    multiple response surface modeling, example 88-91
    PRINT option 94
    spatial uniformity semiconductor processes, example
        94, 97
    testing equality of variances in calibration, example 101
mullet fish data example 87-91
multiple correlation 5
multiple response surface modeling 87-91
multivariate analysis 1-19
    applications 1-2
    basic concepts 2-5
    generating random vectors and matrices 17-19
    sampling from multivariate normal distributions 6-8
    statistics and distributions 8
    testing multivariate normality 9-16
multivariate analysis of experimental data 117-150
    analysis of covariance 145-149
    balanced and unbalanced data 120-123
    blocking 137-139
    fractional factorial experiments 139-145
    one-way classification 123-129
    two-way classification 129-137
multivariate ANOVA partitioning
    See MANOVA partitioning
multivariate data 1
    fractional factorial design 140
    outliers, detecting with plots 50-53
    repeated measures 294-297
    SS&CP matrices 121-123
    unbalanced and unequally spaced 270-274
multivariate data, graphical representation 21-59
    Andrews function plots 33-38
    biplots 38-45
    bivariate normal distribution 53-58
    contour plots 56-58
    outlier detection plots for 50-53
    pdf, plotting 55
    P-P plots 48
    profile plots 31-33
    Q-Q plots 45-51
    SAS/INSIGHT software 58
    scatter plots 22-31
multivariate normal distribution 5, 21
    See also bivariate normal distribution
    Q-Q plots to assess 45-50
    testing for 9-16
multivariate normality test, example 102-107
multivariate normal population, sampling 6-8
multivariate regression 61-116
    ANOVA partitioning 64-65

fish data example 73-80
    general linear hypotheses 91-98
    least squares estimation 63
    regression diagnostics 102-116
    simultaneous confidence intervals 84-87
    statistical background 62
    testing linear hypotheses 66-83
    variance and bias analyses for calibration problems 98-101
multivariate repeated measures data 294-297
    dental data example 295-297
multivariate skewness and kurtosis
    CALIS procedure to calculate, example 11-13
    Mardia's measures for 10
    outliers, detecting 53
multivariate tests
    crossover design, example 239-242
    Hotelling T^2 statistic for cork data example 69-73
    multivariate regression for fish data example 73-80

N

_N_ automatic variable 46
NCOL function 307
negatively correlated variables in biplots 45
NLIN procedure 80
    maximum likelihood estimates for growth models 232
    MAXITER= option 233
    PARMS= option 233
NOINT option
    MODEL statement 94
    REG procedure 230
NOITER option, PARMS statement 263
NOM option, REPEATED statement 207
    testing sphericity of orthogonal contrasts, example 163
nonlinear regression model for growth 231-236
normal distribution, multivariate
    See bivariate normal distribution
    See multivariate normal distribution
NORMAL function 18, 312-313
normality assessment, example 102-107
normal random numbers 18, 312-313
NOUNI option
    MANOVA statement 186
    MODEL statement 70, 72
NOU option, REPEATED statement 207
NROW function 307

O

observations, plotting with variables 38-45
one-way classification models 123-129
    laboratories classification data example 123-126
onions in diet, example 239-242

optimization of uniformity and selectivity in etching process, example 132-137
ORPOL function, IML procedure 165-170, 224
orthogonal contrasts, testing sphericity 161
orthogonal Latin-square crossover designs 243-246
orthogonal transformation 164
outliers
    detecting in multivariate normal distribution 110-111
    detecting with plots 50-53
    influential observations, compared with 111-116
OUT= option, OUTPUT statement 102
OUTPUT statement 35, 102
OVERLAY option, PLOT statement 46

P

PAGESIZE= option 23
parallel profiles, k populations 179-186
PARMS= option, NLIN procedure 233
PARMS statement, MIXED procedure
    testing circular covariance structures, example 260
    testing prespecified covariance structures, example 263
partial tests, MTEST statement for 75
pdf (probability density function) of bivariate normal distribution 53-56
Pearson's correlation coefficient 4-5
percentiles, computing 313
pharmaceutical stability study example 288-291
Pillai's trace statistic 65, 67
    MTEST statement to calculate, example 91
PLOT statement
    LEVELS option 57
    OVERLAY option 46
    two-dimensional scatter plots 22-26
plotting symbol, specifying 23
polynomial curve fitting 165-170
polynomial growth model 219-220
POLYNOMIAL option, REPEATED statement
    sphericity of orthogonal contrasts 164
    subject-specific covariates, example 214
    three-factor experiment, two repeated measures, example 203
    time trend studies 194-195
    univariate analysis of k populations, example 191
population coefficient of determination 5
populations
    k populations 176-195
    multivariate normal population, sampling 6-8
    single population analysis 152-176
    three-population study, example 178
positively correlated variables in biplots 45

P-P plots 48
PPPLOT statement, CAPABILITY procedure 48
PREFIX option, MANOVA statement 186
prespecified known variance-covariance matrix 262-264
PRINCOMP procedure 45-46
    STD option 46
PRINTE option, MANOVA statement 73
    balanced two-way classification, example 131
    one-way balanced MANOVA partitioning, example 124, 126
    two-way factorial experiment, example 171
PRINTE option, REPEATED statement, example 163
PRINTH option, MANOVA statement 73
    balanced two-way classification, example 131
    one-way balanced MANOVA partitioning, example 124, 126
PRINTH option, REPEATED statement, example 163
printing matrices 306
PRINT option, MTEST statement 75, 94
probability density function (pdf) 21
    bivariate normal distribution 53-56
probability distribution 4
probability-probability plots 48
PROBF function 217
PROBPLOT statement, CAPABILITY procedure 48
profile analysis
    coincidental profiles, k populations 186
    horizontal profiles, k populations 187
    k populations 178-188
    parallel profiles, k populations 179-186
    single population, example 152-156
profile plots 31-33
    cubic growth model, example 221
projection (hat) matrix 113
    test performance data example 114-116
PS= option 23

Q

Q-Q plots 45-50
    detecting outliers 51
    multivariate normality test, example 105-106
    outlier detection in multivariate normal distribution 110-111
QQPLOT statement, CAPABILITY procedure 47
QR decomposition 38
quality control for car door panels example 262-264
quality improvement of mullet fish example 87-91
quantile-quantile plots
    See Q-Q plots

R

ramus height data example 291-294
random coefficient model 288-294
  pharmaceutical stability study example 288-291
  ramus height data example 291-294
random effects parameters 254
random matrix generation, example 17-19
random number generator 18, 312-313
RANDOM statement
  GLM procedure, univariate analysis of k populations
    example 189
  MIXED procedure, random coefficient model example
    292
random vectors 2-5
  generating 17
Rao-Khatri reduction 220-228
  choosing covariates to improve efficiency, example
    225-228
  cubic growth model, example 221-225
Rao's F statistic 68
  mixed effects model 250
RDSQ (robust squared distances) 52-53
reference cell model 123
REG procedure
  See also MTEST statement, REG procedure
  MTEST statement 75
  multivariate tests as options 68-69
  obtaining growth curve estimate, example 228
  simultaneous confidence intervals, fish data example 84
  spatial uniformity semiconductor processes, example
    92-98
  two-level fractional factorial experiment, example 140-145
  variance and bias analysis 98
regression, multivariate
  See multivariate regression
regression, univariate 61
regression analysis 61
regression coefficients, homogeneity tests 228-231
regression diagnostics 102-116
  homogeneity of dispersion, assessing 107-110
  influential observations 111-116
  multivariate normality test, example 102-107
  outliers, detecting 110-111
REML estimator 250, 267-270
  MIXED procedure for 252
  random coefficient model, example 291
repeated measures analysis 151-246
  analysis in presence of covariates 207-219
  crossover designs 236-246
  factorial designs 195-207
  growth curve models 219-236
  k populations 176-195

  multivariate data 294-297
  single population 152-176
  treatment combinations/conditions 170-176
  with fixed effects only 265-270
  with fixed effects only, example 266-270
repeated measures analysis using mixed models 247-297
  analysis in presence of covariates 274-287
  fixed effects only 265-274
  mixed effects linear model 248-252
  multivariate repeated measures 294-297
  random coefficient model 288-294
  statistical tests for covariance structures 255-265
repeated measures variables 201
REPEATED statement, GLM procedure
  See also POLYNOMIAL option, REPEATED statement
    TYPE= option, REPEATED statement
  CONTRAST option 163
  HELMERT option 164
  NOM option 163, 207
  NOU option 207
  PRINTE option 163
  PRINTH option 163
  profile analysis of single population 154
  sphericity tests 156
  subject-specific covariates, example 214
  three-factor experiment, two repeated measures,
    example 203
  two-factor experiment, both repeated measures,
    example 201
  two-way factorial experiment, example 174, 200
  univariate analysis of k populations, example 191
  within-subject hypotheses 219
REPEATED statement, IML procedure
  LDATA= option 260
  testing circular covariance structure, example 260-262
REPEATED statement, MIXED procedure 252
  random coefficient model, example 291
  repeated measures with time-varying covariates,
    example 287
  R option 294
  SUBJECT= option 268
  testing covariance structure 256-257
  testing covariance structure, example 259
response surface modeling 87-91
response variables 1
restricted maximum linear estimator
  See REML estimator
robust squared distances (RDSQ) 52-53
ROOT function 17-18, 308, 309
R option, REPEATED statement 294
rows in matrix determining 307
Roy's maximum root criterion 67

S

sample statistics and distributions 8-9
sample variance-covariance matrix 6-8
SAS data sets, converting from/into matrices 312
SAS/INSIGHT software 58
scalars, specifying 305
scatter plot matrices 28-31
scatter plots 22-31
    three-dimensional 26-28
    two-dimensional 22-26
SCATTER statement, G3D procedure 26-28
Schwartz's Bayesian criterion (BIC) 253, 264
second-degree polynomial
    fitting, example 165-169
    mice data example 227-228
semiconductor processes example 92-98
sequential sums of squares 120
SET statement 266
sheep data example 280-287
simultaneous confidence intervals 84-87
single population analysis 152-176
    fitting polynomial curve 165-170
    profile analysis, example 152-156
    testing for covariance structures 156-162
    treatment combinations/conditions 170-176
    univariate analysis 162-164
singular value decomposition (SVD) 39, 309
skewness, multivariate
    Mardia's measures for 10
S option, MODEL statement 292
spatial power covariance structure
    See Markov covariance structure
spatial uniformity in semiconductor processes, example
    92-98
sphericity of orthogonal contrasts 161
sphericity tests 156
split plot design
    F test validity 189
    time trends analysis 195
    univariate analysis, repeated measures data 162
    univariate approach, with covariates 215-219
squared Mahalanobis distance
    approximately with biplots 39
    assessing multivariate normality 45
    outlier detection 50, 110-111
    sample version 10
square of multiple correlation coefficient 5
square root of symmetric nonnegative definite matrix 308
square roots of matrices 4

SS&CP matrices 121-123
    analysis of covariances 147-149
    blocking, corn varieties comparison example 137-139
    pairwise comparisons of different types 138
    profile analysis of k populations, example 186
    spatial uniformity semiconductor processes, example 97
    total sum of squares in ANOVA 64
    two-way classification models 129-130
    two-way classification models, example 131
    two-way factorial experiment of treatment combinations, example 173
    two-way unbalanced, etching process, example 133
standardized test performance data, example
    detection of outliers 110-111
    influential observations, detecting 114-116
    multivariate normality test 102-107
standardizing variables 31
STANDARD procedure 31
    obtaining standardized response variables 87
STD option, PRINCOMP procedure 46
stepdown analysis 80-83
    fish data example 81-83
Student-Newman-Keuls test 99
SUBJECT= option, REPEATED statement 268
subject-specific covariates 208-214
    fixed over time, example 275-278
sums of squares, Type I through Type IV 120-123
sums of squares and crossproducts
    See SS&CP matrices
SVD (singular value decomposition) 39, 309
SVD subroutine 309
symmetric matrices, eigenvalues and eigenvectors 307
symmetric nonnegative definite matrix
    square root 308
    symmetric square root 309
symmetric square root of symmetric nonnegative definite matrix 309
symmetry of distribution
    compound, assumption of 162
    compound, testing for 157-158
    scatter plots to examine 23-26

T

t distribution 313
testability of linear hypotheses 119
TEST option, RANDOM statement 190
test performance data, example
    detection of outliers 110-111
    influential observations, detecting 114-116
    multivariate normality test 102-107

TEST statement, fish data example 82
test statistics and distributions 8-9
T function 306
thermocouples calibration, example 99-101
third-degree polynomial
    dog response time example 223-225
    fitting, example 169-170
three-population study, example 178
time-fixed covariates 275-278
time trend studies 194-195
time-varying covariates 278-287
    general linear model approach 283-287
    sheep data example 280-287
tire wear data example 196-200
TITLE statement, J= option 28
Toeplitz covariance structure 161, 256
total variance 3, 63
trace, matrix 63, 307
TRACE function 307
transpose, matrix 306
TRANSPOSE procedure 31
treatment combinations/conditions
    k populations 178
    repeated measures design, dietary treatment example
        174-176
    repeated measures design, dog data example 170-174
two bands Toeplitz covariance structure 256
two-dimensional scatter plots 22-26
two-way classification 129-137
    two-way balanced MANOVA, mice weight loss example
        130-132
    two-way factorial experiment 196
    two-way factorial experiment, example 170-174
    two-way unbalanced MANOVA, etching process example
        132-137
Type H structure/condition 161
    testing for, example 163-164
    time trend analysis 195
    treatment combinations 170
    univariate analysis of k populations, example 189
TYPE= option, RANDOM statement 252
TYPE= option, REPEATED statement 252, 270
    fitting Markov covariance structure, example 271
    repeated measures with time-varying covariates, example
        287
    testing covariance structure, example 259
    testing linear covariance structure 257
Type I through IV SS&CP matrices 121-122
Type I through IV sums of squares 120-121, 272

unbalanced data 120-123
    one-way classification, diabetic patients study example
        126-129
    two-way classification, etching process example 132-137
    unequally spaced 270-274
univariate analysis
    crossover design analysis 237-239
    k populations 189-193
    single population, repeated measures 162-164
    split plot design with covariates 215-219
univariate normal density 5
univariate regression 61
    influential observations, detecting 113
    influential observations, example 114-116
unstructured covariance structure 256
USS function 46

variables
    blocking variables 69, 137-139
    confounded and fractional factorial experiments 140-145
    correlations, in biplots 45
    plotting with observations (biplots) 38-45
    repeated measures variables 201
    standardizing 31
variable-time covariates 278-287
    general linear model approach 283-287
    sheep data example 280-287
variance and bias analysis for calibration problems 98-101
variance-covariance matrices 3
    circular covariance testing 161
    circular pattern for, testing 158-161
    confidence intervals for parameters 254
    prespecified, testing for 262-264
    sample 6-8
    testing covariance structure 259
variances 3-5
    MIVQUE0 (minimum variance quadratic unbiased
        estimator) 252
    testing equality of, example 99-101
    total variance 63
Von Bertalanffy models 231-236

Wald's statistic, mixed effects model 250
weight loss in mice example 130-132
Welsh-Kuh type statistic 113
    influential observations, example 114-116
whole plot model 215

U

V

W

Wilks' ratio 65, 67-68
   MTEST statement to calculate, example 91
   obtaining Hotelling T^2 from, example 71
Williams' designs for crossover analysis 244
Wishart distribution 6-8
   generating random matrix 17-19
within-subject hypotheses 178
   REPEATED statement 219

Special Characters

@ operator for Kronecker product 309
// operator for arranging matrices 310
' (leading quote) for transpose 306
|| operator for arranging matrices 310

Books from SAS Institute's
Books by Users Press

Advanced Log-Linear Models Using SAS®
by **Daniel Zelterman**

Analysis of Clinical Trials Using SAS®: A Practical Guide
by **Alex Dmitrienko, Walter Offen, Christy Chuang-Stein, and Geert Molenbergs**

Annotate: Simply the Basics
by **Art Carpenter**

Applied Multivariate Statistics with SAS® Software, Second Edition
by **Ravindra Khattree and Dayanand N. Naik**

Applied Statistics and the SAS® Programming Language, Fourth Edition
by **Ronald P. Cody and Jeffrey K. Smith**

An Array of Challenges — Test Your SAS® Skills
by **Robert Virgile**

Carpenter's Complete Guide to the SAS® Macro Language, Second Edition
by **Art Carpenter**

The Cartoon Guide to Statistics
by **Larry Gonick and Woollcott Smith**

Categorical Data Analysis Using the SAS® System, Second Edition
by **Maura E. Stokes, Charles S. Davis, and Gary G. Koch**

Cody's Data Cleaning Techniques Using SAS® Software
by **Ron Cody**

Common Statistical Methods for Clinical Research with SAS® Examples, Second Edition
by **Glenn A. Walker**

Debugging SAS® Programs: A Handbook of Tools and Techniques
by **Michele M. Burlew**

Efficiency: Improving the Performance of Your SAS® Applications
by **Robert Virgile**

The Essential PROC SQL Handbook for SAS® Users
by **Katherine Prairie**

Fixed Effects Regression Methods for Longitudinal Data Using SAS®
by **Paul D. Allison**

Genetic Analysis of Complex Traits Using SAS®
Edited by **Arnold M. Saxton**

A Handbook of Statistical Analyses Using SAS®, Second Edition
by **B.S. Everitt and G. Der**

Health Care Data and the SAS® System
by **Marge Scerbo, Craig Dickstein, and Alan Wilson**

The How-To Book for SAS/GRAPH® Software
by **Thomas Miron**

In the Know ... SAS® Tips and Techniques From Around the Globe
by **Phil Mason**

Instant ODS: Style Templates for the Output Delivery System
by **Bernadette Johnson**

Integrating Results through Meta-Analytic Review Using SAS® Software
by **Morgan C. Wang and Brad J. Bushman**

Learning SAS® in the Computer Lab, Second Edition
by **Rebecca J. Elliott**

The Little SAS® Book: A Primer
by **Lora D. Delwiche and Susan J. Slaughter**

The Little SAS® Book: A Primer, Second Edition
by **Lora D. Delwiche and Susan J. Slaughter**
(updated to include Version 7 features)

The Little SAS® Book: A Primer, Third Edition
by **Lora D. Delwiche and Susan J. Slaughter**
(updated to include SAS 9.1 features)

Logistic Regression Using the SAS® System: Theory and Application
by **Paul D. Allison**

Longitudinal Data and SAS®: A Programmer's Guide
by **Ron Cody**

Maps Made Easy Using SAS®
by **Mike Zdeb**

Models for Discrete Data
by **Daniel Zelterman**

Multiple Comparisons and Multiple Tests Using SAS®
Text and Workbook Set
(books in this set also sold separately)
by **Peter H. Westfall, Randall D. Tobias,**
Dror Rom, Russell D. Wolfinger,
and **Yosef Hochberg**

Multiple-Plot Displays: Simplified with Macros
by **Perry Watts**

Multivariate Data Reduction and Discrimination with
SAS® Software
by **Ravindra Khattree**
and **Dayanand N. Naik**

Output Delivery System: The Basics
by **Lauren E. Haworth**

Painless Windows: A Handbook for SAS® Users, Third Edition
by **Jodie Gilmore**
(updated to include Version 8 and SAS 9.1 features)

PROC TABULATE by Example
by **Lauren E. Haworth**

Professional SAS® Programming Shortcuts
by **Rick Aster**

Quick Results with SAS/GRAPH® Software
by **Arthur L. Carpenter**
and **Charles E. Shipp**

Quick Results with the Output Delivery System
by **Sunil K. Gupta**

Quick Start to Data Analysis with SAS®
by **Frank C. Dilorio**
and **Kenneth A. Hardy**

Reading External Data Files Using SAS®: Examples Handbook
by **Michele M. Burlew**

Regression and ANOVA: An Integrated Approach Using
SAS® Software
by **Keith E. Muller**
and **Bethel A. Fetterman**

SAS®Applications Programming: A Gentle Introduction
by **Frank C. Dilorio**

SAS® for Forecasting Time Series, Second Edition
by **John C. Brocklebank**
and **David A. Dickey**

SAS® for Linear Models, Fourth Edition
by **Ramon C. Littell, Walter W. Stroup,**
and **Rudolf J. Freund**

SAS® for Monte Carlo Studies: A Guide for Quantitative
Researchers
by **Xitao Fan, Ákos Felsővályi, Stephen A. Sivo,**
and **Sean C. Keenan**

SAS® Functions by Example
by **Ron Cody**

SAS® Macro Programming Made Easy
by **Michele M. Burlew**

SAS® Programming by Example
by **Ron Cody**
and **Ray Pass**

SAS® Programming for Researchers and Social Scientists,
Second Edition
by **Paul E. Spector**

SAS® Survival Analysis Techniques for Medical Research,
Second Edition
by **Alan B. Cantor**

SAS® System for Elementary Statistical Analysis,
Second Edition
by **Sandra D. Schlotzhauer**
and **Ramon C. Littell**

SAS® System for Mixed Models
by **Ramon C. Littell, George A. Milliken, Walter W. Stroup,**
and **Russell D. Wolfinger**

SAS® System for Regression, Third Edition
by **Rudolf J. Freund**
and **Ramon C. Littell**

SAS® System for Statistical Graphics, First Edition
by **Michael Friendly**

The SAS® Workbook and *Solutions* Set
(books in this set also sold separately)
by **Ron Cody**

Selecting Statistical Techniques for Social Science Data:
A Guide for SAS® Users
by **Frank M. Andrews, Laura Klem, Patrick M. O'Malley,**
Willard L. Rodgers, Kathleen B. Welch,
and **Terrence N. Davidson**

Statistical Quality Control Using the SAS® System
by **Dennis W. King**

A Step-by-Step Approach to Using the SAS® System
for Factor Analysis and Structural Equation Modeling
by **Larry Hatcher**

A Step-by-Step Approach to Using the SAS® System
for Univariate and Multivariate Statistics, Second Edition
by **Norm O'Rourke, Larry Hatcher,**
and **Edward J. Stepanski**

Step-by-Step Basic Statistics Using SAS®: Student Guide
and *Exercises*
(books in this set also sold separately)
by **Larry Hatcher**

Survival Analysis Using the SAS® System:
A Practical Guide
by **Paul D. Allison**

Tuning SAS® Applications in the OS/390 and z/OS
Environments, Second Edition
by **Michael A. Raithel**

Univariate and Multivariate General Linear Models:
Theory and Applications Using SAS® Software
by **Neil H. Timm**
and **Tammy A. Mieczkowski**

Using SAS® in Financial Research
by **Ekkehart Boehmer, John Paul Broussard,**
and **Juha-Pekka Kallunki**

Using the SAS® Windowing Environment: A Quick Tutorial
by **Larry Hatcher**

Visualizing Categorical Data
by **Michael Friendly**

Web Development with SAS® by Example
by **Frederick Pratter**

Your Guide to Survey Research Using the SAS® System
by **Archer Gravely**

JMP® Books

JMP® for Basic Univariate and Multivariate Statistics: A Step-by-Step Guide
by **Ann Lehman, Norm O'Rourke, Larry Hatcher,**
and **Edward J. Stepanski**

JMP® Start Statistics, Third Edition
by **John Sall, Ann Lehman,**
and **Lee Creighton**

Regression Using JMP®
by **Rudolf J. Freund, Ramon C. Littell,**
and **Lee Creighton**

WILEY SERIES IN PROBABILITY AND STATISTICS
ESTABLISHED BY WALTER A. SHEWHART AND SAMUEL S. WILKS

Editors: *David J. Balding, Peter Bloomfield, Noel A. C. Cressie, Nicholas I. Fisher, Iain M. Johnstone, J. B. Kadane, Louise M. Ryan, David W. Scott, Adrian F. M. Smith, Jozef L. Teugels*
Editors Emeriti: *Vic Barnett, J. Stuart Hunter, David G. Kendall*

The *Wiley Series in Probability and Statistics* is well established and authoritative. It covers many topics of current research interest in both pure and applied statistics and probability theory. Written by leading statisticians and institutions, the titles span both state-of-the-art developments in the field and classical methods.

Reflecting the wide range of current research in statistics, the series encompasses applied, methodological and theoretical statistics, ranging from applications and new techniques made possible by advances in computerized practice to rigorous treatment of theoretical approaches.

This series provides essential and invaluable reading for all statisticians, whether in academia, industry, government, or research.

ABRAHAM and LEDOLTER · Statistical Methods for Forecasting
AGRESTI · Analysis of Ordinal Categorical Data
AGRESTI · An Introduction to Categorical Data Analysis
AGRESTI · Categorical Data Analysis, *Second Edition*
ANDĚL · Mathematics of Chance
ANDERSON · An Introduction to Multivariate Statistical Analysis, *Second Edition*
*ANDERSON · The Statistical Analysis of Time Series
ANDERSON, AUQUIER, HAUCK, OAKES, VANDAELE, and WEISBERG · Statistical Methods for Comparative Studies
ANDERSON and LOYNES · The Teaching of Practical Statistics
ARMITAGE and DAVID (editors) · Advances in Biometry
ARNOLD, BALAKRISHNAN, and NAGARAJA · Records
*ARTHANARI and DODGE · Mathematical Programming in Statistics
*BAILEY · The Elements of Stochastic Processes with Applications to the Natural Sciences
BALAKRISHNAN and KOUTRAS · Runs and Scans with Applications
BARNETT · Comparative Statistical Inference, *Third Edition*
BARNETT and LEWIS · Outliers in Statistical Data, *Third Edition*
BARTOSZYNSKI and NIEWIADOMSKA-BUGAJ · Probability and Statistical Inference
BASILEVSKY · Statistical Factor Analysis and Related Methods: Theory and Applications
BASU and RIGDON · Statistical Methods for the Reliability of Repairable Systems
BATES and WATTS · Nonlinear Regression Analysis and Its Applications
BECHHOFER, SANTNER, and GOLDSMAN · Design and Analysis of Experiments for Statistical Selection, Screening, and Multiple Comparisons
BELSLEY · Conditioning Diagnostics: Collinearity and Weak Data in Regression
BELSLEY, KUH, and WELSCH · Regression Diagnostics: Identifying Influential Data and Sources of Collinearity
BENDAT and PIERSOL · Random Data: Analysis and Measurement Procedures, *Third Edition*
BERRY, CHALONER, and GEWEKE · Bayesian Analysis in Statistics and Econometrics: Essays in Honor of Arnold Zellner
BERNARDO and SMITH · Bayesian Theory
BHAT and MILLER · Elements of Applied Stochastic Processes, *Third Edition*
BHATTACHARYA and JOHNSON · Statistical Concepts and Methods
BHATTACHARYA and WAYMIRE · Stochastic Processes with Applications
BILLINGSLEY · Convergence of Probability Measures, *Second Edition*
BILLINGSLEY · Probability and Measure, *Third Edition*
BIRKES and DODGE · Alternative Methods of Regression
BLISCHKE AND MURTHY (editors) · Case Studies in Reliability and Maintenance
BLISCHKE AND MURTHY · Reliability: Modeling, Prediction, and Optimization
BLOOMFIELD · Fourier Analysis of Time Series: An Introduction, *Second Edition*
BOLLEN · Structural Equations with Latent Variables
BOROVKOV · Ergodicity and Stability of Stochastic Processes
BOULEAU · Numerical Methods for Stochastic Processes
BOX · Bayesian Inference in Statistical Analysis
BOX · R. A. Fisher, the Life of a Scientist
BOX and DRAPER · Empirical Model-Building and Response Surfaces
*BOX and DRAPER · Evolutionary Operation: A Statistical Method for Process Improvement
BOX, HUNTER, and HUNTER · Statistics for Experimenters: An Introduction to Design, Data Analysis, and Model Building
BOX and LUCEÑO · Statistical Control by Monitoring and Feedback Adjustment
BRANDIMARTE · Numerical Methods in Finance: A MATLAB-Based Introduction
BROWN and HOLLANDER · Statistics: A Biomedical Introduction

*Now available in a lower priced paperback edition in the Wiley Classics Library.

BRUNNER, DOMHOF, and LANGER · Nonparametric Analysis of Longitudinal Data in Factorial Experiments

BUCKLEW · Large Deviation Techniques in Decision, Simulation, and Estimation

CAIROLI and DALANG · Sequential Stochastic Optimization

CHAN · Time Series: Applications to Finance

CHATTERJEE and HADI · Sensitivity Analysis in Linear Regression

CHATTERJEE and PRICE · Regression Analysis by Example, *Third Edition*

CHERNICK · Bootstrap Methods: A Practitioner's Guide

CHERNICK and FRIIS · Introductory Biostatistics for the Health Sciences

CHILÈS and DELFINER · Geostatistics: Modeling Spatial Uncertainty

CHOW and LIU · Design and Analysis of Clinical Trials: Concepts and Methodologies

CLARKE and DISNEY · Probability and Random Processes: A First Course with Applications, *Second Edition*

*COCHRAN and COX · Experimental Designs, *Second Edition*

CONGDON · Bayesian Statistical Modelling

CONOVER · Practical Nonparametric Statistics, *Second Edition*

COOK · Regression Graphics

COOK and WEISBERG · Applied Regression Including Computing and Graphics

COOK and WEISBERG · An Introduction to Regression Graphics

CORNELL · Experiments with Mixtures, Designs, Models, and the Analysis of Mixture Data, *Third Edition*

COVER and THOMAS · Elements of Information Theory

COX · A Handbook of Introductory Statistical Methods

*COX · Planning of Experiments

CRESSIE · Statistics for Spatial Data, *Revised Edition*

CSÖRGŐ and HORVÁTH · Limit Theorems in Change Point Analysis

DANIEL · Applications of Statistics to Industrial Experimentation

DANIEL · Biostatistics: A Foundation for Analysis in the Health Sciences, *Sixth Edition*

*DANIEL · Fitting Equations to Data: Computer Analysis of Multifactor Data, *Second Edition*

DASU and JOHNSON · Exploratory Data Mining and Data Cleaning

DAVID · Order Statistics, *Second Edition*

*DEGROOT, FIENBERG, and KADANE · Statistics and the Law

DEL CASTILLO · Statistical Process Adjustment for Quality Control

DETTE and STUDDEN · The Theory of Canonical Moments with Applications in Statistics, Probability, and Analysis

DEY and MUKERJEE · Fractional Factorial Plans

DILLON and GOLDSTEIN · Multivariate Analysis: Methods and Applications

DODGE · Alternative Methods of Regression

*DODGE and ROMIG · Sampling Inspection Tables, *Second Edition*

*DOOB · Stochastic Processes

DOWDY and WEARDEN · Statistics for Research, *Second Edition*

DRAPER and SMITH · Applied Regression Analysis, *Third Edition*

DRYDEN and MARDIA · Statistical Shape Analysis

DUDEWICZ and MISHRA · Modern Mathematical Statistics

DUNN and CLARK · Applied Statistics: Analysis of Variance and Regression, *Second Edition*

DUNN and CLARK · Basic Statistics: A Primer for the Biomedical Sciences, *Third Edition*

DUPUIS and ELLIS · A Weak Convergence Approach to the Theory of Large Deviations

*ELANDT-JOHNSON and JOHNSON · Survival Models and Data Analysis

ENDERS · Applied Econometric Time Series

ETHIER and KURTZ · Markov Processes: Characterization and Convergence

EVANS, HASTINGS, and PEACOCK · Statistical Distributions, *Third Edition*

FELLER · An Introduction to Probability Theory and Its Applications, Volume I, *Third Edition,* Revised; Volume II, *Second Edition*

FISHER and VAN BELLE · Biostatistics: A Methodology for the Health Sciences

*FLEISS · The Design and Analysis of Clinical Experiments

FLEISS · Statistical Methods for Rates and Proportions, *Second Edition*

FLEMING and HARRINGTON · Counting Processes and Survival Analysis

FULLER · Introduction to Statistical Time Series, *Second Edition*

FULLER · Measurement Error Models

GALLANT · Nonlinear Statistical Models

GHOSH, MUKHOPADHYAY, and SEN · Sequential Estimation

GIFI · Nonlinear Multivariate Analysis

GLASSERMAN and YAO · Monotone Structure in Discrete-Event Systems

GNANADESIKAN · Methods for Statistical Data Analysis of Multivariate Observations, *Second Edition*

GOLDSTEIN and LEWIS · Assessment: Problems, Development, and Statistical Issues

GREENWOOD and NIKULIN · A Guide to Chi-Squared Testing

GROSS and HARRIS · Fundamentals of Queueing Theory, *Third Edition*

*HAHN and SHAPIRO · Statistical Models in Engineering

HAHN and MEEKER · Statistical Intervals: A Guide for Practitioners

HALD · A History of Probability and Statistics and their Applications Before 1750

*Now available in a lower priced paperback edition in the Wiley Classics Library.

HALD · A History of Mathematical Statistics from 1750 to 1930
HAMPEL · Robust Statistics: The Approach Based on Influence Functions
HANNAN and DEISTLER · The Statistical Theory of Linear Systems
HEIBERGER · Computation for the Analysis of Designed Experiments
HEDAYAT and SINHA · Design and Inference in Finite Population Sampling
HELLER · MACSYMA for Statisticians
HINKELMAN and KEMPTHORNE: · Design and Analysis of Experiments, Volume 1: Introduction to Experimental Design
HOAGLIN, MOSTELLER, and TUKEY · Exploratory Approach to Analysis of Variance
HOAGLIN, MOSTELLER, and TUKEY · Exploring Data Tables, Trends and Shapes
*HOAGLIN, MOSTELLER, and TUKEY · Understanding Robust and Exploratory Data Analysis
HOCHBERG and TAMHANE · Multiple Comparison Procedures
HOCKING · Methods and Applications of Linear Models: Regression and the Analysis of Variance, *Second Edition*
HOEL · Introduction to Mathematical Statistics, *Fifth Edition*
HOGG and KLUGMAN · Loss Distributions
HOLLANDER and WOLFE · Nonparametric Statistical Methods, *Second Edition*
HOSMER and LEMESHOW · Applied Logistic Regression, *Second Edition*
HOSMER and LEMESHOW · Applied Survival Analysis: Regression Modeling of Time to Event Data
HØYLAND and RAUSAND · System Reliability Theory: Models and Statistical Methods
HUBER · Robust Statistics
HUBERTY · Applied Discriminant Analysis
HUNT and KENNEDY · Financial Derivatives in Theory and Practice
HUSKOVA, BERAN, and DUPAC · Collected Works of Jaroslav Hajek—with Commentary
IMAN and CONOVER · A Modern Approach to Statistics
JACKSON · A User's Guide to Principle Components
JOHN · Statistical Methods in Engineering and Quality Assurance
JOHNSON · Multivariate Statistical Simulation
JOHNSON and BALAKRISHNAN · Advances in the Theory and Practice of Statistics: A Volume in Honor of Samuel Kotz
JUDGE, GRIFFITHS, HILL, LÜTKEPOHL, and LEE · The Theory and Practice of Econometrics, *Second Edition*
JOHNSON and KOTZ · Distributions in Statistics
JOHNSON and KOTZ (editors) · Leading Personalities in Statistical Sciences: From the Seventeenth Century to the Present
JOHNSON, KOTZ, and BALAKRISHNAN · Continuous Univariate Distributions, Volume 1, *Second Edition*
JOHNSON, KOTZ, and BALAKRISHNAN · Continuous Univariate Distributions, Volume 2, *Second Edition*
JOHNSON, KOTZ, and BALAKRISHNAN · Discrete Multivariate Distributions
JOHNSON, KOTZ, and KEMP · Univariate Discrete Distributions, *Second Edition*
JUREČKOVÁ and SEN · Robust Statistical Procedures: Aymptotics and Interrelations
JUREK and MASON · Operator-Limit Distributions in Probability Theory
KADANE · Bayesian Methods and Ethics in a Clinical Trial Design
KADANE AND SCHUM · A Probabilistic Analysis of the Sacco and Vanzetti Evidence
KALBFLEISCH and PRENTICE · The Statistical Analysis of Failure Time Data, *Second Edition*
KASS and VOS · Geometrical Foundations of Asymptotic Inference
KAUFMAN and ROUSSEEUW · Finding Groups in Data: An Introduction to Cluster Analysis
KEDEM and FOKIANOS · Regression Models for Time Series Analysis
KENDALL, BARDEN, CARNE, and LE · Shape and Shape Theory
KHURI · Advanced Calculus with Applications in Statistics, *Second Edition*
KHURI, MATHEW, and SINHA · Statistical Tests for Mixed Linear Models
KLUGMAN, PANJER, and WILLMOT · Loss Models: From Data to Decisions
KLUGMAN, PANJER, and WILLMOT · Solutions Manual to Accompany Loss Models: From Data to Decisions
KOTZ, BALAKRISHNAN, and JOHNSON · Continuous Multivariate Distributions, Volume 1, *Second Edition*
KOTZ and JOHNSON (editors) · Encyclopedia of Statistical Sciences: Volumes 1 to 9 with Index
KOTZ and JOHNSON (editors) · Encyclopedia of Statistical Sciences: Supplement Volume
KOTZ, READ, and BANKS (editors) · Encyclopedia of Statistical Sciences: Update Volume 1
KOTZ, READ, and BANKS (editors) · Encyclopedia of Statistical Sciences: Update Volume 2
KOVALENKO, KUZNETZOV, and PEGG · Mathematical Theory of Reliability of Time-Dependent Systems with Practical Applications
LACHIN · Biostatistical Methods: The Assessment of Relative Risks
LAD · Operational Subjective Statistical Methods: A Mathematical, Philosophical, and Historical Introduction
LAMPERTI · Probability: A Survey of the Mathematical Theory, *Second Edition*
LANGE, RYAN, BILLARD, BRILLINGER, CONQUEST, and GREENHOUSE · Case Studies in Biometry
LARSON · Introduction to Probability Theory and Statistical Inference, *Third Edition*
LAWLESS · Statistical Models and Methods for Lifetime Data, *Second Edition*
LAWSON · Statistical Methods in Spatial Epidemiology
LE · Applied Categorical Data Analysis
LE · Applied Survival Analysis
LEE and WANG · Statistical Methods for Survival Data Analysis, *Third Edition*
LePAGE and BILLARD · Exploring the Limits of Bootstrap
LEYLAND and GOLDSTEIN (editors) · Multilevel Modelling of Health Statistics
LIAO · Statistical Group Comparison

*Now available in a lower priced paperback edition in the Wiley Classics Library.

LINDVALL · Lectures on the Coupling Method

LINHART and ZUCCHINI · Model Selection

LITTLE and RUBIN · Statistical Analysis with Missing Data, *Second Edition*

LLOYD · The Statistical Analysis of Categorical Data

MAGNUS and NEUDECKER · Matrix Differential Calculus with Applications in Statistics and Econometrics, *Revised Edition*

MALLER and ZHOU · Survival Analysis with Long Term Survivors

MALLOWS · Design, Data, and Analysis by Some Friends of Cuthbert Daniel

MANN, SCHAFER, and SINGPURWALLA · Methods for Statistical Analysis of Reliability and Life Data

MANTON, WOODBURY, and TOLLEY · Statistical Applications Using Fuzzy Sets

MARDIA and JUPP · Directional Statistics

MASON, GUNST, and HESS · Statistical Design and Analysis of Experiments with Applications to Engineering and Science, *Second Edition*

McCULLOCH and SEARLE · Generalized, Linear, and Mixed Models

McFADDEN · Management of Data in Clinical Trials

McLACHLAN · Discriminant Analysis and Statistical Pattern Recognition

McLACHLAN and KRISHNAN · The EM Algorithm and Extensions

McLACHLAN and PEEL · Finite Mixture Models

McNEIL · Epidemiological Research Methods

MEEKER and ESCOBAR · Statistical Methods for Reliability Data

MEERSCHAERT and SCHEFFLER · Limit Distributions for Sums of Independent Random Vectors: Heavy Tails in Theory and Practice

*MILLER · Survival Analysis, *Second Edition*

MONTGOMERY, PECK, and VINING · Introduction to Linear Regression Analysis, *Third Edition*

MORGENTHALER and TUKEY · Configural Polysampling: A Route to Practical Robustness

MUIRHEAD · Aspects of Multivariate Statistical Theory

MURRAY · X-STAT 2.0 Statistical Experimentation, Design Data Analysis, and Nonlinear Optimization

MYERS and MONTGOMERY · Response Surface Methodology: Process and Product Optimization Using Designed Experiments, *Second Edition*

MYERS, MONTGOMERY, and VINING · Generalized Linear Models. With Applications in Engineering and the Sciences

NELSON · Accelerated Testing, Statistical Models, Test Plans, and Data Analyses

NELSON · Applied Life Data Analysis

NEWMAN · Biostatistical Methods in Epidemiology

OCHI · Applied Probability and Stochastic Processes in Engineering and Physical Sciences

OKABE, BOOTS, SUGIHARA, and CHIU · Spatial Tesselations: Concepts and Applications of Voronoi Diagrams, *Second Edition*

OLIVER and SMITH · Influence Diagrams, Belief Nets and Decision Analysis

PANKRATZ · Forecasting with Dynamic Regression Models

PANKRATZ · Forecasting with Univariate Box-Jenkins Models: Concepts and Cases

*PARZEN · Modern Probability Theory and Its Applications

PEÑA, TIAO, and TSAY · A Course in Time Series Analysis

PIANTADOSI · Clinical Trials: A Methodologic Perspective

PORT · Theoretical Probability for Applications

POURAHMADI · Foundations of Time Series Analysis and Prediction Theory

PRESS · Bayesian Statistics: Principles, Models, and Applications

PRESS · Subjective and Objective Bayesian Statistics, *Second Edition*

PRESS and TANUR · The Subjectivity of Scientists and the Bayesian Approach

PUKELSHEIM · Optimal Experimental Design

PURI, VILAPLANA, and WERTZ · New Perspectives in Theoretical and Applied Statistics

PUTERMAN · Markov Decision Processes: Discrete Stochastic Dynamic Programming

*RAO · Linear Statistical Inference and Its Applications, *Second Edition*

RENCHER · Linear Models in Statistics

RENCHER · Methods of Multivariate Analysis, *Second Edition*

RENCHER · Multivariate Statistical Inference with Applications

RIPLEY · Spatial Statistics

RIPLEY · Stochastic Simulation

ROBINSON · Practical Strategies for Experimenting

ROHATGI and SALEH · An Introduction to Probability and Statistics, *Second Edition*

ROLSKI, SCHMIDLI, SCHMIDT, and TEUGELS · Stochastic Processes for Insurance and Finance

ROSENBERGER and LACHIN · Randomization in Clinical Trials: Theory and Practice

ROSS · Introduction to Probability and Statistics for Engineers and Scientists

ROUSSEEUW and LEROY · Robust Regression and Outlier Detection

RUBIN · Multiple Imputation for Nonresponse in Surveys

RUBINSTEIN · Simulation and the Monte Carlo Method

RUBINSTEIN and MELAMED · Modern Simulation and Modeling

RYAN · Modern Regression Methods

RYAN · Statistical Methods for Quality Improvement, *Second Edition*

SALTELLI, CHAN, and SCOTT (editors) · Sensitivity Analysis

*SCHEFFE · The Analysis of Variance

*Now available in a lower priced paperback edition in the Wiley Classics Library.

SCHIMEK · Smoothing and Regression: Approaches, Computation, and Application

SCHOTT · Matrix Analysis for Statistics

SCHUSS · Theory and Applications of Stochastic Differential Equations

SCOTT · Multivariate Density Estimation: Theory, Practice, and Visualization

*SEARLE · Linear Models

SEARLE · Linear Models for Unbalanced Data

SEARLE · Matrix Algebra Useful for Statistics

SEARLE, CASELLA, and McCULLOCH · Variance Components

SEARLE and WILLETT · Matrix Algebra for Applied Economics

SEBER and LEE · Linear Regression Analysis, *Second Edition*

SEBER · Multivariate Observations

SEBER and WILD · Nonlinear Regression

SENNOTT · Stochastic Dynamic Programming and the Control of Queueing Systems

*SERFLING · Approximation Theorems of Mathematical Statistics

SHAFER and VOVK · Probability and Finance: It's Only a Game!

SMALL and McLEISH · Hilbert Space Methods in Probability and Statistical Inference

SRIVASTAVA · Methods of Multivariate Statistics

STAPLETON · Linear Statistical Models

STAUDTE and SHEATHER · Robust Estimation and Testing

STOYAN, KENDALL, and MECKE · Stochastic Geometry and Its Applications, *Second Edition*

STOYAN and STOYAN · Fractals, Random Shapes and Point Fields: Methods of Geometrical Statistics

STYAN · The Collected Papers of T. W. Anderson: 1943–1985

SUTTON, ABRAMS, JONES, SHELDON, and SONG · Methods for Meta-Analysis in Medical Research

TANAKA · Time Series Analysis: Nonstationary and Noninvertible Distribution Theory

THOMPSON · Empirical Model Building

THOMPSON · Sampling, *Second Edition*

THOMPSON · Simulation: A Modeler's Approach

THOMPSON and SEBER · Adaptive Sampling

THOMPSON, WILLIAMS, and FINDLAY · Models for Investors in Real World Markets

TIAO, BISGAARD, HILL, PEÑA, and STIGLER (editors) · Box on Quality and Discovery: with Design, Control, and Robustness

TIERNEY · LISP-STAT: An Object-Oriented Environment for Statistical Computing and Dynamic Graphics

TSAY · Analysis of Financial Time Series

UPTON and FINGLETON · Spatial Data Analysis by Example, Volume II: Categorical and Directional Data

VAN BELLE · Statistical Rules of Thumb

VIDAKOVIC · Statistical Modeling by Wavelets

WEISBERG · Applied Linear Regression, *Second Edition*

WELSH · Aspects of Statistical Inference

WESTFALL and YOUNG · Resampling-Based Multiple Testing: Examples and Methods for *p*-Value Adjustment

WHITTAKER · Graphical Models in Applied Multivariate Statistics

WINKER · Optimization Heuristics in Economics: Applications of Threshold Accepting

WONNACOTT and WONNACOTT · Econometrics, *Second Edition*

WOODING · Planning Pharmaceutical Clinical Trials: Basic Statistical Principles

WOOLSON and CLARKE · Statistical Methods for the Analysis of Biomedical Data, *Second Edition*

WU and HAMADA · Experiments: Planning, Analysis, and Parameter Design Optimization

YANG · The Construction Theory of Denumerable Markov Processes

*ZELLNER · An Introduction to Bayesian Inference in Econometrics

ZHOU, OBUCHOWSKI, and McCLISH · Statistical Methods in Diagnostic Medicine